VOLUME SEVENTY TWO

ADVANCES IN CARBOHYDRATE CHEMISTRY AND BIOCHEMISTRY

BOARD OF ADVISORS

VOLUME SEVENTY TWO

ADVANCES IN CARBOHYDRATE CHEMISTRY AND BIOCHEMISTRY

Editors

DAVID C. BAKER
University of Tennessee, Knoxville, TN

DEREK HORTON
Ohio State University, Columbus, Ohio American University, Washington, DC

AMSTERDAM • BOSTON • HEIDELBERG • LONDON
NEW YORK • OXFORD • PARIS • SAN DIEGO
SAN FRANCISCO • SINGAPORE • SYDNEY • TOKYO
Academic Press is an imprint of Elsevier

Academic Press is an imprint of Elsevier
225 Wyman Street, Waltham, MA 02451, USA
525 B Street, Suite 1800, San Diego, CA 92101–4495, USA
The Boulevard, Langford Lane, Kidlington, Oxford OX5 1GB, UK
125 London Wall, London, EC2Y 5AS, UK

First edition 2015

ISBN: 978-0-12-802141-5
ISSN: 0065-2318

For information on all Academic Press publications
visit our website at http://store.elsevier.com/

CONTENTS

Contributors *vii*
Preface *ix*

1. ROBERT JOHN (ROBIN) FERRIER 1932–2013 **3**
Richard Furneaux, Ralf Schwörer, and Sarah Wilcox

2. Synthetic Approaches to L-Iduronic Acid and L-Idose: Key Building Blocks for the Preparation of Glycosaminoglycan Oligosaccharides **21**
Shifaza Mohamed and Vito Ferro

1. Introduction 23
2. Epimerization at C-5 of D-Glucose Derivatives 25
3. Homologation of Tetroses and Pentoses 36
4. Isomerization of Unsaturated Sugars 42
5. Miscellaneous Methods 48
6. Conclusions 52

Acknowledgments 52
References 52

3. Glycosylation of Cellulases: Engineering Better Enzymes for Biofuels **63**
Eric R. Greene, Michael E. Himmel, Gregg T. Beckham, and Zhongping Tan

1. Introduction 65
2. Glycosylation of Cellulose-Degrading Enzymes 66
3. Recombinant Expression of Fungal Cellulases 82
4. Modifications by Glycan-Trimming Enzymes 88
5. Summary and Future Perspectives 96

Acknowledgments 96
Appendix 1 Molecular Dynamics Simulation of a Linker Interacting with Crystalline Cellulose 97
References 97

4. Human Milk Oligosaccharides (HMOS): Structure, Function, and Enzyme-Catalyzed Synthesis **113**

Xi Chen

1. Introduction 115
2. Structures of HMOS 117
3. Biosynthesis of HMOS 151
4. Functions of HMOS 152
5. Production of HMOS by Enzyme-Catalyzed Processes 158
6. Perspectives 169

Acknowledgments 170
References 170

Author Index *191*
Subject Index *223*

CONTRIBUTORS

Gregg T. Beckham
National Bioenergy Center, National Renewable Energy Laboratory, Golden, Colorado, USA

Xi Chen
Department of Chemistry, University of California, Davis, California, USA

Vito Ferro
School of Chemistry and Molecular Biosciences, The University of Queensland, Brisbane, Queensland, Australia

Richard Furneaux
Ferrier Research Institute, Victoria University of Wellington, Petone 5046, New Zealand

Eric R. Greene
Department of Chemistry and Biochemistry and BioFrontiers Institute, University of Colorado, Boulder, Colorado, USA

Michael E. Himmel
Biosciences Center, National Renewable Energy Laboratory, Golden, Colorado, USA

Shifaza Mohamed
School of Chemistry and Molecular Biosciences, The University of Queensland, Brisbane, Queensland, Australia

Ralf Schwörer
Ferrier Research Institute, Victoria University of Wellington, Petone 5046, New Zealand

Zhongping Tan
Department of Chemistry and Biochemistry and BioFrontiers Institute, University of Colorado, Boulder, Colorado, USA

Sarah Wilcox
Ferrier Research Institute, Victoria University of Wellington, Petone 5046, New Zealand

PREFACE

The publication of this volume of *Advances in Carbohydrate Chemistry and Biochemistry*, Vol. 72, marks a transition, the passing of Prof. Derek Horton, who had been associated with editing the series since Vol. 24 (publ. date 1969) when he served as Assistant Editor with Melville L. Wolfrom and R. Stuart Tipson. After nearly two decades as Coeditor with R.S.T., he assumed sole editorship with the publication of Vol. 49 in 1991. (For a detailed history of *Advances*, see Vol. 70.[1]) Prof. Horton worked to maintain the outstanding rigor and quality of articles marked by the series from the appearance of Vol. 1 in 1945. In the carbohydrate community of scientists, he was *Advances*' most visible promoter and ardent advocate, and his name had become virtually synonymous with that of the series, defining its place among journals and other publications in the carbohydrate field. His expertise and leadership are sorely missed. The authors of the articles contained herein had been recruited by Prof. Horton, with the writing in various stages of development at the time of his death.

Shifaza Mohamed and Vito Ferro of the University of Queensland, Brisbane, Australia have examined the various pathways for definitively producing and L-idose, L-iduronic acid, and related derivatives, which are important components of the glycosaminoglycans (GAGs, including heparin, heparan sulfate, and dermatan sulfate). While as students of carbohydrate chemistry we have all marveled at Emil Fischer's accomplishments in the synthesis of key compounds in the L-ido series over 125 years ago, we also realize that straightforward, definitive routes to these compounds still remain a synthetic challenge, especially to produce the products in workable amounts to serve the serious synthetic chemist. The authors group the chemistry into three main areas: (1) the epimerization at C-5 of D-glucose derivatives, (2) homologation of tetroses and pentoses, and (3) isomerization of unsaturated sugars, as well as a number of miscellaneous methods. By seriously considering the chemistry presented, one realizes that a great many processes bode well for the development of synthetic chemistry to GAG and GAG-like structures, which will enhance our understanding of important structure–activity relationships among these complex compounds.

The team of Eric R. Greene, Michael E. Himmel, Gregg T. Beckham, and Zhongping Tan of the University of Colorado in Boulder, CO and the National Renewable Energy Laboratory in Golden, CO, have collaborated to address the issues of improving the cellulose breakdown process by engineering better enzymes. Cellulases, the principal enzymes of saccharification—the hydrolysis of cellulose into simpler, soluble sugars—lend themselves to both N- and O-glycosylation, processes that are not directly genetically controlled, with attendant influences on activity and structure that are the subject of much research and study. The authors address the critical issues of glycosylation and its control, especially with the industrially important cellobiohydrolase from *Trichoderma reesei*, with an eye to exploit fungal enzymes through expression host engineering methods for more efficient biofuels production.

Xi Chen of the University of California, Davis, CA surveys the structure, function, and enzyme-catalyzed synthesis of human milk oligosaccharides and discusses their importance in establishing and maintaining the health of infants. While over 100 human milk oligosaccharides (HMOS) are known, far less is known about the precise function of individual HMOS, mainly due to the limited quantities of these compounds that have been isolated or synthesized. The author concentrates on both whole-cell and living-cell enzymatic methods to synthesize complex HMOS (only cursory mention is made of synthetic chemical methods) with the aim of providing inspiration for chemists and biochemists to delve into the synthesis of even larger, more complex and branched-chain HMOS with as yet undefined properties.

Richard Furneaux, Ralf Schwörer, and Sarah Wilcox of the Ferrier Research Institute, Victoria University of Wellington, New Zealand, provide a an insightful obituary for Prof. Robert John (Robin) Ferrier, a well-known and much appreciated carbohydrate chemist. Trained as a polysaccharide chemist at the University of Edinburgh in Scotland, he later focused on monosaccharide chemistry and served as a world leader in developing rigorous organic chemistry methods with and for monosaccharides. He is especially remembered for his work on 2,3-unsaturated glycosides synthesized from glycals, a reaction that, in the community of chemists, universally bears his name. For those of us who knew him on the world stage, his likable demeanor, clear and even-tempered lectures, and sometimes wry sense of humor will be missed. A true gentleman among scientists he was.

David C. Baker
University of Tennessee
Knoxville, Tennessee, USA
August 2015

REFERENCE

1. Horton, D. Seven Decades of "Advances". *Adv. Carbohydr. Chem. Biochem.* **2013**, *70*, 13–209.

ROBERT JOHN (ROBIN) FERRIER
1932–2013

In 1954, at the outset of my post-graduate experience in the Department of Chemistry in Edinburgh…I was introduced to carbohydrate chemistry, and with amazing lack of imagination I have never strayed too far from the subject since—but it has taken me to some undreamed of areas of chemistry and also of the world.

Robin Ferrier c. 2000

Robert John Ferrier was born in Edinburgh on the 7th of August 1932 to Edward, a policemen who would later become head of Edinburgh Criminal Investigation Department, and his wife Sophia. Although he was named Robert John, he was always called Robin, following the idiosyncratic family naming tradition, where his father Edward was called William, his mother Sophia was known as Rita, and Robin's twin sister Barbara was known as Ray.

Apart from a short stay at Traquair to where he was evacuated during the war, Robin attended George Heriot's School, one of the historic landmarks of the city of Edinburgh. This independent school dates back to its foundation in 1628 and boasts a long list of notable alumni. As a 17-year-old schoolboy, Robin came upon organic chemistry, a discovery he described as a major piece of good fortune, and decided this was the subject he wanted to pursue. He enrolled in undergraduate chemistry studies at the University of Edinburgh and obtained a Bachelor of Science with first-class honors in 1954. He joined the group of Prof. Gerald Aspinall for his Ph.D., training in the field of polysaccharide chemistry, and was awarded the degree in 1957. Several publications came out of this work, ranging from "The Preparation of Cyclic Ketals by Use of Trifluoroacetic Anhydride" to "The Constitution of Barley Husk Hemicellulose."[1–6]

One of his favorite anecdotes was that he told of his conscientiously dehusking a sack of barley by hand, only to later realize that the hard-won product was actually an inexpensive by-product of milling and available by the bagful. He was pleased, however, with the insight that led to the facile monitoring by UV absorbance of the progress of periodate oxidation reactions, then an essential tool for polysaccharide structural analysis.[3]

Advances in Carbohydrate Chemistry and Biochemistry, Volume 72
ISSN 0065-2318
http://dx.doi.org/10.1016/bs.accb.2015.08.003

During his first teaching position as Assistant Lecturer at Birkbeck College, University of London, Robin changed his research interest from polysaccharides to monosaccharides.[7–29] He became one of the pioneering chemists who treated carbohydrates as ordinary organic compounds rather than as a separate field of research. This approach was made possible by new methods and laboratory tools that became available during that time, allowing the structure, reactions, and mechanisms of carbohydrates to be studied in the same way as other organic molecules.

Robin's early work investigated the formation of esters between hexopyranosides and phenylboronic acid, and he published a paper on the reaction of methyl α-D-glucopyranoside with phenylboronic acid.[10] In subsequent years, this work was extended to xylopyranosides[21] and xylofuranosides,[26] as well as to other hexopyranosides[27] and pentopyranosides.[28] The use of phenylboronic acid as a protecting group[29] and as an aid in configurational analysis[19] was also investigated.

In 1961, Robin took a position as NATO Postdoctoral Fellow in Professor Melvin Calvin's laboratory at the University of California, Berkeley for 15 months during a leave of absence from Birkbeck College. His time in California had a significant impact on him in many ways. Calvin was awarded the Nobel Prize in Chemistry in 1961 while Robin was there, for work on the light-independent reactions of photosynthesis, now known as the Calvin cycle. From this time onward, Robin relished speculating about potential Nobel nominees and likely recipients. He also discovered a love of fresh fruit, which continued throughout his life.

But the most important encounter was meeting his U.S.-born, future wife, Carolyn Tompkins. They were both living at the student hostel, International House, where she was the in-house nurse. They were married in Edinburgh on 25 July 1962 and had two children together, which were both born in London—Alison in 1964 and Duncan in 1968.

Robin's research with Calvin focussed on the characterization of a photosynthesis product, which was believed to be a branched hexonic acid diphosphate. While the synthesis of the desired 2-*C*-hydroxymethylpentonic acids[13] had to be abandoned due to the complexity of the ether mixtures that were produced, studies on the trimethylsilyl ethers of carbohydrates and their gas- or paper-chromatographic characteristics were based on this work.[9,15,16,18] The desired pentuloses were then made by enzymatic synthesis,[11] and the photosynthetic product was finally characterized as a 2-keto-L-gulonic acid phosphate.[12] An impressive number of seven publications resulted from his short time in California.

Returning to the United Kingdom, Robin spent the next eight years in London,[30–65] apart from a short period of teaching chemistry at the University of Ife in Nigeria in 1966 and a memorable two-week visit to Professor Nicolai Kochetkov's laboratory in Moscow in 1968, sponsored by the British Council. While in Russia, he carried out early mass spectrometric work on carbohydrates.[58] The experience and skills Robin had gained in working with radiochemicals at Berkeley enabled him to contribute to elucidating the mechanism of free-sugar alcoholysis (Fischer glycosidation) and thiolysis.[42–44] In another, rather unexpected publication in 1964, Robin described an alternative design of a rotary evaporator that had been developed for use in his department.[23]

While his work on the chemistry of carbohydrates and boronic acids continued, Robin developed another major research interest in unsaturated sugars, especially 2-deoxy glycosides.[14,25,32–37,40,45,50–54] Based on products observed in reactions of 3,4,6-tri-*O*-acetyl-D-glucal with phenols under reflux[14] or with other nucleophiles in the absence of protic acids,[34] the reaction of alcohols with 3,4,6-tri-*O*-acetyl-D-glucal in presence of boron trifluoride was developed.[50] This scheme predominantly produced 2,3-dideoxy-α-D-glycosides after nucleophilic substitution, combined with an allylic shift. The reaction would later be known as the Ferrier rearrangement.

A desire to progress his career and give his family a better quality of life led to Robin's decision to move to New Zealand. He was appointed Victoria University of Wellington's first Chair of Organic Chemistry in 1970. While his research remained in carbohydrate chemistry, its focus was gradually shifted to the use of monosaccharides in the synthesis of functionalized cyclohexanes and cyclopentanes.[66–194] Of particular interest were synthetic routes to compounds of medicinal significance, such as aminoglycosides, prostaglandins, and anthracyclinones.

Although Robin's research group remained small compared with those in Europe or America, it was probably an appropriate size for Australasia in the 1970s and 1980s, and never exceeded a handful of graduate research students and an occasional postdoctoral researcher. Nevertheless, the group collectively produced a remarkable volume of research. A new α-glycosidation procedure using mercury(II) ion activated thioglycosides was developed,[72] alkylthio groups were introduced into carbohydrates to produce polydeoxy-polythiosugars,[66,69,74] and a considerable effort went into the study of photobromination reactions of carbohydrates and their usefulness in chemical synthesis, for example, in the synthesis of vitamin C,[92] L-idose,[107,108] and unsaturated sugars.[114]

The synthesis of functionalized cyclopentanes and cyclohexanes was a major focus, and Robin and his coworkers developed the syntheses of cyclopentanes as prostaglandin synthons—compounds that at the time were of great interest to the pharmaceutical industry. The group elaborated the necessary carbon–carbon bond-forming chemistries—either dipolar cycloadditions or photochemical [2+2] cycloadditions—that were compatible with the sensitive carbohydrate functionality and went on to publish a number of papers on the subject.[116,119–123,136]

Many of Robin's most significant discoveries were made by following up unexpected chemical observations, which often led him into uncharted territory. His second "name" reaction, the Ferrier carbocyclization,[102] was the result of this approach. While on sabbatical leave in 1977 as a Royal Society Research Fellow at Heriot–Watt University in Edinburgh, he set up a reaction to run overnight and the next morning found his reaction flask full of a crystalline product, but not the one he expected. Upon investigation, he found he had discovered a new carbon–carbon bond-forming reaction that allowed cyclitols to be made from free sugars—something nature does, but chemists had struggled to achieve. (The 270-MHz proton NMR spectrum for the characterization of the reaction product was run by Professor Hans Paulsen in Hamburg, as there was no instrument with such a high field available at Victoria University.) The related Petasis–Ferrier rearrangement, a Lewis acid-catalyzed carbon–carbon bond-forming cyclization reaction, is also named in his honor.

The Ferrier reactions have become useful tools for chemists in their work of piecing together synthetic routes to complex molecules for research and industry. Robin himself would never refer to the reactions using his own name, however, as he believed that would have been boastful. In a review[170] in 1993 on the synthesis of cyclohexanes and cyclopentanes from carbohydrates he wrote:

> *This reaction (…) has become well recognized, and there has been a tendency to refer to it as a name reaction after one of us (RJF). An entirely unconnected reaction—the conversion of acylated glycals into 2,3-unsaturated glycopyranosides—has, however, also acquired the same name, and in the interest of avoiding confusion, we do not use it.*

At Victoria University, Robin supervised 10 successful Ph.D. students, many of whom went on to make significant contributions in New Zealand and overseas. Three of them, Regine Blattner, Peter Tyler, and one of the authors (RHF), were founding members of the carbohydrate chemistry team at New Zealand's Department of Scientific and Industrial Research

(DSIR) in 1985. Robin had high expectations of his students' progress and a rigor that demanded all experimental results be fully validated before they could be believed. This firm approach was softened by a quick wit and mischievous sense of humor. Due to funding restraints and the isolation of New Zealand at the time, many of the chemical reagents used in his lab, which would nowadays be purchased, were made by the students themselves. This led to a deeper appreciation of their origins and properties. His Ph.D. students recall that with some fear they could hear the sound of his footsteps down the long third-floor corridor of the Easterfield Building each evening as he was coming for an update on the day's progress. A visit was particularly terrifying if the day had been frittered away, but wonderful if there was a sniff of new knowledge or a problem to be solved—student and teacher went straight to the chalkboard for an inspirational discourse.

Robin also found opportunities where his chemistry could serve the nation. Notable were his service on the Toxic Substances Board in the 1980s and the leadership of the Royal Society of New Zealand report *Lead in the Environment* that in 1986 confirmed the toxic effects of lead and began the phase-out of leaded petrol in New Zealand.[143,144] In other applied work, he and Regine Blattner created a model system (the oxidation of cellobiitol) that elucidated the chemical reactions that cause nail sickness—an iron-catalyzed oxidative degradation of the cellulosic component of wood.[135] With Anton Erasmuson, he made haptens based on synthetic fragments of the fungal toxophore, sporodesmin, that were used in attempts to create a vaccine against facial eczema in sheep.[101] Robin also conducted model studies that elucidated the mechanisms involved in a high-temperature and pressure biomass liquefaction process with Wayne Severn.[164,166] With Anna-Karin Tidèn, he made the cyclitolamine moiety required for a 23-step total synthesis of the insecticidal natural product allosamidin,[172] which is still highly sought after as an insecticidal standard.

In his 50-year career, Robin published 180 papers, reviews, and books and gave 10 invited plenary lectures at international symposia. His reviews, notably as Senior Reporter for the Specialist Periodical Reports, *Carbohydrate Chemistry*, published annually by the Royal Society of Chemistry (UK) from 1966 to 2000, were of particular benefit to the chemical community. Perhaps the most valuable of these contributions was the book *Monosaccharide Chemistry*[68] written with Dr. Peter Collins in 1972 and majorly updated as *Monosaccharides: Their Chemistry and their Roles in Natural Products*[174] in 1995.

Robin was elected Fellow of the Royal Society of New Zealand (1977) and the New Zealand Institute of Chemistry (1972) and was awarded a D.Sc. (London, 1968). To his credit, by the time his of retirement, most of the world's top chemists had engaged with the challenges and opportunities offered by the chemical synthesis and modification of carbohydrates.

Robin loved sport. He played for the school cricket team from age 14, was an East of Scotland doubles champion in tennis, and sported a nose rearranged by rugby. He spent days in the sun (burning to a crisp) at the Basin Reserve watching New Zealand play test cricket, and later in life he took up golf, one of his few concessions to retirement.

Robin and Carolyn were unfailingly hospitable to visiting chemists and loved to entertain people from all over the world. They were also loyal and regular members of St. John's in the City Presbyterian Church; Robin became an Elder in 1974 and served as Session Clerk and on various committees throughout his life. He was also the Presbytery representative on the Board of Governors at Scots College for 13 years.

After his retirement from Victoria University in 1998, Robin entered what he referred to as his "supposed retirement," working with the carbohydrate chemists at Industrial Research Ltd. Here he continued to foster the next generation of carbohydrate chemists in New Zealand—his "grandchildren," instilling his rigorous approach to chemistry with mentoring and assistance with the group's publications.

In August 2012, Robin celebrated his 80th birthday and retired a second time. Later that year, the Ferrier Trust Fund was set up in his honor, to bring a highly regarded scientist to New Zealand each year to engage with chemistry students and lecture. Peppi Prasit, a Ferrier Ph.D. graduate and founder of Amira Pharmaceuticals and Inception Sciences in the United States, was the trust's foundation donor. Despite Robin's failing health and memory, it was a delight to have him attend the inaugural Ferrier Lecture in March 2013.

In January 2014, the Ferrier Research Institute was founded at Victoria University of Wellington and named in Robin's honor. The 27-strong group was previously the carbohydrate chemistry team at Callaghan Innovation and Industrial Research Ltd (IRL), which was formed when the DSIR was dissolved in 1992.

Richard Furneaux
Ralf Schwörer
Sarah Wilcox

BIBLIOGRAPHY

1. Aspinall, G. O.; Ferrier, R. J. The Constitution of Barley-Husk Hemicellulose. *J. Chem. Soc.* **1957**, 4188–4194.
2. Ferrier, R. J.; Tedder, J. M. Preparation of Cyclic Ketones by Use of Trifluoroacetic Anhydride. *J. Chem. Soc.* **1957**, 1435–1437.
3. Aspinall, G. O.; Ferrier, R. J. Spectrophotometric Method for the Determination of Periodate Consumed During the Oxidation of Carbohydrates. *Chem. Ind. (London)* **1957**, 1216.
4. Aspinall, G. O.; Ferrier, R. J. The Synthesis of 2-*O*-β-D-Xylopyranosyl-L-arabinose. *Chem. Ind. (London)* **1957**, *819*.
5. Aspinall, G. O.; Ferrier, R. J. Cereal Gums. III. The Constitution of an Arabinoxylan from Barley Flour. *J. Chem. Soc.* **1958**, 638–642.
6. Aspinall, G. O.; Ferrier, R. J. Synthesis of 2-[*O*-(β-D-Xylopyranosyl)]-L-arabinose and Its Isolation from the Partial Hydrolysis of Esparto Hemicellulose. *J. Chem. Soc.* **1958**, 1501–1505.
7. Ferrier, R. J.; Overend, W. G. Newer Aspects of the Stereochemistry of Carbohydrates. *Q. Rev., Chem. Soc.* **1959**, *13*, 265–286.
8. Ferrier, R. J.; Overend, W. G. Periodate Oxidation of 2-Deoxyglycosides: Structure of 1-(2-Deoxy-D-galactosyl)benzimidazole. *J. Chem. Soc.* **1959**, 3638–3639.
9. Ferrier, R. J. The Gas–Liquid Partition Chromatography of the Tetra-*O*-acetylpentopyranoses. *Chem. Ind. (London)* **1961**, 831–832.
10. Ferrier, R. J. Interaction of Phenylboronic Acid with Hexosides. *J. Chem. Soc.* **1961**, 2325–2330.
11. Moses, V.; Ferrier, R. J. Biochemical Preparation of D-Xylulose and L-Ribulose. Details of the Action of *Acetobacter suboxydans* on D-Arabitol, Ribitol, and Other Polyhydroxy Compounds. *Biochem. J.* **1962**, *83*, 8–14.
12. Moses, V.; Ferrier, R. J.; Calvin, M. Characterization of the Photosynthetically Synthesized γ-Keto Acid Phosphate as a Diphosphate Ester of 2-Keto-L-gulonic Acid. *Proc. Natl. Acad. Sci. U.S.A.* **1962**, *48*, 1644–1650.
13. Ferrier, R. J. The Configuration of the 2-*C*-Hydroxymethylpentonic Acids. *J. Chem. Soc.* **1962**, 3544–3549.
14. Ferrier, R. J.; Overend, W. G.; Ryan, A. E. The Reaction Between 3,4,6-Tri-*O*-Acetyl-D-glucal and *p*-Nitrophenol. *J. Chem. Soc.* **1962**, 3667–3670.
15. Ferrier, R. J. Structure and Chromatographic Properties of Carbohydrates. II. The Liquid–Liquid Partition Mobilities of Aldono-γ-lactones. *J. Chromatogr.* **1962**, *9*, 251–252.
16. Ferrier, R. J. Structure and Chromatographic Properties of Carbohydrates. III. Gas–Liquid Partition Mobilities of Pyranose Derivatives. *Tetrahedron* **1962**, *18*, 1149–1154.
17. Ferrier, R. J.; Overend, W. G.; Ryan, A. E. Structure and Reactivity of Anhydro Sugars. IV. Action of Alkali on 2-Deoxy-D-*arabino*-hexose: Structure of iso-Glucal. *J. Chem. Soc.* **1962**, 1488–1490.
18. Ferrier, R. J.; Singleton, M. F. The Trimethylsilylation of D-Xylose and the Nuclear Magnetic Resonance Spectra of the Major Products. *Tetrahedron* **1962**, *18*, 1143–1148.
19. Ferrier, R. J.; Overend, W. G.; Rafferty, G. A.; Wall, H. M.; Williams, N. R. Determination of the Configuration of Branched-Chain Sugars. *Proc. Chem. Soc. (London)* **1963**, 133.
20. Ferrier, R. J.; Williams, N. R. Applications of Instrumental Techniques in Structural Monosaccharide Chemistry. *Chem. Ind. (London)* **1964**, 1696–1707.
21. Ferrier, R. J.; Prasad, D.; Rudowski, A.; Sangster, I. Boric Acid Derivatives as Reagents in Carbohydrate Chemistry. Part II. The Interaction of Phenylboronic Acid with Methyl Xylopyranosides. *J. Chem. Soc.* **1964**, 3330–3334.

22. Ferrier, R. J.; Prasad, D.; Rudowski, A. Isolation of the Four Methyl Xylosides. *Chem. Ind. (London)* **1964**, 1260–1261.
23. Ferrier, R. J.; Marsh, R. W. Ring-Sealed Rotary Evaporator. *Chem. Ind. (London)* **1964**, 1559–1560.
24. Ferrier, R. J. The Role of Physical Methods in the Elucidation of Organic Structures: Essay Review. *Chem. Ind. (London)* **1964**, 64–66.
25. Ferrier, R. J. Unsaturated Carbohydrates. II. Three Reactions Leading to Unsaturated Glycopyranosides. *J. Chem. Soc.* **1964**, 5443–5449.
26. Ferrier, R. J.; Prasad, D.; Rudowski, A. Boric Acid Derivatives as Reagents in Carbohydrate Chemistry. III. The Interaction of Phenylboronic Acid with Methyl Xylofuranosides and the Isolation of the Four Methyl Xylosides. *J. Chem. Soc.* **1965**, 858–863.
27. Ferrier, R. J.; Hannaford, A. J.; Overend, W. G.; Smith, B. C. Boric Acid Derivatives as Reagents in Carbohydrate Chemistry. IV. The Interaction of Phenylboronic Acid with Hexopyranoid Compounds. *Carbohydr. Res.* **1965**, *1*, 38–43.
28. Ferrier, R. J.; Prasad, D. Boric Acid Derivatives as Reagents in Carbohydrate Chemistry. V. The Interaction of Phenylboronic Acid with Methyl Pentopyranosides. Syntheses of 3- and 2,4-Substituted Ribose Derivatives. *J. Chem. Soc.* **1965**, 7425–7428.
29. Ferrier, R. J.; Prasad, D. Boric Acid Derivatives as Reagents in Carbohydrate Chemistry. VI. Phenylboronic Acid as a Protecting Group in Disaccharide Synthesis. *J. Chem. Soc.* **1965**, 7429–7432.
30. Williams, D. T.; Jones, J. K. N.; Dennis, N. J.; Ferrier, R. J.; Overend, W. G. The Oxidation of Sugar Acetals and Thioacetals by *Acetobacter suboxydans*. *Can. J. Chem.* **1965**, *43*, 955–959.
31. Ferrier, R. J.; Overend, W. G.; Ryan, A. E. Reactions at Position 1 of Carbohydrates. VII. The Action of Potassium Hydroxide on Aryl 2-Deoxy-α-D-*arabino*-hexopyranosides. *J. Chem. Soc.* **1965**, 3484–3486.
32. Ferrier, R. J.; Overend, W. G.; Sankey, G. H. Unsaturated Carbohydrates. III. A Rearrangement Reaction in the 2-Hydroxyglycal Series: Further Exceptions to Hudson's Isorotation Rule. *J. Chem. Soc.* **1965**, 2830–2836.
33. Ferrier, R. J. Unsaturated Sugars. *Adv. Carbohydr. Chem.* **1965**, *20*, 67–137.
34. Ciment, D. M.; Ferrier, R. J. Unsaturated Carbohydrates. IV. Allylic Rearrangement Reactions of 3,4,6-Tri-*O*-acetyl-D-galactal. *J. Chem. Soc. C* **1966**, 441–445.
35. Ciment, D. M.; Ferrier, R. J.; Overend, W. G. Unsaturated Carbohydrates. V. Displacements of 4- and 6-Sulfonyloxy Groups from Methyl 2,3-Didehydro-2,3-dideoxyhexopyranoside Derivatives. *J. Chem. Soc. C* **1966**, 446–449.
36. Ferrier, R. J.; Sankey, G. H. Unsaturated Carbohydrates. VI. A Modified Synthesis of 2-hydroxyglycal Esters, and Their Conversion into Esters of 2,3-Didehydro-3-deoxyaldoses. *J. Chem. Soc. C* **1966**, 2339–2345.
37. Ferrier, R. J.; Sankey, G. H. Unsaturated Carbohydrates. VII. The Preference Shown by Allylic Ester Groupings on Pyranoid Rings for the *quasi*-Axial Orientation. *J. Chem. Soc. C* **1966**, 2345–2349.
38. Ferrier, R. J.; Hatton, L. R. The Acid-Catalysed Condensation of D-Xylose with Benzaldehyde in the Presence of Alcohols. Two Diastereoisomeric 1,2:3,5-Di-*O*-Benzaldehyde-α-D-xylofuranoses. *Carbohydr. Res.* **1967**, *5*, 132–139.
39. Ferrier, R. J.; Prasad, N. Conformational Free Energies of Allylic Hydroxy, Acetoxy, and Methoxy Groups in Cyclohexenes. *J. Chem. Soc. C* **1967**, 1417–1420.
40. Ferrier, R. J.; Prasad, N. The Application of Unsaturated Carbohydrates to Glycoside Syntheses: 6-*O*-α-D-Mannopyranosyl-, 6-*O*-α-D-Altropyranosyl-, 6-*O*-(3,6-Anhydro-α-D-glucopyranosyl)-D-galactose. *Chem. Commun.* **1968**, 476–477.
41. Ferrier, R. J.; Overend, W. G.; Rafferty, G. A.; Wall, H. W.; Williams, N. R. Branched-Chain Sugars. VIII. A Contribution to the Chemistry of 2-*C*-Methyl-L-Arabinose and -Ribose. *J. Chem. Soc. C* **1968**, 1091–1095.

42. Ferrier, R. J.; Hatton, L. R. Radioactive Sugars. I. Aspects of the Alcoholysis of D-Xylose and D-Glucose. Role of the Acyclic Acetals. *Carbohydr. Res.* **1968**, *6*, 75–86.
43. Ferrier, R. J.; Hatton, L. R.; Overend, W. G. Radioactive Sugars. II. Ethanethiolysis of D-Xylose. *Carbohydr. Res.* **1968**, *6*, 87–96.
44. Ferrier, R. J.; Hatton, L. R.; Overend, W. G. Studies with Radioactive Sugars. III. The Mechanism of the Anomerisation of Ethyl α- and β-D-Xylopyranoside. *Carbohydr. Res.* **1968**, *8*, 56–60.
45. Ferrier, R. J.; Prasad, N.; Sankey, G. H. Unsaturated Carbohydrates. VIII. Intramolecular Allylic Isomerisations of 1-Deoxyald-1-enopyranose (2-Hydroxyglycal) Esters. *J. Chem. Soc. C* **1968**, 974–977.
46. Guthrie, R. D.; Ferrier, R. J.; How, M. J. *Carbohydrate Chemistry*, Vol. 1. Specialist Periodical Reports; The Chemical Society: London, 1968; 294 pages.
47. Ferrier, R. J.; Williams, D. T. Attempted Intramolecular Sulfoxide Oxidations of Alcohols; the Diastereoisomeric 1-Deoxy-1-(ethylsulfinyl)-L-galactitols. *Carbohydr. Res.* **1969**, *10*, 157–160.
48. Ferrier, R. J. Configurational Analysis in Carbohydrate Chemistry. *Progr. Stereochem.* **1969**, *4*, 43–96.
49. Ferrier, R. J. Monosaccharide Chemistry. *Chem. Br.* **1969**, *5*, 15–21.
50. Ferrier, R. J.; Prasad, N. Unsaturated Carbohydrates. IX. Synthesis of 2,3-Dideoxy-α-D *erythro* hex 2 enopyranosides from Tri O acetyl D glucal. *J. Chem. Soc. C* **1969**, 570–575.
51. Ferrier, R. J.; Prasad, N. Unsaturated Carbohydrates. X. Epoxidations and Hydroxylations of 2,3-Dideoxy-α-D-hex-2-enopyranosides. Four Methyl 4,6-Di-*O*-acetyl-2,3-anhydro-α-D-hexopyranosides. *J. Chem. Soc. C* **1969**, 575–580.
52. Ferrier, R. J.; Prasad, N. Unsaturated Carbohydrates. XI. Isomerization and Dimerization of Tri-*O*-acetyl-D-glucal. *J. Chem. Soc. C* **1969**, 581–586.
53. Ferrier, R. J.; Prasad, N.; Sankey, G. H. Unsaturated Carbohydrates. XII. Synthesis and Properties of 3-Deoxy-α- and -β-D-*erythro*-hex-2-enopyranoside Esters. *J. Chem. Soc. C* **1969**, 587–591.
54. Ferrier, R. J. Unsaturated Sugars. *Adv. Carbohydr. Chem. Biochem.* **1969**, *24*, 199–266.
55. Guthrie, R. D.; Ferrier, R. J.; How, M. J.; Somers, P. J. *Carbohydrate Chemistry*, Vol. 2. Specialist Periodical Reports; The Chemical Society: London, 1969; 309 pages.
56. Ferrier, R. J.; Vethaviyasar, N. Allylic Isomerisations of Unsaturated Carbohydrate Derivatives. New Approach to the Synthesis of Amino, Thio, and Branched-Chain Sugars. *J. Chem. Soc. D Chem. Commun.* **1970**, 1385–1387.
57. Ferrier, R. J. Newer Observations on the Synthesis of O-Glycosides. *Fortschr. Chem. Forsch.* **1970**, *14*, 389–429.
58. Ferrier, R. J.; Vethaviyasar, N.; Chizhov, O. S.; Kadentsev, V. I.; Zolotarev, B. M. Unsaturated Carbohydrates. XIII. Mass Spectrometry of Hex-2- and Hex-3-enopyranosyl Derivatives. *Carbohydr. Res.* **1970**, *13*, 269–280.
59. Guthrie, R. D.; Ferrier, R. J.; Inch, T. D.; Somers, P. J. *Carbohydrate Chemistry*, Vol. 3. Specialist Periodical Reports; The Chemical Society: London, 1970; 297 pages.
60. Ferrier, R. J. *From Mysteries to Dilemmas: An Inaugural Address Delivered on 9 September 1971*; Victoria University of Wellington: Wellington, 1971.
61. Ferrier, R. J.; Ponpipom, M. M. Unsaturated Carbohydrates. XIV. Isomerization of (2-Hexenopyranosyl)purine Nucleoside Derivatives. *J. Chem. Soc. C* **1971**, 553–559.
62. Ferrier, R. J.; Ponpipom, M. M. Unsaturated Carbohydrates. XV. Synthesis of (3′-Deoxy-2′-hexenopyranosyl)purine Derivatives. *J. Chem. Soc. C* **1971**, 560–562.
63. Ferrier, R. J.; Vethaviyasar, N. Unsaturated Carbohydrates. XVI. Isomerization of Allylic 4-Azides and 4-Thiocyanates and the Subsequent Synthesis of 2-Acetamido-2-deoxyhexopyranoside Derivatives. *J. Chem. Soc. C* **1971**, 1907–1913.

64. Brimacombe, J. S.; Ferrier, R. J.; Guthrie, R. D.; Inch, T. D.; Kennedy, J. F. *Carbohydrate Chemistry*, Vol. 4. Specialist Periodical Reports; The Chemical Society: London, 1971; 278 pages.
65. Ferrier, R. J. Applications of Phenylboronic Acid in Carbohydrate Chemistry. *Methods Carbohydr. Chem.* **1972**, *6*, 419–426.
66. Bethell, G. S.; Ferrier, R. J. Direct Conversion of 3,5,6-Tri-*O*-benzoyl-1,2-*O*-isopropylidene-α-D-glucose into 4,5,6-Tri-*O*-benzoyl-2,3-di-*S*-ethyl-2,3-dithio-D-allose Diethyl Dithioacetal. *J. Chem. Soc., Perkin Trans. 1* **1972**, 1033–1037.
67. Ferrier, R. J. Modified Synthesis of 1-Deoxyald-1-enopyranose (2-Hydroxyglycal) Esters. Conversion of 1,2-Unsaturated Pyranoid Compounds into 2,3-Unsaturated Glycopyranosyl Derivatives. *Methods Carbohydr. Chem.* **1972**, *6*, 307–311.
68. Ferrier, R. J.; Collins, P. M. *Monosaccharide Chemistry*; Penguin Books: London, 1972. 318 pages.
69. Bethell, G. S.; Ferrier, R. J. Path of the Conversion of 3,5,6-Tri-*O*-benzoyl-1,2-*O*-isopropylidene-α-D-glucose into 4,5,6-Tri-*O*-benzoyl-2,3-di-*S*-ethyl-2,3-dithio-D-allose Diethyl Dithioacetal. *J. Chem. Soc., Perkin Trans. 1* **1972**, 2873–2878.
70. Brimacombe, J. S.; Ferrier, R. J.; Guthrie, R. D.; Inch, T. D.; Kennedy, J. F.; Sturgeon, R. J. *Carbohydrate Chemistry*, Vol. 5. Specialist Periodical Reports; The Chemical Society: London, 1972; 434 pages.
71. Brimacombe, J. S.; Da'aboul, I.; Tucker, L. C. N.; Calvert, N.; Ferrier, R. J. Intramolecular Cyclization of 3,6-Anhydro-D-glucal. *Carbohydr. Res.* **1973**, *27*, 254–256.
72. Ferrier, R. J.; Hay, R. W.; Vethaviyasar, N. Potentially Versatile Synthesis of Glycosides. *Carbohydr. Res.* **1973**, *27*, 55–61.
73. Bethell, G. S.; Ferrier, R. J. Radioactive Sugars. IV. Methanolysis of D-Fructose and L-Sorbose. *Carbohydr. Res.* **1973**, *31*, 69–80.
74. Bethell, G. S.; Ferrier, R. J. The Relayed Introduction of Alkylthio-Groups into Carbohydrate Derivatives: A Novel Synthesis of Amicetose. *J. Chem. Soc., Perkin Trans. 1* **1973**, 1400–1405.
75. Ferrier, R. J.; Vethaviyasar, N. Unsaturated Carbohydrates. XVII. Synthesis of Branched-Chain Sugar Derivatives by the Claisen Rearrangement. *J. Chem. Soc., Perkin Trans. 1* **1973**, 1791–1793.
76. Brimacombe, J. S.; Ferrier, R. J.; Guthrie, R. D.; Inch, T. D.; Kennedy, J. F.; Sturgeon, R. J. *Carbohydrate Chemistry*, Vol. 6. Specialist Periodical Reports; The Chemical Society: London, 1973; 620 pages.
77. Boolieris, D. S.; Ferrier, R. J.; Branda, L. A. Metal-Assisted Synthesis of Glycofuranosylamine Derivatives. *Carbohydr. Res.* **1974**, *35*, 131–139.
78. Bethell, G. S.; Ferrier, R. J. Selective C-1 Ethanethiolysis and Deuteration in *keto*-Hexulose Pentacetates. *Carbohydr. Res.* **1974**, *34*, 194–199.
79. Ferrier, R. J.; Hurford, J. R. Unsaturated Carbohydrates. XVIII. Esters of 1,4-Anhydro-L-*erythro*-pent-1-enitol (a Furanoid 2-Hydroxyglycal). *Carbohydr. Res.* **1974**, *38*, 125–131.
80. Ferrier, R. J. Polyhydric Alcohols (Alditols). In: *Supplements to the 2nd Edition of Rodd's Chemistry of Carbon Compounds*; Ansell, M. F. Ed.; Elsevier: Amsterdam, 1975; pp 1–18. Chapter 22.
81. Ferrier, R. J. Monosaccharides and Their Derivatives. In: *Supplements to the 2nd Edition of Rodd's Chemistry of Carbon Compounds*; Ansell, M. F. Ed.; Elsevier: Amsterdam, 1975; pp 19–229. Chapter 23.
82. Brimacombe, J. S.; Ferrier, R. J.; Guthrie, R. D.; Hughes, N. A.; Kennedy, J. F.; Marshall, R. D.; Sturgeon, R. J. *Carbohydrate Chemistry*, Vol. 7. Specialist Periodical Reports; The Chemical Society: London, 1975; 612 pages.
83. Ferrier, R. J. Displacement, Elimination and Rearrangement Reactions. In: *Carbohydrates*; Aspinall, G. O., Ed.; International Review of Science, Organic Chemistry, Series Two, Vol. 7; Butterworth: London, 1976; pp 35–88. Chapter 2.

84. Ferrier, R. J. The Sea as a Source of Fine Chemicals. *Chem. N. Z.* **1976**, *40*, 106–109.
85. Ferrier, R. J.; Furneaux, R. H. Synthesis of 1,2-*trans*-Related 1-Thioglycoside Esters. *Carbohydr. Res.* **1976**, *52*, 63–68.
86. Brimacombe, J. S.; Ferrier, R. J.; Hughes, N. A.; Kennedy, J. F.; Marshall, R. D.; Sturgeon, R. J.; Williams, N. R. *Carbohydrate Chemistry*, Vol. 8. Specialist Periodical Reports; The Chemical Society: London, 1976; 486 pages.
87. McLennan, T. J.; Robinson, W. T.; Bethell, G. S.; Ferrier, R. J. 4,5,6-Tri-O-benzoyl-2,3-di-*S*-ethyl-2,3-dithio-D-allose Diethyl Dithioacetal. *Acta Crystallogr.* **1977**, *B33*, 2888–2891.
88. Erasmuson, A. F.; Ferrier, R. J.; Franca, N. C.; Gottlieb, H. E.; Wenkert, E. ^{13}C Nuclear Magnetic Resonance Spectroscopy of Vanillin Derivatives. *J. Chem. Soc., Perkin Trans. 1* **1977**, 492–494.
89. Blattner, R.; Ferrier, R. J.; Tyler, P. C. Aspects of the Tautomerism of 2-(D-*galacto*-1,2,3,4,5-pentahydroxypentyl)benzothiazoline. *Carbohydr. Res.* **1977**, *54*, 199–208.
90. Ferrier, R. J.; Furneaux, R. H. Carbon-5 Bromination of Some Glucopyranuronic Acid Derivatives. *J. Chem. Soc., Perkin Trans. 1* **1977**, 1996–2000.
91. Ferrier, R. J.; Furneaux, R. H. The Chemistry of Some 1-Mercury(II)-thio-D-glucose Compounds, a New Synthesis of 1-Thio Sugars. *Carbohydr. Res.* **1977**, *57*, 73–83.
92. Ferrier, R. J.; Furneaux, R. H. A New Route to L-Ascorbic Acid (Vitamin C). *J. Chem. Soc., Chem. Commun.* **1977**, 332–333.
93. Ferrier, R. J.; Furneaux, R. H.; Tyler, P. C. Observations on the Possible Application of Glycosyl Disulfides, Sulfenic Esters, and Sulfones in the Synthesis of Glycosides. *Carbohydr. Res.* **1977**, *58*, 397–404.
94. Ferrier, R. J.; Srivastava, V. K. The Synthesis of Some Methyl 6-Deoxy-2-*O*-*p*-tolylsulfonyl-D-*xylo*-hexofuranosid-5-ulose Derivatives and Their Reactions With Bases. *Carbohydr. Res.* **1977**, *59*, 333–341.
95. Ferrier, R. J.; Furneaux, R. H. Unsaturated Carbohydrates. Part 20. Direct Conversion of Phenyl 1-Thiohexoside Esters into Phenyl 1-Thiohex-1-enopyranosid-3-ulose Esters. *J. Chem. Soc., Perkin Trans. 1* **1977**, 1993–1996.
96. Ferrier, R. J.; Vethaviyasar, N. Unsaturated Carbohydrates: Part XIX. Isomerization, with Allylic Rearrangement, of Hex-2-enopyranoside 4-Xanthate Esters. *Carbohydr. Res.* **1977**, *58*, 481–483.
97. Brimacombe, J. S.; Ferrier, R. J.; Hughes, N. A.; Kennedy, J. F.; Sturgeon, R. J.; Williams, N. R. *Carbohydrate Chemistry*, Vol. 9. Specialist Periodical Reports; The Chemical Society: London, 1977; 486 pages.
98. Ferrier, R. J. Carbohydrate Boronates. *Adv. Carbohydr. Chem. Biochem.* **1978**, *35*, 31–80.
99. Ferrier, R. J.; Tyler, P. C. Introduction, Substitution, and Elimination of Bromine at C-5 of Aldopyranose Peresters. *J. Chem. Soc., Chem. Commun.* **1978**, 1019–1020.
100. Brimacombe, J. S.; Catley, B. J.; Ferrier, R. J.; Kennedy, J. F.; Sturgeon, R. J.; Williams, J. M.; Williams, N. R. *Carbohydrate Chemistry*, Vol. 10. Specialist Periodical Reports; The Chemical Society: London, 1978; 524 pages.
101. Blunt, J. W.; Erasmuson, A. F.; Ferrier, R. J.; Munro, M. H. G. Syntheses of Haptens Related to the Benzenoid and Indole Portions of Sporidesmin A; ^{13}C NMR Spectra of Indole Derivatives. *Aust. J. Chem.* **1979**, *32*, 1045–1054.
102. Ferrier, R. J. Unsaturated Carbohydrates. Part 21. A Carbocyclic Ring Closure of a Hex-5-enopyranoside Derivative. *J. Chem. Soc., Perkin Trans. 1* **1979**, 1455–1458.
103. Brimacombe, J. S.; Catley, B. J.; Ferrier, R. J.; Kennedy, J. F.; Sturgeon, R. J.; Williams, J. M.; Williams, N. R. *Carbohydrate Chemistry*, Vol. 11. Specialist Periodical Reports; The Chemical Society: London, 1979; 546 pages.

104. Ferrier, R. J.; Furneaux, R. H. 1,2-*trans*-1-Thioglycosides. *Methods Carbohydr. Chem.* **1980**, *8*, 251–253.
105. Blattner, R.; Ferrier, R. J.; Prasit, P. New Approach to Aminoglycoside Antibiotics. *J. Chem. Soc., Chem. Commun.* **1980**, 944–945.
106. Blattner, R.; Ferrier, R. J. Photobromination of Carbohydrate Derivatives. Part 2. Penta-*O*-acetyl-β-D-glucopyranose; The 5-Bromo Derivative and Products of Further Bromination. *J. Chem. Soc., Perkin Trans. 1* **1980**, 1523–1527.
107. Ferrier, R. J.; Tyler, P. C. Photobromination of Carbohydrate Derivatives. Part 3. C-5 Bromination of Penta-*O*-benzoyl-α- and -β-D-glucopyranose; a Route to D-Xylohexos-5-ulose Derivatives and α-L-Idopyranosides. *J. Chem. Soc., Perkin Trans. 1* **1980**, 1528–1534.
108. Ferrier, R. J.; Tyler, P. C. Photobromination of Carbohydrate Derivatives. Part 4. Observations on Some Glucopyranoside Esters; a Simple Route to Aryl α-L-Idopyranosides. *J. Chem. Soc., Perkin Trans. 1* **1980**, 2762–2767.
109. Ferrier, R. J.; Tyler, P. C. Photobromination of Carbohydrate Derivatives. Part 5. Preparation and Reactions of (5*S*)-1,2,3,4-Tetra-*O*-acetyl-5-bromo-β-D-xylopyranose; a New Type of 'Double-Headed' Nucleoside. *J. Chem. Soc., Perkin Trans. 1* **1980**, 2767–2773.
110. Ferrier, R. J.; Furneaux, R. H. Photobromination of Carbohydrate Derivatives. VI. Functionalization at C6 of 2,3,4-Tri-*O*-acetyl-1,6-anhydro-β-D-glucopyranose. *Aust. J. Chem.* **1980**, *33*, 1025–1036.
111. Ferrier, R. J. *Physical Methods for Structural Analysis. III. Polarimetry*; Academic Press: New York, 1980; pp. 1354–1375.
112. Ferrier, R. J. *Physical Methods for Structural Analysis. VI. X-Ray and Neutron Diffraction*; Academic Press: New York, 1980; pp. 1437–1444.
113. Ferrier, R. J.; Prasit, P. Unsaturated Carbohydrates. 23. The Methoxymercuration of a 6-Deoxyhex-5-enopyranoside Derivative and Reactions of the Products. *Carbohydr. Res.* **1980**, *82*, 263–272.
114. Blattner, R.; Ferrier, R. J.; Tyler, P. C. Unsaturated Carbohydrates. Part 22. Alkenes from 5-Bromohexopyranose Derivatives. *J. Chem. Soc., Perkin Trans. 1* **1980**, 1535–1539.
115. Ferrier, R. J. Unsaturated Sugars. In: Pigman, W., Horton, D., Eds.; *The Carbohydrates,* Vol. 1B, 2nd ed.; Academic Press: New York, 1980; pp 843–879. Chapter 19.
116. Ferrier, R. J.; Prasit, P. A General Route to Optically Pure Prostaglandins from a D-Glucose Derivative. *J. Chem. Soc., Chem. Commun.* **1981**, 983–985.
117. Kennedy, J. F.; Williams, N. R.; Davidson, B.; Morrison, I. M.; Ferrier, R. J.; Sturgeon, C. M.; Sturgeon, R. J. *Carbohydrate Chemistry*, Vol. 12. Specialist Periodical Reports; The Royal Society of Chemistry: London, 1981; 624 pages.
118. Kennedy, J. F.; Williams, N. R.; Davidson, B.; Morrison, I. M.; Ferrier, R. J.; Richardson, A. C.; Sturgeon, C. M.; Sturgeon, R. J. *Carbohydrate Chemistry*, Vol. 13. Specialist Periodical Reports; The Royal Society of Chemistry: London, 1982; 746 pages.
119. Ferrier, R. J.; Furneaux, R. H.; Prasit, P.; Tyler, P. C.; Brown, K. L.; Gainsford, G. J.; Diehl, J. W. Functionalised Carbocycles from Carbohydrates. Part 2. The Synthesis of 3-Oxa-2-azabicyclo[3.3.0]octanes. X-Ray Crystal Structure of (1*R*,5*S*)-6-*exo*,7-*endo*,8-*exo*-Triacetoxy-*N*-methyl-4-*endo*-phenylthio-3-oxa-2-azabicyclo[3.3.0] octane. *J. Chem. Soc., Perkin Trans. 1* **1983**, 1621–1628.
120. Ferrier, R. J.; Prasit, P.; Gainsford, G. J. Functionalized Carbocycles from Carbohydrates. Part 3. The Synthesis of the Epoxy Lactone Prostaglandin Intermediate via an Isoxazolidine Derivative. X-Ray Crystal Structure of (1*R*,5*R*)-6-*exo*,7-*endo*-Bis (benzoyloxy)-8-*exo*-iodo-3-oxo-2-oxabicyclo[3.3.0]octane. *J. Chem. Soc., Perkin Trans. 1* **1983**, 1629–1634.

121. Ferrier, R. J.; Prasit, P.; Gainsford, G. J.; Le, P. Y. Functionalized Carbocycles from Carbohydrates. Part 4. The Synthesis of the Epoxy Lactone Prostaglandin Intermediate via Bicyclo[3.2.0]heptane Derivatives. X-Ray Crystal Structure of (1*R*)-5-*endo*-Acetyl-2-*exo*,3-*endo*-bis(benzoyloxy)bicyclo[2.2.1]heptane. *J. Chem. Soc., Perkin Trans. 1* **1983**, 1635–1640.
122. Ferrier, R. J.; Prasit, P.; Tyler, P. C. Functionalized Carbocycles from Carbohydrates. Part 5. The Synthesis of the Epoxybicyclo[3.2.0]heptanone Ethylene Acetal Prostaglandin Intermediate. *J. Chem. Soc., Perkin Trans. 1* **1983**, 1641–1644.
123. Ferrier, R. J.; Prasit, P. Routes to Prostaglandins from Sugars. *Pure Appl. Chem.* **1983**, *55*, 565–576.
124. Ferrier, R. J.; Prasit, P. Unsaturated Carbohydrates. Part 25. Abbreviated Synthesis of the Insect Pheromone (+)-*exo*-Brevicomin from a Nona-3,8-dienulose Derivative. *J. Chem. Soc., Perkin Trans. 1* **1983**, 1645–1647.
125. Williams, N. R.; Davidson, B.; Ferrier, R. J.; Richardson, A. C. *Carbohydrate Chemistry*, Vol. 14, *Part I: Monosaccharides, disaccharides and specific oligosaccharides*. Specialist Periodical Reports; The Royal Society of Chemistry: London, 1983; 236 pages.
126. Williams, N. R.; Davidson, B.; Ferrier, R. J.; Furneaux, R. H. *Carbohydrate Chemistry*, Vol. 15, *Part I: Monosaccharides, disaccharides and specific oligosaccharides*. Specialist Periodical Reports; The Royal Society of Chemistry: London, 1983; 277 pages.
127. Chew, S.; Ferrier, R. J. Application of Ultrasound in Carbohydrate Chemistry. Synthesis of Optically Pure Functionalized Hexahydro-Anthracenes and -Naphthacenes. *J. Chem. Soc., Chem. Commun.* **1984**, 911–912.
128. Prasad, D.; Prasad, N.; Prasad, R. M.; Ferrier, R. J.; Milgate, S. M. ^{13}C N.M.R. Examination of Ethyl Cyano(arylhydrazono)acetates. *J. Chem. Soc., Perkin Trans. 1* **1984**, 1397–1399.
129. Ferrier, R. J.; Haines, S. R. Functionalized Carbocycles from carbohydrates. Part 6. A Route to Functionalized Cyclopentanes from 6-Deoxyhex-5-enopyranoside Derivatives. *Carbohydr. Res.* **1984**, *130*, 135–146.
130. Ferrier, R. J.; Haines, S. R. Photobromination of Carbohydrate Derivatives. Part 7. Reaction of Furanose Derivatives with Bromine: 4′-Bromo- and 4′-Fluoro-Aldofuranose and -Nucleoside Esters. *J. Chem. Soc., Perkin Trans. 1* **1984**, 1675–1681.
131. Ferrier, R. J.; Haines, S. R.; Gainsford, G. J.; Gabe, E. J. Photobromination of Carbohydrate Derivatives. Part 8. Reaction of Furanose Derivatives with *N*-Bromosuccinimide. X-Ray Molecular Structure of 1-*O*-Acetyl-2,5,6-tri-*O*-benzoyl-4-hydroxy-3,4-*O*-(α-succinimidobenzylidene)-β-D-galactofuranose. *J. Chem. Soc., Perkin Trans. 1* **1984**, 1683–1687.
132. Ferrier, R. J.; Haines, S. R. A Synthesis of 1,2-*trans*-Related Glycofuranosyl Acetates. *Carbohydr. Res.* **1984**, *127*, 157–161.
133. Ferrier, R. J.; Haines, S. R. Unsaturated Carbohydrates. Part 26. Alkenes from 4-Bromohexofuranose Esters. Reactions of 5-Deoxyald-4-enofuranose Derivatives in the Presence of Mercury(II) Ions. *J. Chem. Soc., Perkin Trans. 1* **1984**, 1689–1692.
134. Williams, N. R.; Davidson, B.; Ferrier, R. J.; Furneaux, R. H.; Khan, R. *Carbohydrate Chemistry*, Vol. 16, *Part I: Monosaccharides, disaccharides and specific oligosaccharides*. Specialist Periodical Reports; The Royal Society of Chemistry: London, 1984; 286 pages.
135. Blattner, R.; Ferrier, R. J. Effects of Iron, Copper, and Chromate Ions on the Oxidative Degradation of Cellulose Model Compounds. *Carbohydr. Res.* **1985**, *138*, 73–82.
136. Ferrier, R. J.; Tyler, P. C.; Gainsford, G. J. Functionalized Carbocycles from Carbohydrates. Part 7. A Route to Carbacyclin from a D-Glucose Derivative. X-Ray Crystal Structure of 3-*endo*-Benzoyloxy-2-*exo*-(1,3-diphenyl-1,3,2-diazaphospholan-2-yloxymethyl)-6-oxobicyclo[3.3.0]octane. *J. Chem. Soc., Perkin Trans. 1* **1985**, 295–300.

137. Chew, S.; Ferrier, R. J.; Prasit, P.; Tyler, P. C. Routes to Carbocyclic Compounds of Pharmaceutical Importance from Carbohydrates. In: *Organic Synthesis: An Interdisciplinary Challenge*; Streith, J., Prinzbach, H., Schill, G., Eds.; Blackwell Scientific Inc: Oxford, 1985; pp 247–253.
138. Ferrier, R. J.; Schmidt, P.; Tyler, P. C. Unsaturated Carbohydrates. Part 27. Synthesis of (−)-*exo*-Brevicomin from a Nona-3,8-dienulose Derivative. *J. Chem. Soc., Perkin Trans. 1* **1985**, 301–303.
139. Ferrier, R. J.; Tyler, P. C. Synthesis and Photolysis of Some Carbohydrate 1,6-Dienes. *Carbohydr. Res.* **1985**, *136*, 249–258.
140. Blattner, R.; Ferrier, R. J.; Haines, S. R. Unsaturated Carbohydrates. Part 28. Observations on the Conversion of 6-Deoxyhex-5-enopyranosyl Compounds into 2-Deoxyinosose Derivatives. *J. Chem. Soc., Perkin Trans. 1* **1985**, 2413–2416.
141. Williams, N. R.; Davidson, B.; Ferrier, R. J.; Furneaux, R. H. *Carbohydrate Chemistry*, Vol. 17, *Part I: Monosaccharides, disaccharides and specific oligosaccharides*. Specialist Periodical Reports; The Royal Society of Chemistry: London, 1985; 275 pages.
142. Blattner, R.; Ferrier, R. J. Functionalized Carbocycles From Carbohydrates. Part 9. Direct Synthesis of 6-Oxabicyclo[3.2.1]octane Derivatives from Deoxyinoses. *Carbohydr. Res.* **1986**, *150*, 151–162.
143. Ferrier, R. J.; Davies, R. B.; Edwards, I. R.; Fergusson, D. M.; Reeves, R. D.; Scott, P. J.; Stevenson, C. D. *Lead in the Environment in New Zealand*. The Royal Society of New Zealand Miscellaneous Series 14, The Royal Society of New Zealand: Wellington, 1986, 130 pages.
144. Ferrier, R. J. Lead in the Environment in New Zealand. *Chem. N. Z.* **1986**, 107–111.
145. Williams, N. R.; Davidson, B.; Ferrier, R. J.; Furneaux, R. H.; Wightman, R. H. *Carbohydrate Chemistry*, Vol. 18, *Part I: Monosaccharides, disaccharides and specific oligosaccharides*. Specialist Periodical Reports; The Royal Society of Chemistry: London, 1986; 284 pages.
146. Blattner, R.; Ferrier, R. J.; Renner, R. Chain Extensions from C-1 and C-5 of D-Xylopyranose Derivatives. *J. Chem. Soc., Chem. Commun.* **1987**, 1007–1008.
147. Blattner, R.; Ferrier, R. J. Crystalline Pseudo-α-D-glucopyranose. *J. Chem. Soc., Chem. Commun.* **1987**, 1008–1009.
148. De Raadt, A.; Ferrier, R. J. A Simple, Efficient Synthesis of Vinyl β-D-Glucopyranosides. *J. Chem. Soc., Chem. Commun.* **1987**, 1009–1010.
149. Williams, N. R.; Davidson, B.; Ferrier, R. J.; Furneaux, R. H.; Wightman, R. H. *Carbohydrate Chemistry*, Vol. 19, *Part I: Monosaccharides, disaccharides and specific oligosaccharides*. Specialist Periodical Reports; The Royal Society of Chemistry: London, 1987; 292 pages.
150. Chew, S.; Ferrier, R. J.; Sinnwell, V. Functionalized Carbocycles from Carbohydrates. Part 10. An Approach to the Pyranonaphthoquinones. *Carbohydr. Res.* **1988**, *174*, 161–168.
151. Ferrier., R. J. The Synthesis and Reactions of Monosaccharide Derivatives. In: *Carbohydrate Chemistry*, Kennedy, J. F., Ed.; Oxford University Press: Oxford, 1988; pp 443–499. Chapter 11.
152. Williams, N. R.; Davidson, B.; Ferrier, R. J.; Furneaux, R. H.; Wightman, R. H. *Carbohydrate Chemistry*, Vol. 20, *Part I: Monosaccharides, disaccharides and specific oligosaccharides*. Specialist Periodical Reports; The Royal Society of Chemistry: London, 1988; 301 pages.
153. Ferrier, R. J.; Petersen, P. M.; Taylor, M. A. Radical Cyclization Reactions Leading to Doubly Branched Carbohydrates and 6- and 8-Oxygenated 2,9-Dioxabicyclo[4.3.0]-nonane Derivatives. *J. Chem. Soc., Chem. Commun.* **1989**, 1247–1248.
154. Williams, N. R.; Davidson, B.; Ferrier, R. J.; Furneaux, R. H.; Tyler, P. C.; Wightman, R. H. *Carbohydrate Chemistry*, Vol. 21, *Part I: Monosaccharides, disaccharides*

and specific oligosaccharides. Specialist Periodical Reports; The Royal Society of Chemistry: Cambridge, 1989; 297 pages.
155. Ferrier, R. J.; Stuetz, A. E. Functionalized Carbocycles from Carbohydrates. 11. Two Routes to Enantiomerically Pure 3-Aminoinosose Derivatives. *Carbohydr. Res.* **1990**, *200*, 237–245.
156. Ferrier, R. J. Synthetic Carbohydrate Chemistry of the 1980s. *Carbohydr. Res.* **1990**, *202*, ix–xi.
157. Ferrier, R. J.; Stuetz, A. E. Functionalized Carbocycles from Carbohydrates. Part 12. Hexa-O-benzyl-5-hydroxy-pseudo-α-D-glucopyranose and Its C-5 Epimer. *Carbohydr. Res.* **1990**, *205*, 283–291.
158. Ferrier, R. J.; Petersen, P. M. Unsaturated Carbohydrates. Part 29. Their Application to the Synthesis of Stereospecifically Doubly and Triply Branched Derivatives. *Tetrahedron* **1990**, *46*, 1–11.
159. Ferrier, R. J.; Blattner, R.; Furneaux, R. H.; Tyler, P. C.; Wightman, R. H.; Williams, N. R. *Carbohydrate Chemistry: Monosaccharides, disaccharides and specific oligosaccharides*. Specialist Periodical Reports, Vol. 22; The Royal Society of Chemistry: Cambridge, 1990; 294 pages.
160. Ferrier, R. J.; Lee, C. K.; Wood, T. A. Free Radical Substitutions of Acyloxy Groups in Carbohydrate α-Ketoesters. *J. Chem. Soc., Chem. Commun.* **1991**, 690–691.
161. Somsak, L.; Ferrier, R. J. Radical-Mediated Brominations at Ring Positions of Carbohydrates. *Adv. Carbohydr. Chem. Biochem.* **1991**, *49*, 37–92.
162. De Raadt, A.; Ferrier, R. J. Unsaturated Carbohydrates. Part 30. Syntheses and Reactions of Saturated and 2,3-Unsaturated Vinyl and 1′-Substituted-Vinyl Glycosides. *Carbohydr. Res.* **1991**, *216*, 93–107.
163. Ferrier, R. J.; Blattner, R.; Furneaux, R. H.; Tyler, P. C.; Wightman, R. H.; Williams, N. R. *Carbohydrate Chemistry: Monosaccharides, disaccharides and specific oligosaccharides*. Specialist Periodical Reports, Vol. 23; The Royal Society of Chemistry: Cambridge, 1991; 307 pages.
164. Ferrier, R. J.; Severn, W. B.; Furneaux, R. H.; Miller, I. J. Isotope Studies of the Transfer of the Carbon Atoms of Carbohydrate Derivatives into Aromatic Compounds (Especially Xanthene) Under Degradation Conditions. *Carbohydr. Res.* **1992**, *237*, 87–94.
165. Ferrier, R. J.; Hall, D. W. One-Step Synthesis of Glycosidic Spiroketals from 2,3-Epoxybutyl Glycoside Derivatives. *J. Chem. Soc., Perkin Trans. 1* **1992**, 3029–3034.
166. Ferrier, R. J.; Severn, W. B.; Furneaux, R. H.; Miller, I. J. The Products of the Zinc Chloride-Promoted Decomposition of Cellulose in Aqueous Phenol at 350 °C. *Carbohydr. Res.* **1992**, *237*, 79–86.
167. Ferrier, R. J.; Petersen, P. M. Unsaturated Carbohydrates. Part 31. Trichothecene-Related and Other Branched *C*-Pyranoside Compounds. *J. Chem. Soc., Perkin Trans. 1* **1992**, 2023–2028.
168. Ferrier, R. J.; Blattner, R.; Furneaux, R. H.; Tyler, P. C.; Wightman, R. H.; Williams, N. R. *Carbohydrate Chemistry: Monosaccharides, disaccharides and specific oligosaccharides*. Specialist Periodical Reports, Vol. 24; The Royal Society of Chemistry: Cambridge, 1992; 348 pages.
169. Ferrier, R. J.; Hall, D. W.; Petersen, P. M. Chemical Interaction Between the Aglycon and the Allyl Group of 6-O-Allylhexopyranoside Derivatives. *Carbohydr. Res.* **1993**, *239*, 143–153.
170. Ferrier, R. J.; Middleton, S. The Conversion of Carbohydrate Derivatives into Functionalized Cyclohexanes and Cyclopentanes. *Chem. Rev.* **1993**, *93*, 2779–2831.
171. Ferrier, R. J.; Blattner, R.; Clinch, K.; Furneaux, R. H.; Tyler, P. C.; Wightman, R. H.; Williams, N. R. *Carbohydrate Chemistry: Monosaccharides, disaccharides and specific oligosaccharides*. Specialist Periodical Reports, Vol. 25; The Royal Society of Chemistry: Cambridge, 1993; 369 pages.

172. Blattner, R.; Furneaux, R. H.; Kemmitt, T.; Tyler, P. C.; Ferrier, R. J.; Tiden, A.-K. Syntheses of the Fungicidal and Insecticidal Allosamidin and a Structural Isomer. *J. Chem. Soc., Perkin Trans. 1* **1994**, 3411–3421.
173. Ferrier, R. J.; Blattner, R.; Clinch, K.; Furneaux, R. H.; Gallagher, T. C.; Tyler, P. C.; Wightman, R. H.; Williams, N. R. *Carbohydrate Chemistry: Monosaccharides, disaccharides and specific oligosaccharides*. Specialist Periodical Reports, Vol. 26; The Royal Society of Chemistry: Cambridge, 1994; 356 pages.
174. Collins, P.; Ferrier, R. J. *Monosaccharides: Their Chemistry and Their Roles in Natural Products*; John Wiley & Sons: Chichester, 1995. 574 pages.
175. Ferrier, R. J.; Blattner, R.; Clinch, K.; Furneaux, R. H.; Gallagher, T. C.; Tyler, P. C.; Wightman, R. H.; Williams, N. R. *Carbohydrate Chemistry: Monosaccharides, disaccharides and specific oligosaccharides*. Specialist Periodical Reports, Vol. 27; The Royal Society of Chemistry: Cambridge, 1995; 391 pages.
176. Ferrier, R. J.; Blattner, R.; Clinch, K.; Furneaux, R. H.; Gardiner, J. M.; Tyler, P. C.; Wightman, R. H.; Williams, N. R. *Carbohydrate Chemistry: Monosaccharides, disaccharides and specific oligosaccharides*. Specialist Periodical Reports, Vol. 28; The Royal Society of Chemistry: Cambridge, 1996; 413 pages.
177. Ferrier, R. J. The Conversion of Carbohydrates to Cyclohexane Derivatives. In: *Preparative Carbohydrate Chemistry*; Hanessian, S., Ed.; Marcel Dekker: New York, 1997; pp 569–594. Chapter 26.
178. Ferrier, R. J. Sugars are More Than Sweeteners. Glycoscience—A New Era for Carbohydrates. *Sci. Spectra* **1997**, *9*, 32–39.
179. Ferrier, R. J.; Blattner, R.; Clinch, K.; Furneaux, R. H.; Gardiner, J. M.; Tyler, P. C.; Wightman, R. H. *Carbohydrate Chemistry: Monosaccharides, disaccharides and specific oligosaccharides*. Specialist Periodical Reports, Vol. 29; The Royal Society of Chemistry: Cambridge, 1997; 438 pages.
180. Ferrier, R. J.; Blattner, R.; Clinch, K.; Furneaux, R. H.; Gardiner, J. M.; Tyler, P. C.; Wightman, R. H. *Carbohydrate Chemistry: Monosaccharides, disaccharides and specific oligosaccharides*. Specialist Periodical Reports, Vol. 30; The Royal Society of Chemistry: Cambridge, 1998; 436 pages.
181. Blattner, R.; Ferrier, R. J.; Furneaux, R. H. Intervention of Catalytic Amounts of Water in the Allylic Rearrangements of Glycal Derivatives. *Tetrahedron:Asymmetry* **2000**, *11*, 379–383.
182. Ferrier, R. J.; Holden, S. G.; Gladkikh, O. Proposed Tennis Ball and Basket-and-Lid Routes to C60: Two Relevant C45 Compounds. *J. Chem. Soc., Perkin Trans. 1* **2000**, 3505–3512.
183. Ferrier, R. J.; Blattner, R.; Clinch, K.; Furneaux, R. H.; Gardiner, J. M.; Tyler, P. C.; Wightman, R. H. *Carbohydrate Chemistry: Monosaccharides, disaccharides and specific oligosaccharides*. Specialist Periodical Reports, Vol. 31; The Royal Society of Chemistry: Cambridge, 2000; 417 pages.
184. Ferrier, R. J. Monosaccharides. *Encyclopedia of Life Sciences*; Agrò, A. F., Ed.; John Wiley & Sons: Chichester, 2001. 12 pages.
185. Ferrier, R. J. Substitution-with-Allylic-Rearrangement Reactions of Glycal Derivatives. *Top. Curr. Chem.* **2001**, *215*, 153–175.
186. Ferrier, R. J. Direct Conversion of 5,6-Unsaturated Hexopyranosyl Compounds to Functionalized Cyclohexanones. *Top. Curr. Chem.* **2001**, *215*, 277–291.
187. Ferrier, R. J.; Blattner, R.; Clinch, K.; Field, R. A.; Furneaux, R.H.; Gardiner, J. M.; Kartha, K. P. R.; Tilbrook, D. M. G.; Tyler, P. C.; Wightman, R. H. *Carbohydrate Chemistry: Monosaccharides, disaccharides and specific oligosaccharides*. Specialist Periodical Reports, Vol. 32; The Royal Society of Chemistry: Cambridge, 2001; 433 pages.
188. Ferrier, R. J.; Blattner, R.; Field, R. A.; Furneaux, R.H.; Gardiner, J. M.; Hoberg, J. O.; Kartha, K. P. R.; Tilbrook, D. M. G.; Tyler, P. C.; Wightman, R. H. *Carbohydrate*

Chemistry: Monosaccharides, disaccharides and specific oligosaccharides. Specialist Periodical Reports, Vol. 33; The Royal Society of Chemistry: Cambridge, 2002; 451 pages.

189. Ferrier, R. J.; Hoberg, J. O. Synthesis and Reactions of Unsaturated Sugars. *Adv. Carbohydr. Chem. Biochem.* **2003**, *58*, 55–119.
190. Ferrier, R. J.; Zubkov, O. A. Transformation of Glycals into 2,3-Unsaturated Glycosyl Derivatives. *Org. React.* **2003**, *62*, 569–736.
191. Ferrier, R. J.; Blattner, R.; Field, R. A.; Furneaux, R.H.; Hamilton, C.; Hoberg, J. O.; Kartha, K. P. R.; Tyler, P. C.; Wightman, R. H. *Carbohydrate Chemistry: Monosaccharides, disaccharides and specific oligosaccharides*. Specialist Periodical Reports, Vol. 34; The Royal Society of Chemistry: Cambridge, 2003; 396 pages.
192. Ferrier, R. J. Historical Overview. In: *The Organic Chemistry of Sugars*; Levy, D. E., Fügedi, P., Eds.; CRC Press LLC: Boca Raton, 2006; pp 3–24. Chapter 1.
193. Ferrier, R. J.; Furneaux, R. H. Activation of Sugar Hydroxyl Groups Prior to Glycosylation. *Aust. J. Chem.* **2009**, *62*, 585–589.
194. Ferrier, R. J.; Monteiro, M. A. Gerald Aspinall, 1924–2005. *Adv. Carbohydr. Chem. Biochem.* **2009**, *62*, 2–10.

CHAPTER TWO

Synthetic Approaches to L-Iduronic Acid and L-Idose: Key Building Blocks for the Preparation of Glycosaminoglycan Oligosaccharides

Shifaza Mohamed, Vito Ferro

School of Chemistry and Molecular Biosciences, The University of Queensland, Brisbane, Queensland, Australia

Contents

1. Introduction 23
   1.1 Background 23
2. Epimerization at C-5 of D-Glucose Derivatives 25
   2.1 S_N2 Displacement of Sulfonates 25
   2.2 The Mitsunobu Reaction 32
   2.3 Epimerization via Generation of a C-5 Radical 33
3. Homologation of Tetroses and Pentoses 36
   3.1 The Mukaiyama-Type Aldol Reaction 36
   3.2 Diastereoselective Cyanohydrin Formation 36
   3.3 Addition of Organometallic Reagents 40
   3.4 Homologation Using 2-(Trimethylsilyl)thiazole 41
4. Isomerization of Unsaturated Sugars 42
   4.1 Diastereoselective Hydroboration of exo-Glycals 42
   4.2 From Δ^4-Uronates 46
   4.3 From 4-Deoxypentenosides 47
   4.4 From D-Glucuronic Acid Glycal 47
5. Miscellaneous Methods 48
   5.1 Diastereoselective Tishchenko Reaction 48
   5.2 Homologation with 5,6-Dihydro-1,4-dithiin-2-yl[(4-methoxybenzyl)oxy] methane 49
   5.3 C–H Activation of 6-Deoxy-L-hexoses 50
6. Conclusions 52
Acknowledgments 52
References 52

Advances in Carbohydrate Chemistry and Biochemistry, Volume 72
ISSN 0065-2318
http://dx.doi.org/10.1016/bs.accb.2015.07.001

ABBREVIATIONS

Ac_2O acetic anhydride
AllBr allyl bromide
9-BBN 9-borabicyclo[3.3.1]nonane
BTI bis(trifluoroacetoxy)iodobenzene
β-CD β-cylcodextrin, cyclomaltoheptaose
CSA camphor-10-sulfonic acid
***m*-CPBA** 3-chloroperoxybenzoic acid
DBU 1,8-diazabicyclo[5.4.0]undec-7-ene
DCE dichloroethane
DCM dichloromethane
DDQ 2,3-dichloro-5,6-dicyano-*p*-benzoquinone
DEAD diethyl azodicarboxylate
DIBAL-H diisobutylaluminum hydride
DIPC *N,N*-diisopropyl carbodiimide
DIPEA *N,N*-diisopropylethylamine
DMAP 4-(dimethylamino)pyridine
DMDO dimethyldioxirane
DMF *N,N*-dimethylformamide
DMF·DMA *N,N*-dimethylformamide dimethyl acetal
DMFDNpA *N,N*-dimethylformamide dineopentyl acetal
DMS dimethyl sulfide, methyl sulfide $(Me)_2S$
DMSO dimethyl sulfoxide
DS dermatan sulfate
FmocCl 9-fluorenylmethoxycarbonyl chloride
GAG glycosaminoglycan
GlcA D-glucuronic acid
HS heparan sulfate
IBCF isobutyl chloroformate
IdoA L-iduronic acid
LevOH levulinic acid
LevONa sodium levulinate
Me_4Phen 3,4,7,8-tetramethyl-1,10-phenanthroline
MeCN acetonitrile
NaOPiv sodium pivaloate
NapBr (2-napthyl)methylbromide
NBS *N*-bromosuccinimide
NIS *N*-iodosuccinimide
PivCl pivaloyl chloride
PMB 4-methoxybenzyl ether
TBAF tetra-*n*-butylammonium fluoride
TBAI tetra-*n*-butylammonium iodide
TBDMS *tert*-butyldimethylsilyl
TBDMSOTf *tert*-butyldimethylsilyl trifluoromethanesulfonate
TBDPS *tert*-butyldiphenylsilyl
TEMPO 2,2,6,6-tetramethyl-1-piperidinyloxy
TES triethylsilyl

Tf_2O trifluoromethanesulfonic anhydride
TFA trifluoroacetic acid
TFDO methyl(trifluoromethyl)dioxirane
TfOH trifluoromethanesulfonic acid
THF tetrahydrofuran, oxolane
$TIPDSiCl_2$ 1,3-dichloro-1,1,3,3-tetraisopropyldisiloxane
TMS trimethylsilyl
TrCl triphenylmethyl chloride
TsOH *p*-toluenesulfonic acid
UHP anhydrous urea–hydrogen peroxide complex

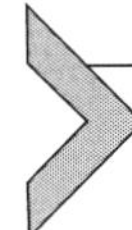

1. INTRODUCTION

1.1 Background

L-Iduronic acid (IdoA) is an important monosaccharide component of the glycosaminoglycans (GAGs) heparin, heparan sulfate (HS), and dermatan sulfate (DS). GAGs are complex, highly sulfated polysaccharides that are composed of repeating disaccharide subunits of L-iduronic acid, (IdoA) or D-glucuronic acid (GlcA), (1→4)-linked to D-glucosamine (for heparin/HS),[1–3] or (1→3)-linked to *N*-acetyl-D-galactosamine (for DS) (Fig. 1).[4,5] GAGs interact with a multitude of diverse proteins such as growth factors, enzymes, morphogens, cell-adhesion molecules, and cytokines[4–9] and in so doing play crucial roles in normal physiological processes—such as embryogenesis and cell adhesion, as well as pathological conditions—including tumor growth, angiogenesis, metastasis, and viral infection.[4–9] These interactions are often facilitated by the remarkable conformational flexibility of IdoA residues, which allow the normally linear GAG chains to kink or bend to better accommodate binding to the protein.[10,11] IdoA can adopt ${}^1C_4$ and ${}^4C_1$ chair conformations as well as the ${}^2S_O$ skew-boat conformation (Fig. 2),[12–14] depending on the adjacent residues in the GAG chain. Heparin itself has been used as an anticoagulant drug for decades,[15] while low molecular weight heparins and fondaparinux, the methyl glycoside of a specific heparin pentasaccharide sequence that binds to antithrombin, are also important anticoagulant drugs.[15,16] The development of new therapeutics for a range of diseases based on inhibition of GAG–protein interactions by specific GAG oligosaccharides or their mimetics is also an area of great current interest.[17–19]

Despite their biomedical importance, the structure–activity relationships of GAGs are poorly understood, owing to their chemical complexity.

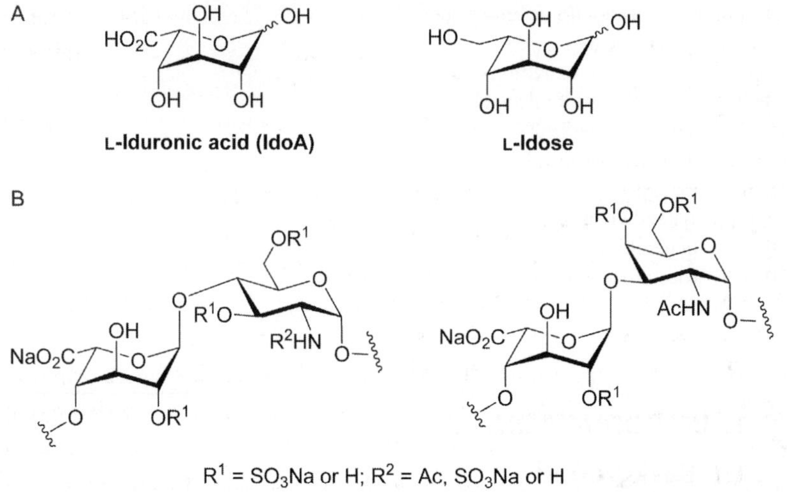

A

L-Iduronic acid (IdoA)

L-Idose

B

R^1 = SO_3Na or H; R^2 = Ac, SO_3Na or H

Figure 1 (A) Structures of L-iduronic acid (IdoA) and L-idose. (B) Structures of the IdoA-containing disaccharide repeating units of HS and heparin (*left*) and DS (*right*). The major disaccharide repeating units containing GlcA are not shown. (Note: IdoA and L-idose are depicted in the $^1C_4$ conformation following common usage in the field. This does not necessarily represent the conformation in solution, which in most cases is an equilibrium between the $^1C_4$ and ^2S_O forms. Throughout this chapter, where there is a doubt or little available evidence about the conformation of an IdoA derivative, the structure will be depicted using the Mills projection.)

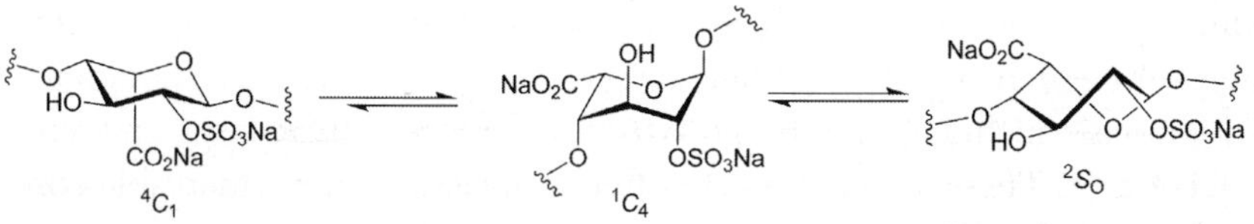

$^4C_1$ $^1C_4$ ^2S_O

Figure 2 Conformational flexibility of 2-*O*-sulfo IdoA residues in GAG oligosaccharides.

Effective structure–activity studies require access to pure, chemically well-defined GAG fragments, which are difficult to obtain from natural sources. Thus, efficient and versatile synthetic protocols are required to gain excess to such GAG oligosaccharides of various lengths with known sulfation patterns. The synthesis of GAG oligosaccharides is the subject of intensive ongoing research and has been reviewed extensively,[20–24] including more recent reviews which cover the latest developments in the past decade or so.[25–28] The main challenge in the synthesis of GAG oligosaccharides is the efficient gram-scale preparation of IdoA building blocks, since neither IdoA nor L-idose is commercially available or readily accessible from natural sources and must therefore be synthesized. Since the early 1980s, there has been much interest in the synthesis of IdoA building blocks and their subsequent incorporation into well-defined GAG oligosaccharides. A range of methodologies for their preparation have been developed, each with its own advantages and disadvantages.[29–32]

Among these, a common approach involves the inversion of configuration at C-5 of the more abundant D-glucose, which differs from L-idose only by the stereochemistry at this carbon. Although some short and efficient routes have been reported recently, the ready availability of adequate amounts of IdoA building blocks still remains one of the major obstacles encountered in the synthesis of GAG oligosaccharides. In this review, the different synthetic approaches for the preparation of IdoA and its derivatives, including L-idose, will be presented and discussed. Derivatives of the latter are often used in GAG synthesis and are elaborated to IdoA via selective oxidation at C-6 after incorporation into a GAG oligosaccharide. Particular focus will be given to the preparation of IdoA building blocks most commonly used for GAG oligosaccharide synthesis, and on the progress made since the last systematic review in this area in 2002.[29] References to some examples of the use of these building blocks in GAG synthesis will be given, but for more examples, the reader is referred to the earlier reviews.[20–28]

2. EPIMERIZATION AT C-5 OF D-GLUCOSE DERIVATIVES

2.1 S_N2 Displacement of Sulfonates

One of the more common strategies to convert D-gluco derivatives into L-ido derivatives is to invert the configuration at C-5 via nucleophilic displacement of a sulfonate leaving group with an oxygen nucleophile. This displacement has been commonly performed[29] on both 1,2-*O*-isopropylidene-α-D-glucofuranurono-6,3-lactone (**1**) and 1,2:5,6-di-*O*-isopropylidene-α-D-glucofuranose (**17**).

Among several routes for this transformation, starting from the commercially available D-glucuronolactone (**1**) provides one of the shortest routes to L-configured sugars. Early attempts to displace the 5-*O*-tosyl derivative of **1** resulted in extensive decomposition[33] or necessitated the initial reduction of the lactone and protection of the resultant lactol as a benzoate prior to displacement, ultimately resulting in unsatisfactory overall yields.[33] Attempted use of the Mitsunobu reaction also led to intractable mixtures. In 1980, Weidmann and coworkers[34] subsequently demonstrated that the 5-*O*-triflyl derivative **2** could be readily displaced by sodium trifluoroacetate in DMF at room temperature. The resultant trifluoroacetate underwent rapid methanolysis to furnish L-iduronolactone **3a** in an overall yield of 78% (Scheme 1A). The triflate could also be cleanly displaced with sodium benzoate; however, deprotection of the resultant benzoate was plagued by eliminations and lactone ring opening. A recent study found that the 5-*O*-mesyl derivative of **1** was unreactive toward sodium trifluoroacetate, while attempts to make the 5-*O*-imidazol-1-yl derivative resulted in formation of a chlorodeoxy compound.[35] The triflate thus remains the leaving group of choice in this strategy.

Linhardt and coworkers reported the synthesis of **3a** in high yield from **1** by a slight modification of the procedure of Weidmann and associates[34] and its subsequent conversion into IdoA synthon **8** (Scheme 1B).[36,37] Lactone **3a** was treated with ethanethiol and concentrated hydrochloric acid resulting in simultaneous removal of the isopropylidene group and C-1 thioacetalization to give dithioacetal **4** in quantitative yield. This was followed by regioselective formation of the 2,4-*O*-*p*-methoxybenzylidene derivative **5** under standard conditions. The lactone ring of **5** was then opened with methylamine in THF to give the L-iduronamide in 91% yield. The formation of the amide was necessary to prevent reformation of the lactone ring, which occurred during attempts to convert **5** into the corresponding methyl ester by methanolysis. Regioselective 5-O-silylation and subsequent 3-O-benzoylation led to the fully protected amide **6**. Regioselective reductive ring opening of the 2,4-*O*-*p*-methoxybenzylidene group with sodium cyanoborohydride–trimethylsilyl chloride, followed by 4-O-benzylation and desilylation, gave alcohol **7**. Finally, deprotection of the dithioacetal group and subsequent pyranose ring closure, followed by acetylation, gave the L-iduronamide **8** in 89% yield.[36] There have been no reports of the use of **8** in GAG oligosaccharide synthesis.

Lassaletta and coworkers[38] also utilized triflate **2** but performed the displacement with sodium pivaloate to give the L-iduronolactone **3b** in 94% yield. Taking advantage of the greater base stability of the pivaloate ester, the selective lactone ring opening was performed with methanolic 1% Et_3N at low temperature. The crude intermediate **9** was immediately benzylated under acidic conditions to prevent lactone ring reformation to give compound **10**. Protection of the 3-OH group was also crucial to prevent reformation of the lactone ring during pyranose ring formation. This is also the limiting step of this strategy as the overall yield over the two steps is 54%. Successive removal of the pivaloyl group (and remethylation of the carboxylate), followed by hydrolysis of the 1,2-*O*-isopropylidene acetal and

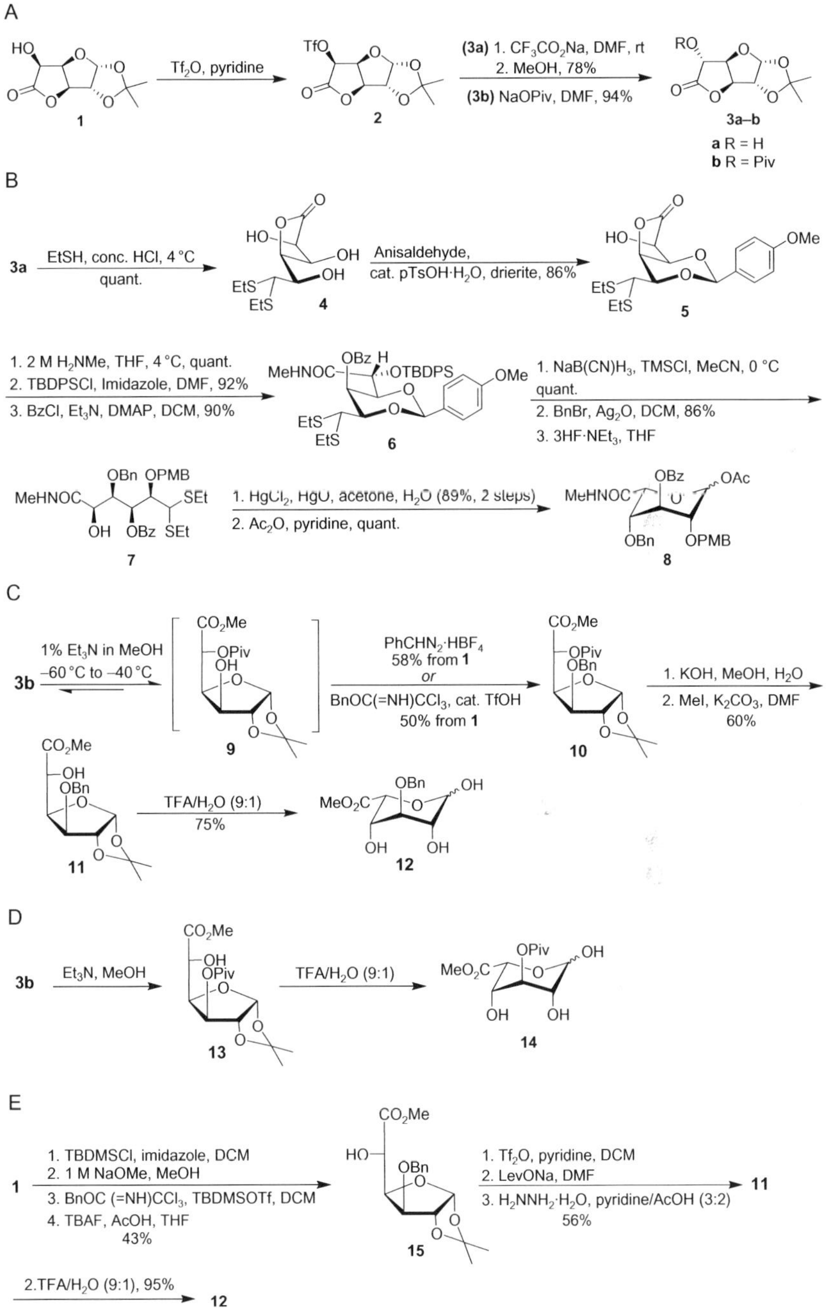

A

1 → Tf_2O, pyridine → 2

2 → (3a) 1. CF_3CO_2Na, DMF, rt 2. MeOH, 78% / (3b) NaOPiv, DMF, 94% → 3a–b

a R = H
b R = Piv

B

3a → EtSH, conc. HCl, 4 °C, quant. → 4

4 → Anisaldehyde, cat. pTsOH·H_2O, drierite, 86% → 5

5 → 1. 2 M H_2NMe, THF, 4 °C, quant. 2. TBDPSCl, Imidazole, DMF, 92% 3. BzCl, Et_3N, DMAP, DCM, 90% → 6

6 → 1. $NaB(CN)H_3$, TMSCl, MeCN, 0 °C quant. 2. BnBr, Ag_2O, DCM, 86% 3. $3HF \cdot NEt_3$, THF → 7

7 → 1. $HgCl_2$, HgO, acetone, H_2O (89%, 2 steps) 2. Ac_2O, pyridine, quant. → 8

C

3b ⇌ 1% Et_3N in MeOH, −60 °C to −40 °C ⇌ [9]

9 → $PhCHN_2 \cdot HBF_4$ 58% from **1** *or* $BnOC(=NH)CCl_3$, cat. TfOH 50% from **1** → 10

10 → 1. KOH, MeOH, H_2O 2. MeI, K_2CO_3, DMF 60% → 11

11 → TFA/H_2O (9:1) 75% → 12

D

3b → Et_3N, MeOH → 13

13 → TFA/H_2O (9:1) → 14

E

1 → 1. TBDMSCl, imidazole, DCM 2. 1 M NaOMe, MeOH 3. BnOC (=NH)CCl_3, TBDMSOTf, DCM 4. TBAF, AcOH, THF 43% → 15

15 → 1. Tf_2O, pyridine, DCM 2. LevONa, DMF 3. $H_2NNH_2 \cdot H_2O$, pyridine/AcOH (3:2) 56% → 11

→ 2.TFA/H_2O (9:1), 95% → 12

Scheme 1 Epimerization of D-glucuronolactone via displacement of a C-5 triflate.

concomitant pyranose formation, gave the known[39] IdoA derivative **12** in moderate overall yield. This method has since been utilized for the synthesis of GAG di- and oligosaccharides.[40–42]

Whitfield and coworkers[43,44] also used a similar strategy for the synthesis of IdoA building blocks, preparing lactone **3b** following the procedure of Lassaletta and coworkers[38] (Scheme 1D). Taking advantage of a 5-O- to 3-O-migration side reaction noted earlier,[38] treatment of **3b** with Et_3N at 0 °C resulted in migration of the pivaloyl group to give **13**, thus eliminating lactone ring reformation. Acid-catalyzed hydrolysis then furnished IdoA derivative **14**, which was utilized in the synthesis of heparin di- and trisaccharides. This route is more convenient than the previous method since it does not require the (low-yielding) benzyl ether protection of the 3-OH group under acidic conditions.

More recently, Suda and coworkers[45] also started from lactone **1** but carried out the displacement of a triflate group at a later stage of the synthetic sequence (Scheme 1E). In this approach, lactone **1** was 5-O-silylated and then subjected to lactone ring opening with sodium methoxide in methanol. Lewis acid-catalyzed 3-O-benzylation and subsequent removal of the silyl protecting group with TBAF then gave the known[46] intermediate **15** in 43% overall yield. Conversion of **15** to the IdoA derivative **12** followed the sequence reported earlier by Seeberger and associates (see below)[46] via formation of the triflate, displacement with sodium levulinate with inversion of configuration, and removal of the levulinate group by treatment with hydrazine hydrate to give the protected L-idofuranose **11** in 56% overall yield. Acid hydrolysis of the isopropylidene acetal then afforded **12** in 95% yield, which was subsequently incorporated into HS disaccharide building blocks.

In 1984, Jacquinet et al.[39] developed a synthetic strategy for the synthesis IdoA derivatives starting from 1,2:5,6-di-*O*-isopropylidene-α-D-glucofuranose (**17**) (Scheme 2A). First, 3-O-benzylation of **17**, followed by chemoselective hydrolysis of the 5,6-*O*-isopropylidene acetal, gave diol **18** in high yield. Treatment of **18** with trityl chloride, followed by acetic anhydride in pyridine, afforded the fully protected intermediate **19**. Compound **19** was then subjected to Jones oxidation to obtain the corresponding acid, which was esterified with diazomethane to give the methyl ester **15** in 51% overall yield. In order to achieve epimerization at C-5, **15** was converted into its triflate derivative, which was then subjected to nucleophilic S_N2 displacement with sodium trifluoroacetate and methanolysis to give **11**. Subsequent acid hydrolysis of the isopropylidene acetal, followed by acetylation, led in 66% yield to the IdoA derivative **20**, which

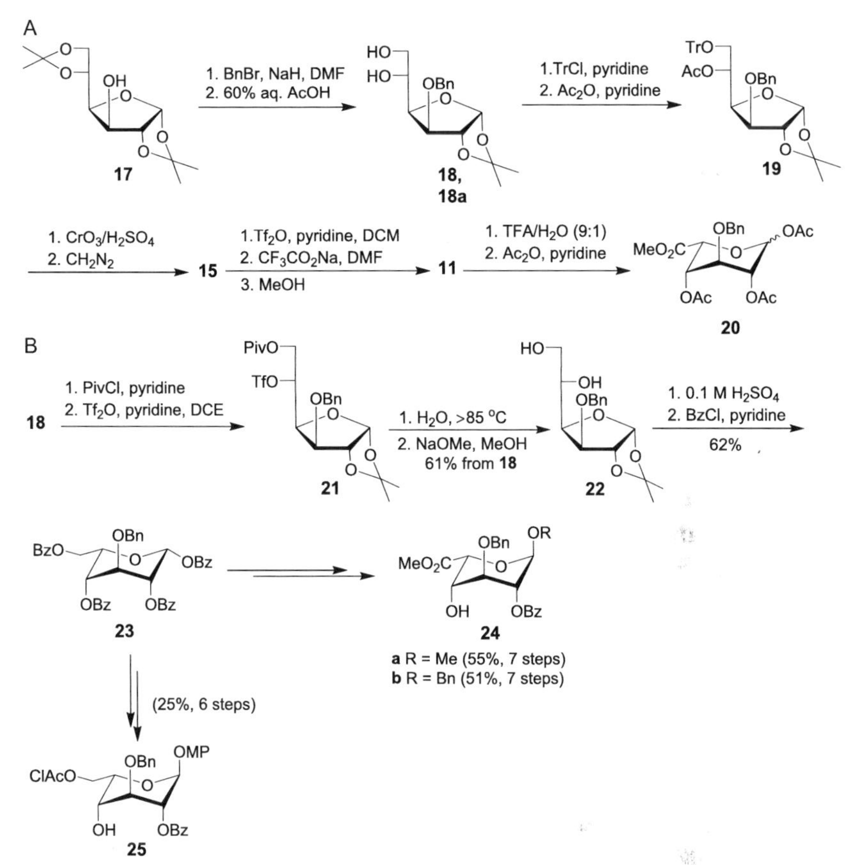

Scheme 2 Epimerization of 1,2-*O*-isopropylidene-α-D-glucofuranose derivatives via displacement of a C-5 triflate.

was used for the first synthesis of the antithrombin III-binding heparin pentasaccharide.[47] This route to **11** was followed by Suda et al.[48] for the synthesis of a heparin disaccharide and by Borbás and coworkers[49,50] who utilized the 3-*O*-methyl analogue of **15** for the preparation of methylated GAG disaccharide analogues. In 2003, Seeberger and coworkers[46] also followed the Jacquinet et al. procedure to intermediate **15**, but then carried out the displacement of the triflate with sodium levulinate. The resultant levulinate ester was then converted into **12** by treatment with hydrazine, followed by acid hydrolysis of the isopropylidene acetal. This sequence was later followed by Suda and coworkers[45] (see Scheme 1E).

In 2000, Barroca and Jacquinet[51] reported a variation to the earlier synthetic strategy. Regioselective pivaloylation at C-6 of diol **18** allowed direct formation of the triflate at C-5 to give compound **21** without prior protection and deprotection of C-5 (Scheme 2B). Treatment with water at 85 °C resulted in epimerization at C-5 to give the L-idofuranose derivative **22**. Subsequent acid hydrolysis, followed by benzoylation, led in 62% yield to L-idopyranoside **23**,[51] which was transformed via a series of steps into the IdoA glycosyl acceptor **24** used for the synthesis of a dermatan sulfate disaccharide. Schwörer et al.[52] prepared **23** via this route before transforming it into an alternative glycosyl acceptor **25** and incorporating it into the synthesis of HS hexa- to dodecasaccharides.

The main drawback of the previously discussed strategies comes in the final cyclization step to form the L-idopyranose derivative(s), which invariably results in the formation of mixtures of pyranose and furanose products. These mixtures often require tedious chromatographic separation in order to avoid complications in the succeeding steps, resulting in poor yields of the desired pyranose product(s). This also makes these routes challenging for large-scale synthesis of IdoA derivatives.

In 1985, van Boeckel et al. reported an alternative synthesis of IdoA derivatives by inversion of configuration at C-5 of diol **18a**.[53] This was achieved by first forming the 5,6-dimesylate, followed by a nucleophilic displacement of the C-6 mesylate with potassium acetate in the presence of a crown ether, to yield **26a** (Scheme 3). Treatment of **26a** with potassium *tert*-butoxide in *tert*-butanol furnished the L-ido-configured epoxide **27a** in excellent yield. The formation of **27a** takes place through formation of the C-6 alkoxide and subsequent intramolecular displacement of the C-5 mesylate with inversion of configuration. Treatment of epoxide **27a** with dilute sulfuric acid results in concomitant hydrolysis of the 1,2-*O*-isopropylidene acetal and opening of the 5,6-epoxide, leading to spontaneous pyranose ring formation. Upon direct acetylation, the acetate **28a** was isolated as a mixture of anomers, which was then converted into a glycosyl donor for incorporation into heparin oligosaccharides, with oxidation to IdoA occurring post-glycosylation.[53] This route involves simple protecting-group manipulations, which proceed in excellent yield with little by-product formation and chromatographic purification, thus making it easily scalable. For these reasons, van Boeckel et al.'s method utilizing **18a** or its congener **18b** has been employed extensively for obtaining gram-scale quantities of IdoA derivatives,[54–65] including those for commercial scale production of the anticoagulant drug fondaparinux.[66]

In 2004, Hung and coworkers reported a variation with fewer synthetic steps of the route developed by van Boeckel and coworkers to obtain the

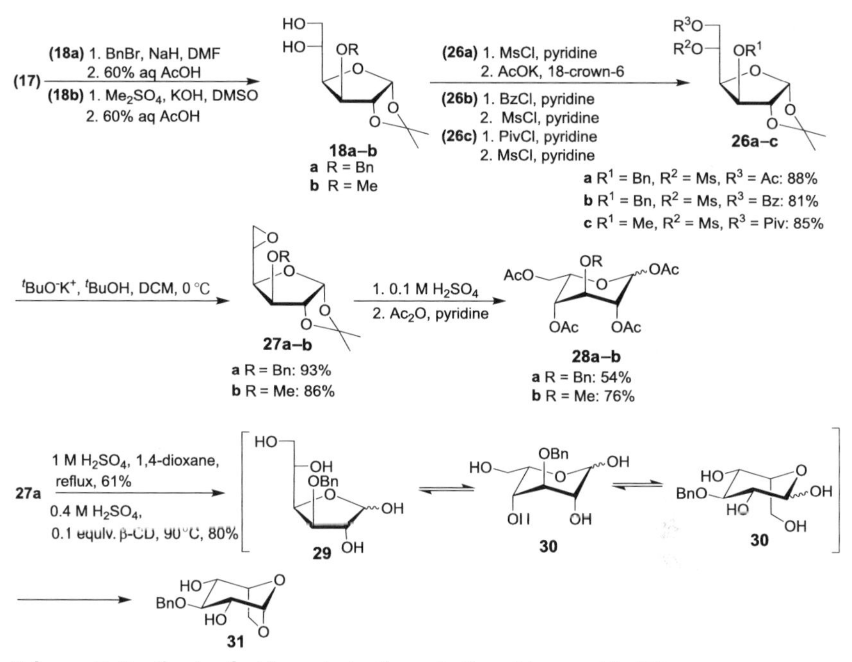

Scheme 3 Synthesis of L-idose derivatives via the L-ido-epoxide **27**.

epoxide **27a**.[67] This involved a regioselective, one-pot benzoylation–mesylation protocol to give the corresponding furanose **26b** as a single isomer in 81% yield. The one-pot synthesis of **26b** was then followed by the formation of epoxide **27a** under the same conditions reported by van Boeckel and coworkers. Furthermore, Hung and coworkers reported the synthesis of 1,6-anhydrosugar **31** from **27a**, which was also independently reported by Fügedi and coworkers.[60,68,69] The preparation of **31** proceeds through a combination of epoxide ring opening and isopropylidene acetal cleavage, leading to an equilibrium between the L-idofuranose **29** and the L-idopyranose form **30**, present as both the $^1C_4$ and the $^4C_1$ conformers. Finally, the elimination of a water molecule at reflux temperature yields **31**. The 1,6-anhydrosugar **31** offers distinct advantages over the pyranose form of L-idose because (1) it eliminates the formation of an anomeric mixture of L-pyranosides, thus avoiding the complicated isomer separation; (2) **31** is crystalline, and thus a simple recrystallization can be used for purification; (3) the number of hydroxyl groups that need to be protected is reduced; (4) the rigid [3.2.1] bicyclic structure forces the 2-, 3-, and 4-OH groups to assume equatorial positions, increasing their reactivity; and (5) the anhydro ring can be readily cleaved and functionalized further as desired.[27] In 2012,

Hung and coworkers further refined this synthetic strategy and reported a one-pot preparation of **31** from diol **18a**, thus obtaining the L-ido sugar from 1,2:5,6-di-*O*-isopropylidene-α-D-glucofuranose (**17**) in only four purification stages and in 43% overall yield.[70] This method has been employed extensively for the preparation of IdoA building blocks for use in the assembly of various GAG oligosaccharides.[60,65,71–81]

Although the 1,6-anhydrosugar method involves fewer steps than the one developed by van Boeckel and coworkers, on a multigram scale, the synthesis of **31** can be problematic. This is because at larger scale, the 6-O-benzoylation is not regioselective, leading to both 6-O- and 5-O-benzoylated products, which have to be separated via chromatography in order to avoid complications later in the synthesis. Furthermore, larger scale preparation of **31** is troublesome due to carbonization of the substrate **27a**, thus considerably lowering the yield. To overcome the issue of selectivity, the much bulkier pivaloyl-protecting group was used instead of benzoyl to obtain compound **26c** in a one-pot process. In this way, **27b** was obtained from **18b** in 86% yield over 5 steps.[82] In order to increase the yield of **31** from **27a**, several phase-transfer catalysts and various acids were screened in order to identify a system that could reduce the carbonization of the substrate **27a**.[83] From this investigation, it was found that the use of cyclomaltoheptaose (β-cyclodextrin) as a phase-transfer catalyst considerably improved the yield of the hydrolysis of **27a** to form **31**, when the reaction was done in 0.4 M H_2SO_4/EtOH as solvent. Thus, this method is a good supplement to Hung and coworkers' preparation of **31**.[83]

2.2 The Mitsunobu Reaction

In 2000, Ikegami and coworkers[84] reported a novel conversion of the D-glucono-1,5-lactone **32a**, derived from D-glucose,[85] into the corresponding L-idose derivative using the Mitsunobu reaction as the key step (Scheme 4). Lactone **32a** was first converted into the δ-hydroxyalkoxamate **33a** by treatment with *O*-benzylhydroxylamine and Me_3Al. Intramolecular cyclization of **33a** under Mitsunobu conditions resulted in O-alkylation leading to compound **35a** with complete inversion of stereochemistry at C-5. The cyclization of δ-hydroxyalkoxamate **33a** is presumably the result of the deprotonation of the amide NH by the reduced diethyl azodicarboxylate (DEAD) anion present in the Mitsunobu reaction mixture, leading to the formation of intermediates **34a** and **34b**. Compound **35a** was converted into the L-idopyranose **37a** in excellent overall yield via acid-catalyzed

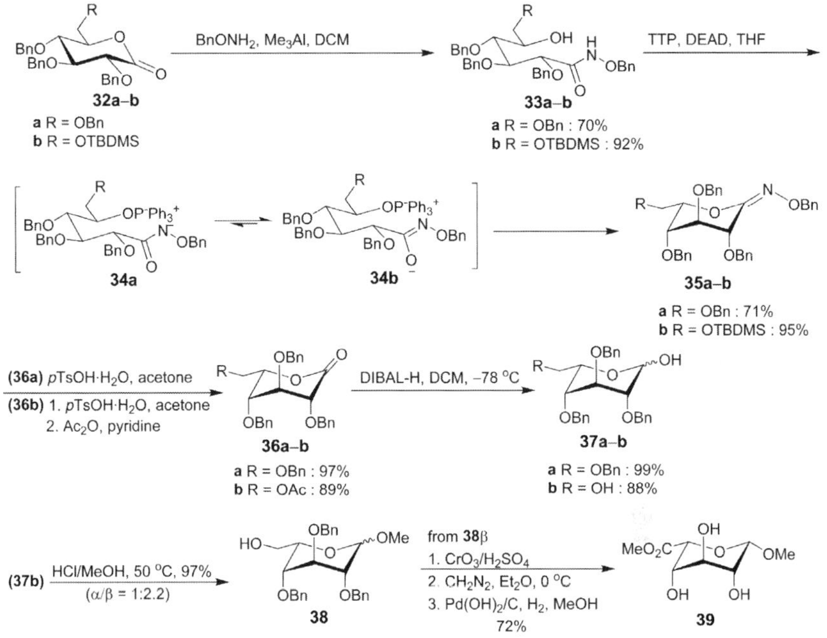

Scheme 4 Conversion of D-glucono-1,5-lactone into IdoA derivatives using the Mitsunobu reaction.

hydrolysis of the oxime to give lactone **36a**, followed by reduction with excess DIBAL-H. Subsequently, the 6-*O-tert*-butyldimethylsilyl (TBDMS) derivative **32b** was utilized in order to allow for selective removal and oxidation to provide access to the IdoA derivative.[86] Acid-catalyzed hydrolysis also removed the TBDMS-protecting group and was followed by acetylation and reduction with DIBAL-H to give **37b**. Subsequently, Fischer glycosidation gave the partially benzyl-protected methyl idoside **38** in 97% yield as a mixture of anomers. Jones oxidation at C-6 of the β-anomer, followed by methylation of the carboxylate and hydrogenolysis of benzyl ether-protecting groups, gave the known[87] IdoA derivative **39** in 72% yield. A drawback of this method is the lack of ready access to lactone **32b**. Consequently, there have thus been no reports of the use of this route for GAG oligosaccharide synthesis.

2.3 Epimerization via Generation of a C-5 Radical

Early attempts at base-promoted epimerization of D-glucuronic acid derivatives resulted in low yields due to degradation and competing β-elimination.[88,89] Thus in 1986 Chiba and Sinaÿ reported on investigations into the synthesis of IdoA derivatives using a radical reduction as a means for C-5 epimerization (Scheme 5A).[90] Commercially available lactone **1** was readily transformed in two steps into the peracetate **40a**. Free-radical bromination[91] of **40a** using *N*-bromosuccinimide (NBS) and UV light in refluxing carbon tetrachloride gave the known,[92] crystalline C-5 bromo derivative **41a**. This is a homolytic process in which the methoxycarbonyl group at C-5 provides stabilization of the intermediate radical leading to the selective bromination of C-5. Tributyltin hydride-mediated reduction of **41a** in toluene at reflux gave the L-idopyranuronate **42a** in 27% yield, along with the D-glucopyranuronate (63%) as the major product. When the β-anomeric substituent was changed from acetyl to methyl, reduction of the C-5 bromo derivative gave an improvement in selectivity to 38% L-ido versus 44.5% D-gluco isomer. This method was subsequently used in the synthesis of HS and DS di- and oligosaccharides.[93–95] Medaković[96] and more recently Wong and coworkers[97] reported that the ratio of L-ido to D-gluco products could be improved to 3:1 when the α-acetate **40b** is used as the starting material and the reduction was carried out at lower temperature in refluxing benzene on the derived bromo compound **41b** to give **42b**. Wong and coworkers also reported a ratio of 2:1 when carrying out the reduction on the β-acetate **41a** under the same conditions.[97]

Glycosyl fluoride **43** has also been used as a substrate for the tributyltin hydride-mediated reduction.[98,99] Interestingly, when carried out in benzene at reflux the reaction led exclusively to the L-idopyranosyl fluoride **44** in

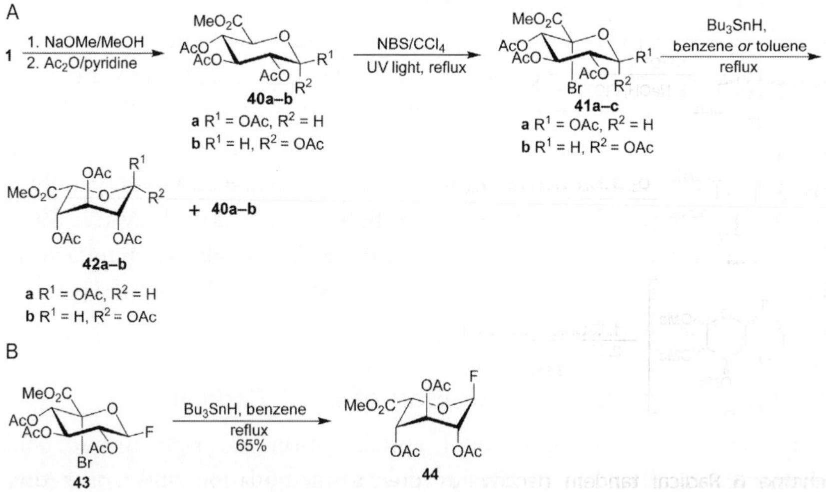

Scheme 5 Synthesis of IdoA derivatives via radical reduction of C-5 bromo compounds.

three isolation steps with an overall yield of 72% (Scheme 5B).[99] Glycosyl fluoride **44** was then utilized as a glycosyl donor for the preparation of fluorogenic substrates for α-L-iduronidase.[98,99] While the improved selectivity in the reduction is a major advance, the above routes are not ideal for multigram-scale synthesis of IdoA derivatives. The major drawbacks include the need for carcinogenic and scarce CCl_4 as the solvent for the free-radical bromination, and the use of toxic tributyltin hydride, which requires tedious chromatography to remove traces of tin by-products. The glycosyl fluoride route also calls for the use of expensive silver fluoride.

Chapleur and coworkers recently reported a new stereoselective strategy to obtain orthogonally protected L-ido derivatives from their D-gluco counterparts using a tandem process involving the formation of a radical at C-5 (Scheme 6).[100] The basis of this method is a kinetically favored 5-exo dig cyclization between a radical generated at C-5 and a suitable tether at O-4 leading exclusively to a 4,5-cis-fused ring system with the L-ido configuration. Methyl α-D-glucopyranoside (**45**) was converted to the 4,6-diol **46** in good yield on a multigram scale via standard transformations. The diol **46** was then subjected to chemoselective oxidation with TEMPO and subsequent Fischer esterification to give the methyl ester **47**. This was followed by the installation of an ethoxypropyne group at C-4 by a transacetalation

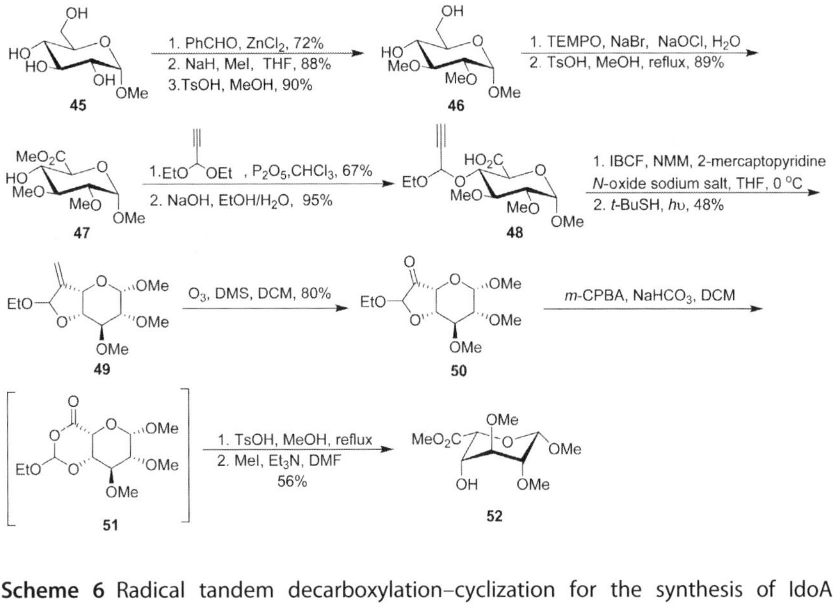

1. PhCHO, $ZnCl_2$, 72%
2. NaH, MeI, THF, 88%
3. TsOH, MeOH, 90%

1. TEMPO, NaBr, NaOCl, H_2O
2. TsOH, MeOH, reflux, 89%

1. EtO OEt, P_2O_5, $CHCl_3$, 67%
2. NaOH, EtOH/H_2O, 95%

1. IBCF, NMM, 2-mercaptopyridine *N*-oxide sodium salt, THF, 0 °C
2. *t*-BuSH, *h*υ, 48%

O_3, DMS, DCM, 80%

m-CPBA, $NaHCO_3$, DCM

1. TsOH, MeOH, reflux
2. MeI, Et_3N, DMF
56%

Scheme 6 Radical tandem decarboxylation–cyclization for the synthesis of IdoA derivatives.

reaction under acidic conditions, followed by saponification to give carboxylic acid **48**. A radical tandem decarboxylation–cyclization via the Barton ester was then performed resulting in the fused-ring compound **49** in 48% yield. The configuration of the product could not be ascertained by NMR spectroscopy due to distortion of the carbohydrate ring, but was confirmed as L-ido by X-ray crystallographic analysis of a derivative. Subsequent ozonolysis of **49** led to ketone **50** in 80% yield, which was then subjected to Baeyer–Villiger oxidation using *m*-CPBA and sodium bicarbonate in DCM, resulting in the unstable lactone **51**, which upon heating in acidic methanol, followed by methylation, gave the IdoA derivative **52**. This route is limited by the moderate yield of the radical decarboxylation–cyclization step. Modifications are also required for the methyl ether-protecting groups and the methyl glycoside for use as a building block for GAG oligosaccharide synthesis.

3. HOMOLOGATION OF TETROSES AND PENTOSES

3.1 The Mukaiyama-Type Aldol Reaction

L-Arabinose (**53**) has been used by Seeberger and coworkers as the starting material for homologation to an IdoA glycosyl donor via cyclization of a suitably protected dithioacetal **56b** (Scheme 7A).[101] The latter was obtained through a Mukaiyama-type aldol reaction between a protected, dithioacetal-containing 4-carbon aldehyde **54** [obtained in 51% overall yield from L-arabinose (**53**)] and TMS–TBDMS-protected enolate **55**. However, the boron trifluoride diethyl etherate-catalyzed condensation gave an equal mixture of three products, which required purification by column chromatography in order to separate the desired dithioacetal **56a** in 31% yield. Finally, a protecting group interconversion to **56b**, followed by *N*-iodosuccinimide (NIS)-promoted activation of the dithioacetal, led to the formation of the IdoA glycosyl donor **57** in quantitative yield. The use of an alternatively protected aldehyde in the aldol reaction and subsequent transformations gave similar results.

3.2 Diastereoselective Cyanohydrin Formation

Due to the poor stereoselectivity of the above aldol reaction and consequent low overall yield (6%), Seeberger and coworkers developed an alternative route using D-xylose **58** as starting material (Scheme 7B).[102,103] First, the fully protected D-xylofuranose **59** was obtained via standard protecting-group

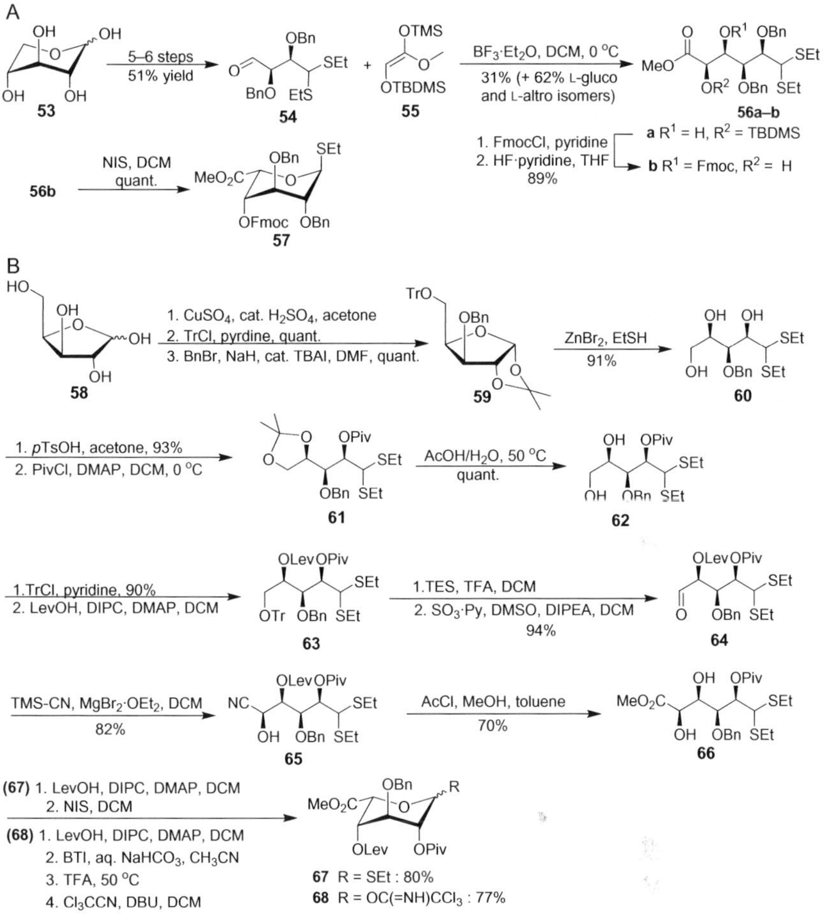

Scheme 7 Synthesis of IdoA derivatives from dithioacetal-containing aldehydes using a Mukaiyama-type aldol reaction (A) and diastereoselective cyanohydrin formation (B).

manipulations in quantitative yield over three steps. Treatment with ethanethiol and $ZnBr_2$ then gave dithioacetal **60** in one pot in 91% yield. Regioselective 5,6-*O*-isopropylidene acetal formation was followed by protection of the 2-OH with pivaloyl chloride to afford the fully protected dithioacetal **61**. Hydrolysis of **61** with aqueous acetic acid then gave the diol **62** in quantitative yield. The primary hydroxyl group of **62** was then selectively protected as the trityl ether, and the secondary hydroxyl group was subjected to esterification with levulinic acid to give the fully protected dithioacetal **63**. Removal of the trityl ether followed by Parikh–Doering oxidation furnished the desired aldehyde **64** in good yield. Next, a chelation-controlled syn-selective cyanation of aldehyde **64** using $MgBr_2{\cdot}OEt_2$ as chelating activator led to the L-ido-configured cyanohydrin **65** with excellent diastereoselectivity (8:1) and good yield (82%). Cyanohydrin **65** underwent a Pinner reaction on treatment with excess hydrochloric acid and methanol in toluene, followed by hydrolysis, to furnish the corresponding methyl ester **66**. Installation of a levulinate-protecting group was followed by NIS-mediated cyclization to give IdoA thioglycoside donor **67** as a mixture of anomers in 80% yield over two steps. Compound **66** could also be converted into an alternative trichloroacetimidate donor **68** by de-thioacetalization with bis(trifluoroacetoxy)iodobenzene (BTI), followed by the reaction of the resultant hemiacetal with DBU and trichloroacetonitrile (77%, 3 steps). The method has been utilized to prepare a range of orthogonally protected IdoA glycosyl donors,[103] which is a major advantage of this route. However, the synthesis is lengthy and suffers from a lack of crystalline intermediates, necessitating a significant number of chromatographic purifications.

In 2009, Gardiner and coworkers published a short route for the synthesis of IdoA derivatives employing a diastereoselective cyanohydrin reaction (Scheme 8).[104] This route involves the oxidative cleavage of diol **18a** with sodium metaperiodate to give the aldehyde **69**, followed by stereoselective addition of a cyano group at room temperature to give the L-ido-configured cyanohydrin **70** with 90% facial selectivity and excellent yield (93%). The diastereoselectivity of the cyanohydrin reaction is dependent on the reaction conditions. Prolonged reaction times favored the formation of L-ido product due to cyanohydrin epimer equilibration coupled with preferential crystallization of the L-ido diastereomer from the reaction mixture. In this manner, the L-ido cyanohydrin **70** was obtained in 76% yield over four steps from **17** without chromatographic purification. The mild conditions and the purification of **70** by crystallization enabled scale-up to >1 mol scale. Compound **70** was then converted into the known IdoA glycosyl acceptor **72** via intermediate **71a** in good yield by a sequence involving isopropylidene acetal hydrolysis with concomitant pyranose formation, 1,2-O-isopropylidenation, oxidation of the nitrile to the amide, and conversion into the methyl ester. Gardiner and coworkers subsequently reported significant improvements to the route resulting in the scalable conversion of **70** into the IdoA acceptor **73** without the need for chromatography.[105] Treatment of **70** with 30% aq HCl effected nitrile hydration and O-isopropylidene hydrolysis leading directly to amide pyranoside **71b**. Compound **71b** was found to undergo a regioselective 1,2-O-diacetylation

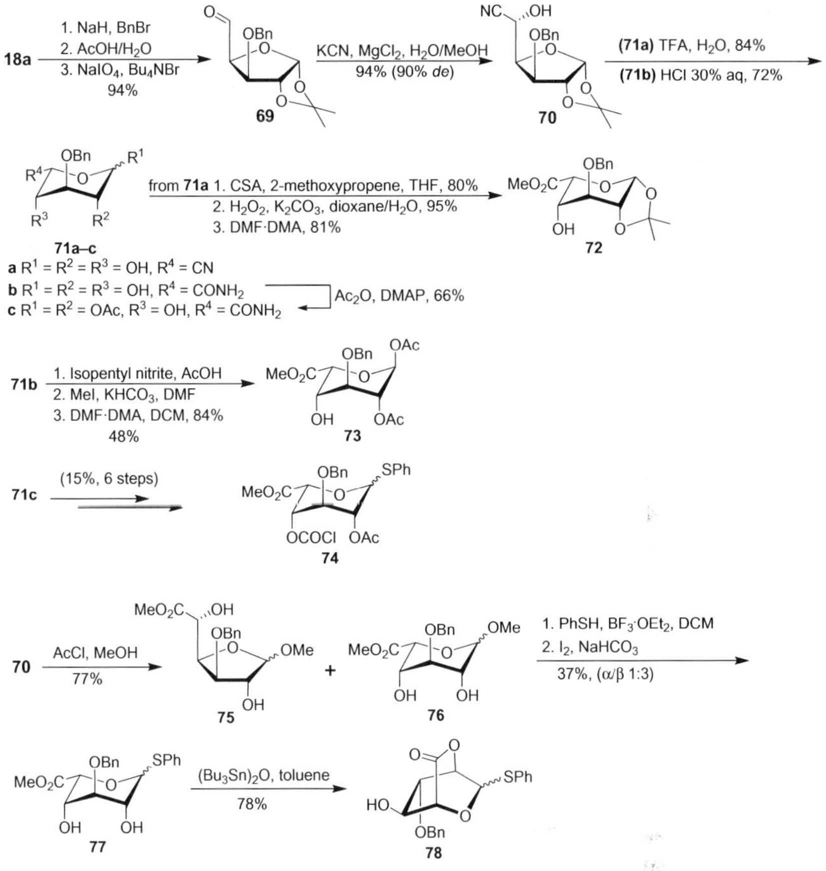

Scheme 8 Diasteroselective cyanohydrin formation for the synthesis of IdoA derivatives.

to give the crystalline diacetate **71c**. Conversion of the amide to the carboxylic acid under mild conditions (heating with isopentyl nitrite in acetic acid) was followed by methylation to give the crystalline acceptor **73** in 48% yield over three steps. Diacetate **71c** was also elaborated into IdoA glycosyl donors of the type **74** in six steps from **70** (15% overall yield) and 10 steps (12% overall yield) from **17**. These building blocks were utilized for the preparation of HS di- and tetrasaccharides.[105] This method has been subsequently utilized for the preparation of multigram amounts of a range of HS oligosaccharides.[105–108] Of particular note is the first example of a gram-scale synthesis of a structurally defined, heparin-related dodecasaccharide.[107] Very recently, Gardiner and coworkers have reported the large-scale synthesis from cyanohydrin **70** of new IdoA thioglycoside lactones[109] as IdoA glycosyl donors, thus adding to the range of IdoA building blocks available. Treatment of **70** with acetyl chloride in methanol can be carried out on a 200-g scale to give the glycoside mixture **75** + **76**. This crude mixture was converted in only three steps into the lactone **78** on a multigram scale. Lactone **78** was shown to be a useful acceptor for the preparation of HS disaccharides. The lactone is substantively disarmed relative to other thioglycosides, but it could be rearmed by subsequent opening of the lactone ring in the derived disaccharide, thereby providing a short route to armed, HS-related, disaccharide thioglycoside donors.[109]

3.3 Addition of Organometallic Reagents

Bonnaffé and coworkers also developed a synthetic route to IdoA derivatives that involved the addition of masked carboxylate nucleophiles to aldehyde **69** (see Schemes 8 and 9).[110] The addition of vinylic organometallic reagents to **69** gave a mixture of both D-gluco- and L-idofuranoses; more specifically, addition of vinylmagnesium bromide yielded an isomeric mixture of **79** and **80** in a ratio of 1:1. The L-idofuranose **80** was then converted into the IdoA derivative **83** via intermediate **81** by a sequence involving ozonolysis, oxidation, and methylation. In order to increase the diastereoselectivity of addition in favor of the L-ido isomer, the addition of different organometallic reagents to **69** was examined. From these studies, it was found that the proportion of L-ido isomer increased with the steric bulk of the nucleophile, and the reaction was completely stereoselective with *tris*-(phenylthio)

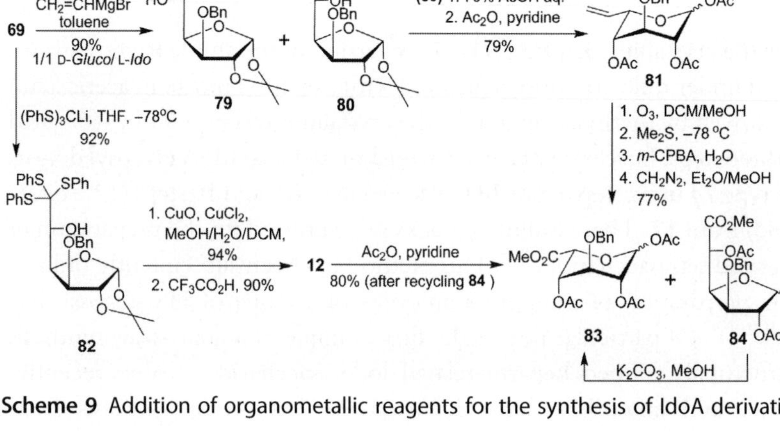

Scheme 9 Addition of organometallic reagents for the synthesis of IdoA derivatives.

methyllithium ((PhS)$_3$CLi), giving compound **82** as the sole product in 92% yield, in a reaction scalable to 100 g. The orthothioester **82** was then converted to the corresponding methyl ester and subjected to acid hydrolysis to give **12** (see Scheme 1), which was subsequently acetylated to give **83** in 80% yield (after recycling the furanose derivative **84**, which also forms in the reaction). The mixture of pyranose **83** and furanose **84** required a tedious chromatographic separation; hence, Bonnaffé developed conditions to trap the crystalline β-pyranose form of **12** as its β-acetate **83β**, which was then isolated by crystallization in 83% yield.[111] This synthetic route has been used to prepare various IdoA glycosyl donors and acceptors via compound **12**.[23,112–117]

3.4 Homologation Using 2-(Trimethylsilyl)thiazole

In 1997, Dondoni et al. reported a short synthetic route to L-idose starting from L-xylose (Scheme 10).[118] This route relies on a highly stereoselective addition of 2-(trimethylsilyl)thiazole and an oxidation/reduction sequence for converting the stereochemistry of the substrate to the required *S* configuration. First, L-xylose **85** was converted into aldehyde **86** via a sequence of dithioacetalation, installation of isopropylidene-protecting groups, and Hg(II)-promoted hydrolysis of the dithioacetal. Addition of the 2-(trimethylsilyl)thiazole to the aldehyde **86**, followed by desilylative workup by treatment with tetrabutylammonium fluoride (TBAF), gave the alcohol **87** in 95% diastereoselectivity and 93% yield. The inversion of configuration at C-2 of **87** was accomplished by an oxidation/reduction sequence using Swern oxidation to obtain the ketone **88**, which was then

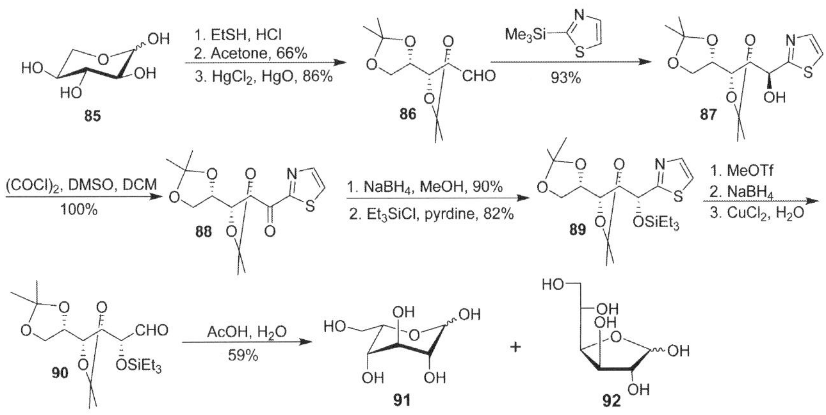

Scheme 10 Synthesis of L-idose via homologation using 2-(trimethylsilyl)thiazole.

subjected to sodium borohydride reduction. The reduction proceeded with excellent stereoselectivity to give the corresponding alcohol in 91% diastereoselectivity, which was protected as the triethylsilyl ether prior to the removal of the thiazole ring. The fully protected compound **89** was then subjected to thiazole to formyl deblocking to obtain the protected aldehyde **90**. Finally, removal of the isopropylidene acetal groups in the presence of aqueous acetic acid afforded L-idose as a mixture of pyranose **91** and furanose **92** forms in 19% overall yield from **85**.[118] L-Idose can be formally converted into various IdoA derivatives for use in GAG synthesis, although there have been no reports of the utilization of the above method for this purpose, perhaps due to the relatively high cost of the L-xylose starting material.

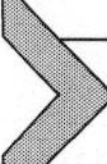

4. ISOMERIZATION OF UNSATURATED SUGARS

4.1 Diastereoselective Hydroboration of exo-Glycals

In 1997, Rochepeau-Jobron and Jacquinet reported the diastereoselective hydroboration of exo-glycal **96** to obtain various L-idose derivatives (Scheme 11).[119] The exo-glycal **96** was obtained from commercially available D-glucose derivative **93**. Selective iodination of the primary hydroxyl group of **93**, followed by acetylation, gave the protected 6-deoxy-6-iodoglucoside **94**. Dehydrohalogenation with 1,8-diazabicyclo[5.4.0]undec-7-ene (DBU) then gave exo-glycal **95** in 74% yield. The protecting groups at C-2 and C-4 were exchanged, and the resulting exo-glycals were subjected to hydroboration

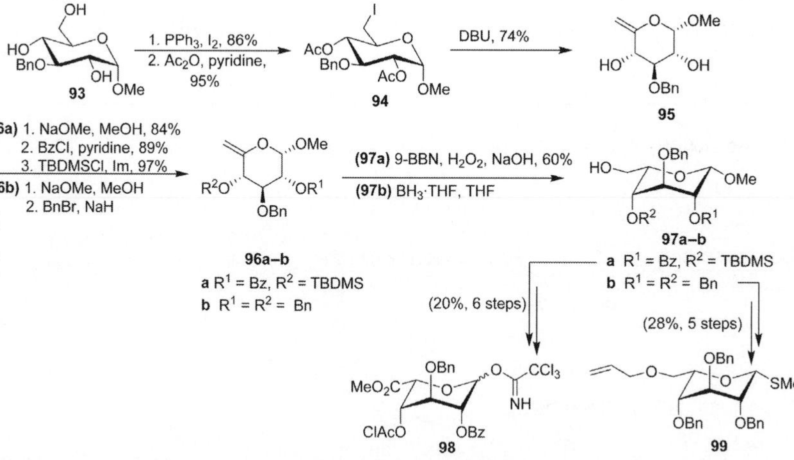

1. PPh_3, I_2, 86%
2. Ac_2O, pyridine, 95%
DBU, 74%
93 94 95
(96a) 1. NaOMe, MeOH, 84%
2. BzCl, pyridine, 89%
3. TBDMSCl, Im, 97%
(96b) 1. NaOMe, MeOH
2. BnBr, NaH
(97a) 9-BBN, H_2O_2, NaOH, 60%
(97b) BH_3·THF, THF
96a–b
a R^1 = Bz, R^2 = TBDMS
b R^1 = R^2 = Bn
97a–b
a R^1 = Bz, R^2 = TBDMS
b R^1 = R^2 = Bn
(20%, 6 steps)
(28%, 5 steps)
98 99

Scheme 11 Synthesis of IdoA derivatives via diastereoselective hydroboration of exo-glycals.

with 9-borabicyclo[3.3.1]nonane (9-BBN). Compound **96a** with *tert*-butyldimethylsilyl- and benzoyl-protecting groups gave the best facial selectivity (9:1, L-ido to D-gluco) to produce the desired L-idoside **97a** in a yield of 60%. Compound **97a** was then converted in moderate yield into a series of IdoA glycosyl donors (e.g., **98**; six steps, 20%). Wong and coworkers reported a facial selectivity of 5:1 L-ido to D-gluco for the hydroboration of the benzyl ether-protected derivative **96b** with $BH_3 \cdot THF$ to give **97b**,[120] which was then converted into the glycosyl donor **99** (28%, 5 steps).

Hinou and coworkers[121,122] reported the hydroboration of α,α-trehalose derivative **102** to obtain L-idose derivative **104** (Scheme 12). In this work, the known ditosylate **100**, available in four steps from trehalose,[123] was converted via the diiodide **101** into the exo-olefin **102**. The hydroboration of **102** was initially carried out using $BH_3 \cdot THF$, which resulted in the formation of the L-ido–L-ido disaccharide **104** (64%) and the L-ido–D-gluco disaccharide **103** (18%). However, employing 9-BBN as the hydroborating reagent gave **104** in higher yield (81%) and selectivity (**103**:**104** = 1:8). Transformation into the corresponding disaccharide uronate **105** was carried

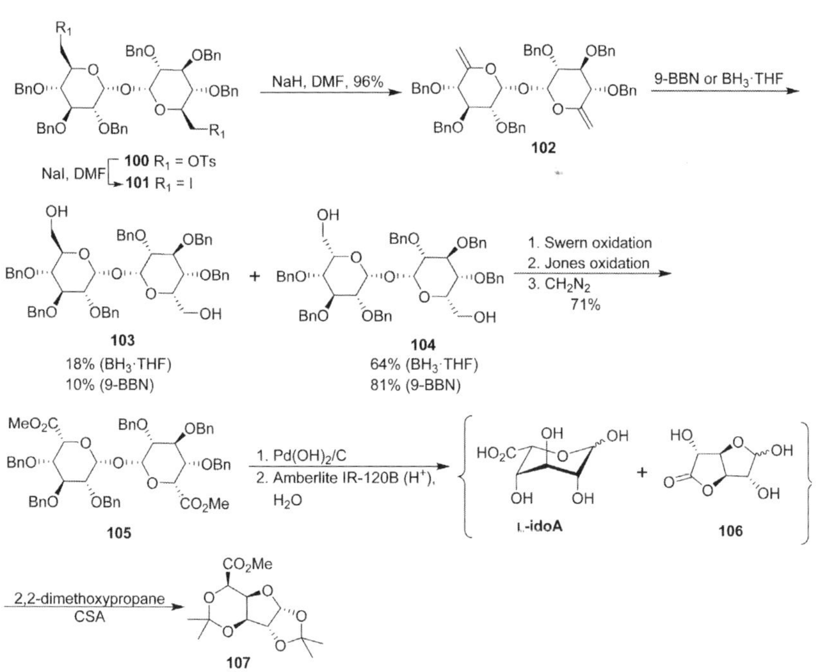

Scheme 12 Synthesis of IdoA via diastereoselective hydroboration of α,α-trehalose exo-glycals.

out by a two-step oxidation of the 6- and 6′-hydroxyl groups of **104** followed by esterification. The esterification and slow reaction rate were essential for preventing epimerization back into the D-gluco configuration. This was then followed by deprotection and hydrolysis of the glycosidic bond to give IdoA. Since IdoA exists in equilibrium with its 3,6-lactone **106** making it difficult to isolate, the mixture was converted to the known[89] isopropylidene derivative **107**.

In 2000, Hung and Chen also reported a synthetic route that utilizes the hydroboration of exo-glycals to obtain IdoA synthons (Scheme 13A),[124] via

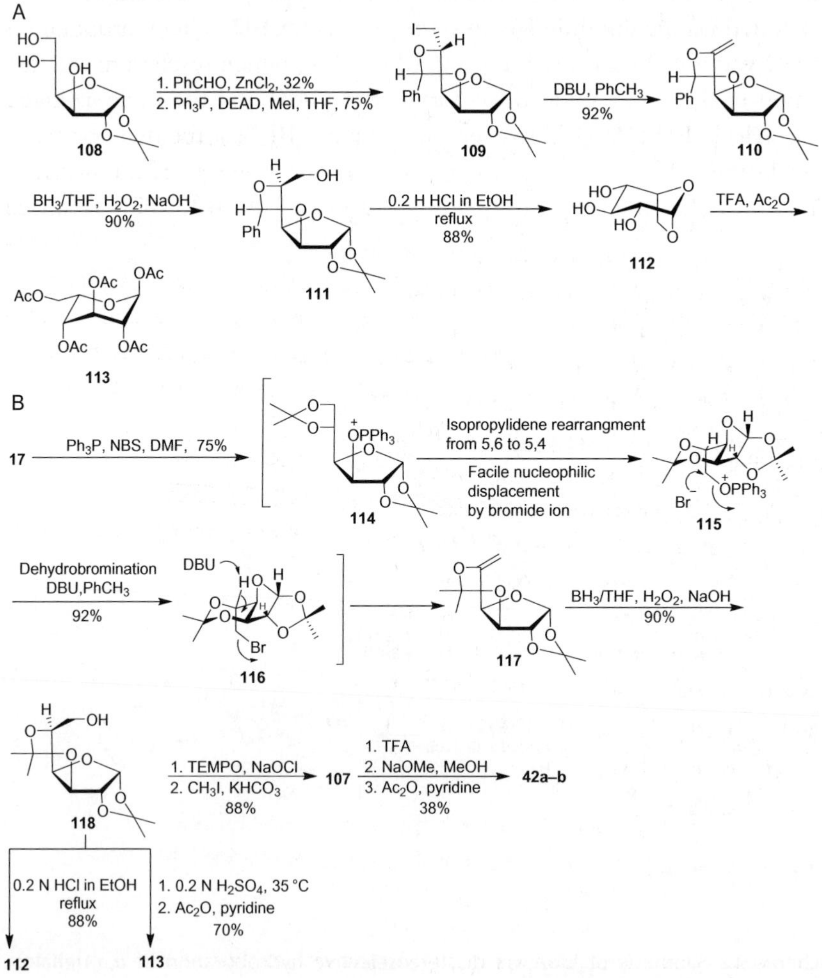

A

108 → 1. PhCHO, $ZnCl_2$, 32%; 2. Ph_3P, DEAD, MeI, THF, 75% → 109 → DBU, $PhCH_3$, 92% → 110

110 → BH_3/THF, H_2O_2, NaOH, 90% → 111 → 0.2 H HCl in EtOH, reflux, 88% → 112 → TFA, Ac_2O → 113

B

17 → Ph_3P, NBS, DMF, 75% → [114 → Isopropylidene rearrangment from 5,6 to 5,4; Facile nucleophilic displacement by bromide ion → 115 → Dehydrobromination DBU,$PhCH_3$, 92% → 116] → 117 → BH_3/THF, H_2O_2, NaOH, 90% → 118

118 → 1. TEMPO, NaOCl; 2. CH_3I, $KHCO_3$, 88% → 107 → 1. TFA; 2. NaOMe, MeOH; 3. Ac_2O, pyridine, 38% → 42a–b

118 → 0.2 N HCl in EtOH, reflux, 88% → 112

118 → 1. 0.2 N H_2SO_4, 35 °C; 2. Ac_2O, pyridine, 70% → 113

Scheme 13 Synthesis of IdoA derivatives via diastereoselective hydroboration of exo-glycal **110**.

the commercially available triol **108** as the starting material. First, **108** was converted into the 6-deoxy-6-iodo derivative **109** via a Mitsunobu-type iodination, and it was then subjected to β-elimination with DBU in refluxing toluene to obtain the enol ether **110** in excellent yield. Hydroboration, followed by oxidative workup, gave the L-idofuranoside **111** as a single diastereomer. The diastereoselectivity of this reaction was attributed to the steric effect of the cis–anti–cis-tricyclic conformation of **110**, resulting in hydroboration of the 5-exo double bond from the less-hindered α-face. This leads to the formation of the substituted group at the equatorial C-5 position forming the L-ido isomer, exclusively. Compound **111** was then converted into the peracetylated idopyranose **113** via 1,6-anhydrosugar **112**.[124] Although this strategy worked well, the disadvantages of this route include: (1) the necessary and tedious purification of each intermediate, (2) the low overall yield (22%) from **108** to **110**, and (3) the starting material **108**, which is more expensive than other common D-glucose derivatives.[125]

Thus to further optimize the synthetic strategy, Hung and coworkers started from inexpensive 1,2:5,6-di-*O*-isopropylidene-α-D-glucofuranose (**17**) to prepare the exo-glycal **117** as the key intermediate.[126–129] Consecutive treatment of **17** with PPh_3, NBS, and DBU in toluene afforded **117** in a one-pot manner via the 6-bromo-6-deoxy compound **116**. In this reaction, PPh_3 first reacts with NBS to generate a phosphonium salt, which is attacked by the 3-hydroxyl group of **17**, resulting in the formation of the alkoxyphosphonium intermediate[130] **114** with a rigid cis-5,5-fused ring system (Scheme 13B). The rigid conformation leads to a 5,6-isopropylidene rearrangement (to give **115**), followed by a facile nucleophilic displacement of the PPh_3 group by bromide ion, thus leading to regioselective bromination at C-6 to give **116**.[127] Dehydrobromination of **116** with DBU furnished **117**, which was then subjected to hydroboration to give the L-idofuranose **118** in two steps with an overall yield of 80%. The acid hydrolysis of **118** at various temperatures to form either compound **112** or **113** was then optimized. To obtain anhydrosugar **112**, compound **118** was heated to reflux in 0.2 M HCl in EtOH. However, suspending **118** in 0.2 M H_2SO_4 at 35 °C, followed by acetylation, gave peracetate **113** in 70% yield.[131] Furthermore, L-idofuranose **118** was converted into the methyl ester **107** via TEMPO oxidation and methylation in good yield, and subsequently into **42a,b** (see Scheme 5) by a three-step sequence of acid hydrolysis, treatment with NaOMe, and acetylation in 38% overall yield.[126]

Overall this method is a short and efficient route to IdoA derivatives which are well suited for the preparation of simple glycosides and α-L-iduronidase substrates. However, for the gram-scale preparation of orthogonally protected IdoA synthons required for the synthesis of GAG oligosaccharides, Hung and coworkers' synthetic routes to anhydrosugar **31** are preferred (see Section 2.1 and Scheme 3).

4.2 From Δ^4-Uronates

Linhardt and coworkers developed a strategy for regioselective conversion of Δ^4-uronates to IdoA derivatives (Scheme 14).[132,133] The glycal **120** was synthesized from the known uronate **119** via β-elimination using DBU. Deacetylation of **120**, followed by silylation with 1,3-dichloro-1,1,3,3-tetraisopropyl-1,3-disiloxane ($TIPDSiCl_2$), led to the Δ^4-uronate **121**. Derivatives of **120** with other protecting groups gave poorer yields and selectivities in the subsequent reactions. Treatment of **121** with NBS in aqueous THF led to the corresponding trans-diaxial bromohydrin **122** together with a small amount of trans-diequatorial bromohydrin **123**. Treatment of **122** with silver oxide (Ag_2O) afforded the corresponding epoxide **124** in good yield. The subsequent acid-catalyzed rearrangement of epoxide **124** using scandium triflate afforded the C-4 keto derivative **125** (63–93%) along with an undesired furanoside by-product. Reduction with sodium borohydride was completely stereoselective and gave the IdoA derivative **126** in 53% yield following acetylation and purification by chromatography. The reducing agent can only approach the C-4 carbon from the

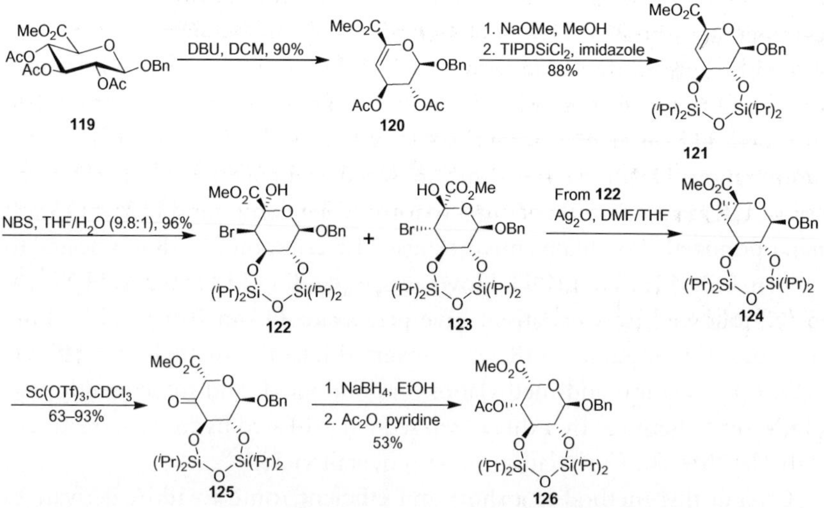

Scheme 14 Regioselective conversion of Δ^4-uronates to IdoA derivatives.

less-hindered α-L face of **125** leading to the formation of the L-ido product. There have been no reports of the use of **126** for GAG oligosaccharide synthesis.

4.3 From 4-Deoxypentenosides

Wei and coworkers have reported a short route to IdoA derivatives using a syn-selective addition of organozinc compounds to 4α-epoxypyranosides generated from methyl β-D-glucopyranoside (**127**) (Scheme 15).[134] The 4-deoxypentenoside **128** was obtained by a two-step oxidation–decarboxylative elimination sequence[135] by treatment with TEMPO followed by *N,N*-dimethylformamide dineopentyl acetal (DMFDNpA). Stereoselective oxidation of **128** with dimethyldioxirane (DMDO) led to the 4α-epoxypyranoside **129** with high facial selectivity (10:1) in quantitative yield. Reaction of **129** with 2-furylzinc bromide at −78 °C resulted in a syn-selective ring opening to give the 2-furyl adduct **130** in 84% yield. Compound **130** was then converted into the IdoA derivative **131** in 73% yield by ozonolysis with a reductive workup. There have been no reports of the use this route for GAG oligosaccharide synthesis.

4.4 From D-Glucuronic Acid Glycal

In 1984, Thiem and Ossowski reported that the treatment of D-glucuronic acid glycal **132a** with 1% sodium methoxide in methanol resulted in epimerization at C-5 to give a mixture of deacetylated glycals **133** and **134a**, with the L-iduronic acid glycal **134a** predominating (4:1) in 54% isolated yield (Scheme 16).[136] This prompted Seeberger and coworkers to further investigate this epimerization for the preparation of differentially protected IdoA building blocks.[137] Treatment of protected glycals **132b** and **132c** with a concentrated solution of sodium methoxide resulted in complete degradation. However, with lower concentrations of base and

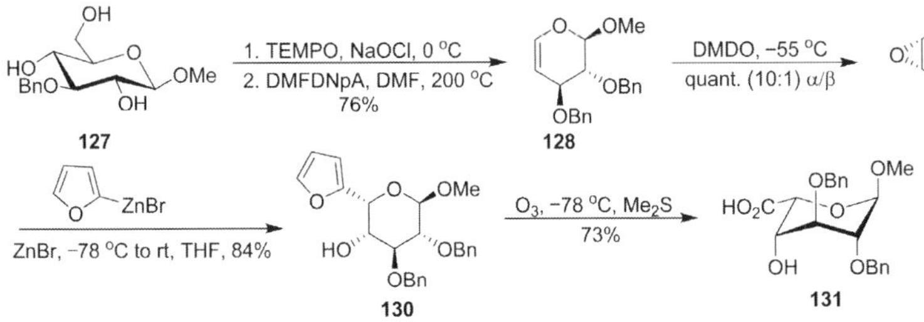

Scheme 15 Synthesis of an IdoA derivative from a 4-deoxypentenoside.

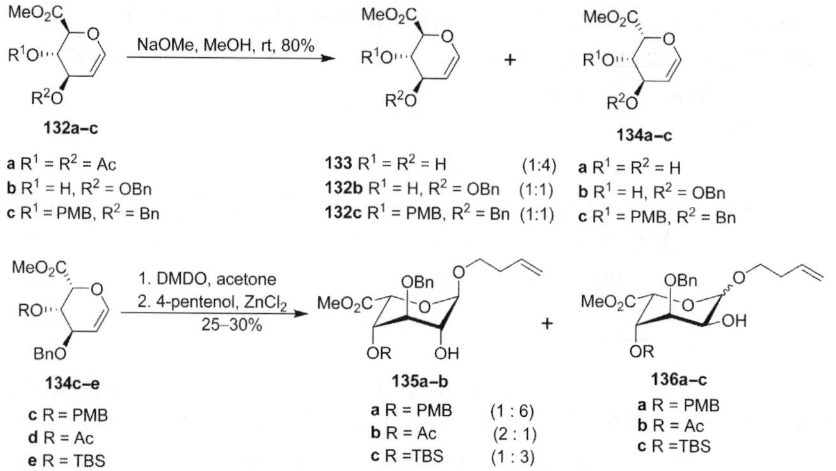

NaOMe, MeOH, rt, 80%

132a–c
a $R^1 = R^2 = Ac$
b $R^1 = H$, $R^2 = OBn$
c $R^1 = PMB$, $R^2 = Bn$

133 $R^1 = R^2 = H$ (1:4)
132b $R^1 = H$, $R^2 = OBn$ (1:1)
132c $R^1 = PMB$, $R^2 = Bn$ (1:1)

134a–c
a $R^1 = R^2 = H$
b $R^1 = H$, $R^2 = OBn$
c $R^1 = PMB$, $R^2 = Bn$

1. DMDO, acetone
2. 4-pentenol, $ZnCl_2$
25–30%

134c–e
c R = PMB
d R = Ac
e R = TBS

135a–b
a R = PMB (1 : 6)
b R = Ac (2 : 1)
c R =TBS (1 : 3)

136a–c
a R = PMB
b R = Ac
c R =TBS

Scheme 16 Conversion to D-glucuronic acid glycal to IdoA derivatives.

shorter reaction times, 1:1 mixtures of **132b/134b** and **132c/134c**, respectively, were obtained. The desired IdoA glycals **134b** and **134c** were isolated by column chromatography. The conversion of **134b** and **134c** to differentially protected IdoA building blocks was next investigated. Epoxidation of **134c–e** with DMDO, followed by conversion to the corresponding *n*-pentyl glycosides, resulted in a mixture of the IdoA glycosides **135a–c** and L-guluronic acid glycosides **136a–c**. However, the yields and selectivities for the desired IdoA glycosides were poor and required extensive chromatographic purifications. This route is also limited by the need for multistep syntheses for the starting glycals and the poor stability of the epoxides derived from IdoA glycals.

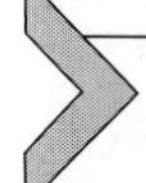

5. MISCELLANEOUS METHODS

5.1 Diastereoselective Tishchenko Reaction

Iadonisi and coworkers developed a novel synthetic route to L-idose (**91**) using the diastereoselective Tishchenko oxidoreduction of hexose-5-ulose **139** induced by t-BuOSmI$_2$ (Scheme 17).[138] The hexose-5-ulose **139** was prepared[139] from methyl glucoside **137** via benzylation, followed by hydrolysis to the hemiacetal and finally $NaBH_4$ reduction, to give alditol **138**. A double Swern oxidation, followed by treatment with t-BuOSmI$_2$ in one pot, generated the *t*-butyl ester **140**, possessing the desired L-ido configuration. An explanation for the observed diastereoselectivity of the Tischenko reaction was proposed[138] based on an earlier model,[140] whereby

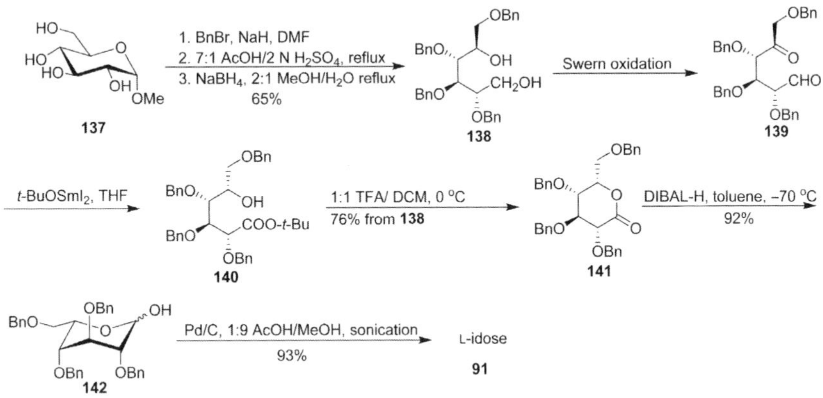

Scheme 17 Diastereoselective Tishchenko reaction of hexose-5-ulose **139**.

the samarium complex leading to the L-ido isomer is favored because of a lack of eclipsing interactions between the 1-*O*-^{t}Bu and 2-*O*-Bn groups. The *tert*-butyl ester **140** was then subjected to lactonization and reduction of lactone **141** with DIBAL-H to give the L-idose derivative **142**. Lastly, hydrogenolysis of the benzyl ether-protecting groups gave the unprotected L-idose (**91**) with an overall yield of 65% from **138**. As noted earlier, L-idose (**91**) can be formally converted into various IdoA derivatives for use in GAG synthesis, but similarly, there have been no reports of the utilization of the above method for this purpose.

5.2 Homologation with 5,6-Dihydro-1,4-dithiin-2-yl [(4-methoxybenzyl)oxy]methane

Guaragna and coworkers[141] developed a novel synthetic route to L-hexoses, which was utilized to gain excess to four rare L-hexopyranoses.[141,142] In 2010, this methodology was extended to include L-idose.[143] 1,2-Bisthioenol ether synthon **143**, prepared from methyl pyruvate,[144] was treated with *n*-butyllithium to generate a lithiated carbanion, which was then coupled with α/β-isopropylidene-L-glycerate **144**, prepared from 5,6-*O*-isopropylidene-L-gulono-1,4-lactone[145] to give the ketone **145** in 98% yield (Scheme 18). Stereoselective reduction of ketone **145** with $NaBH_4$ gave the syn-alcohol **146** in 98% yield. The complete diastereoselectivity of this reaction was due to borohydride attack occurring from the less-hindered face of ketone **145** in its nonchelated conformation. Benzyl protection of the free hydroxyl group, followed by removal of the 4-methoxybenzyl ether (PMB)-protecting group using 2,3-dichloro-5,6-dicyano-*p*-benzoquinone

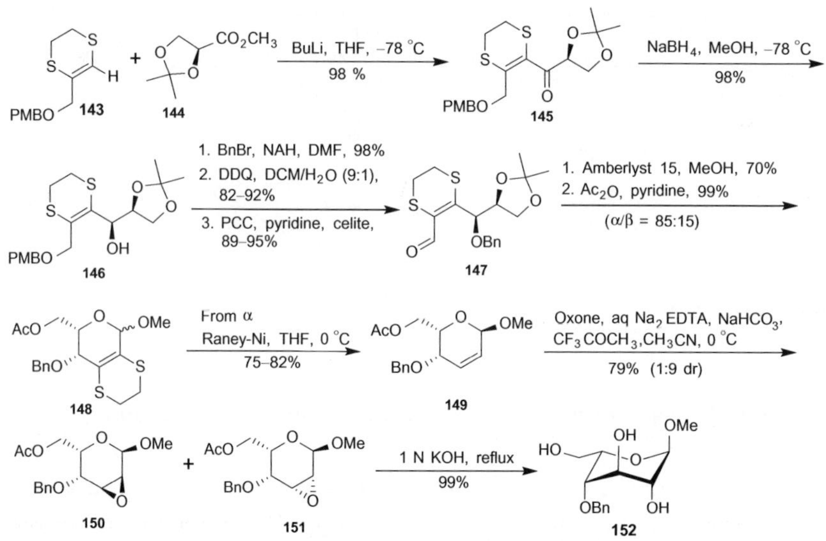

Scheme 18 Synthesis of L-idoside **152** via homologation of methyl 2,3-*O*-isopropylidene-L-glycerate (**144**) with 5,6-dihydro-1,4-dithiin-2-yl[(4-methoxybenzyl) oxy]methane (**143**).

(DDQ) and subsequent oxidation of the resulting primary alcohol, afforded the intermediate **147** in good yield. Synthesis of the intermediate acetal **148** proceeded through cyclization of the main carbon chain of **147** via isopropylidene cleavage followed by direct acetylation. The olefin **149** was then obtained by desulfurization using Raney nickel. Compound **149** was then subjected to epoxidation using in situ generated methyl(trifluoromethyl)dioxirane (TFDO) leading to a diastereomeric mixture (1:9) of L-gulo- and L-talo-epoxides **150** and **151** in a total yield of 79%. Finally, epoxide ring opening of L-talo-epoxide **151** under alkaline conditions afforded the L-idose derivative **152** in 99% yield. The regioselectivity of the epoxide ring opening was attributed to attack of the hydroxide ion from the relatively unhindered C-3 position of the $^{5}H_{o}$ conformer of **151**.

5.3 C–H Activation of 6-Deoxy-L-hexoses

Recently, Bols and coworkers[146] published a new route for the synthesis of all eight L-glycopyranosyl donors, including L-idopyranoside **163**, from their corresponding 6-deoxy L-glycopyranosides (Scheme 19). The 6-deoxy-L-idopyranoside **161** was first synthesized[147] from the commercially available L-rhamnose **153**. First, **153** was converted into its phenyl

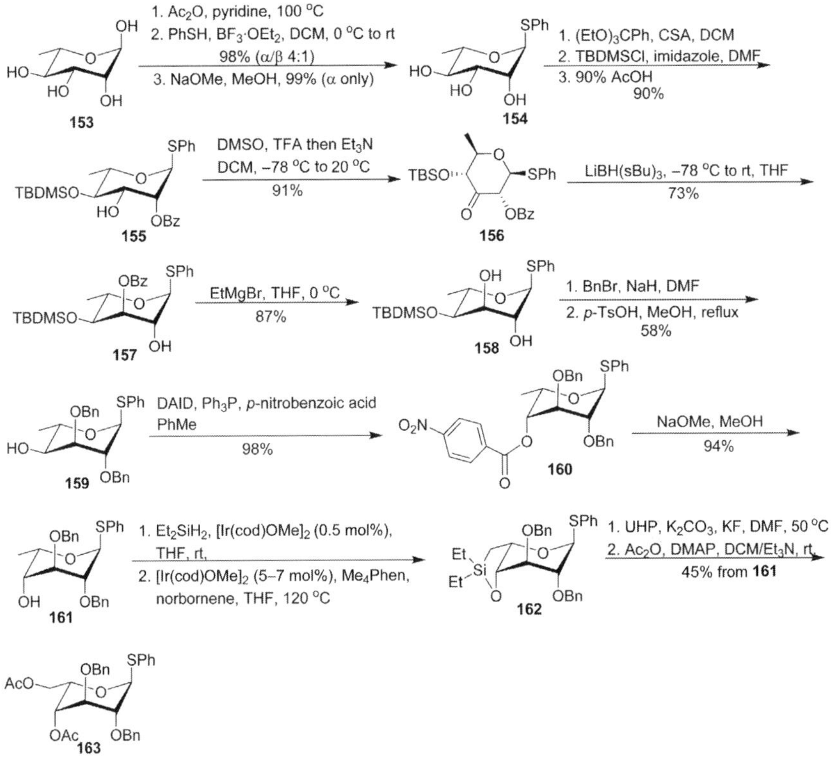

Scheme 19 Synthesis of 1-thio-L-idoglycosides via C–H activation of the 6-deoxy-1-thio-L-hexopyranose **161**.

thioglycoside **154** in three steps in quantitative yield by peracetylation, thioglycosylation, and O-deacetylation. Acid-catalyzed 2,3-O-orthoester formation followed by one-pot silylation of the 4-OH group and opening of the orthoester afforded the intermediate **155** in 90% yield. Compound **155** was then subjected to Swern oxidation to obtain the ketone **156**. Subsequent L-selectride reduction of the ketone accompanied by benzoyl migration from 2-OH to 3-OH gave the C-3 axial epimer **157** in 73% yield. Debenzoylation of **157** with ethylmagnesium bromide (EtMgBr) afforded the diol **158** in 87% yield. Benzylation of diol **158** followed by acid-mediated desilylation gave 6-deoxy-1-thio-L-altropyranoside **159**. This was followed by epimerization of the 4-OH of **159**, which was accomplished via a Mitsunobu reaction to give the *p*-nitrobenzoate **160**. Finally, standard Zemplén deacetylation gave the 6-deoxy-1-thio-L-idopyranoside **161** in 94% yield,[147] which was then converted into its corresponding

L-idopyranoside **163** via an iridium-catalyzed C–H activation of the C-6 methyl group.[146] This involved an $[Ir(cod)OMe]_2$-catalyzed silylation of the 4-OH followed by an intramolecular C–H activation of the methyl group in the γ-position, which is also catalyzed by the same iridium complex in the presence of 3,4,7,8-tetramethyl-1,10-phenanthroline as the ligand, to give the intermediate oxasilolane **162**. Finally, a modified Fleming–Tamao oxidation (UHP, K_2CO_3, KF) of the oxasilolane and acetylation gave the 1-thio-L-idopyranoside **163** in 45% yield from **161**. This method produces an L-idopyranoside as a glycosyl (thioglycoside) donor and is attractive for access to other rare L-sugars. However, the scalability is questionable due to the number of synthetic steps and lack of crystalline intermediates.

6. CONCLUSIONS

The interest in deciphering the molecular details of GAG–protein interactions has continued to grow apace because of the recognition that these interactions mediate a myriad of important biological processes. As a consequence, in recent years there has been a tremendous increase in the development of synthetic strategies for the preparation of a range of well-defined GAG oligosaccharides for biological evaluation. One of the main limitations in the synthesis of GAG oligosaccharides has been access to IdoA (or L-idose) building blocks as glycosyl donors and acceptors. However, advances in the past decade have resulted in a range of diverse methods for this purpose, and a number of options now exist for the large-scale preparation of IdoA building blocks. Of particular note has been the use of these building blocks for the gram-scale preparation of GAG oligosaccharides up to dodecasaccharide, as well as their incorporation into improved syntheses of GAG-based drugs such as fondaparinux. These advances should aid future efforts to better understand GAG–protein interactions and to develop scalable syntheses of potential GAG-based drug candidates.

ACKNOWLEDGMENTS

We thank the University of Queensland and the National MPS Society (USA) for financial support.

REFERENCES

1. Casu, B.; Lindahl, U. Structure and Biological Interactions of Heparin and Heparan Sulfate. *Adv. Carbohydr. Chem. Biochem.* **2001**, *57*, 159–206.
2. Rabenstein, D. L. Heparin and Heparan Sulfate: Structure and Function. *Nat. Prod. Rep.* **2002**, *19*, 312–331.

3. Casu, B.; Naggi, A.; Torri, G. Re-Visiting the Structure of Heparin. *Carbohydr. Res.* **2015**, *403*, 60–68.
4. Malavaki, C.; Mizumoto, S.; Karamanos, N.; Sugahara, K. Recent Advances in the Structural Study of Functional Chondroitin Sulfate and Dermatan Sulfate in Health and Disease. *Connect. Tissue Res.* **2008**, *49*, 133–139.
5. Thelin, M. A.; Bartolini, B.; Axelsson, J.; Gustafsson, R.; Tykesson, E.; Pera, E.; Oldberg, Å.; Maccarana, M.; Malmstrom, A. Biological Functions of Iduronic Acid in Chondroitin/Dermatan Sulfate. *FEBS J.* **2013**, *280*, 2431–2446.
6. Capila, I.; Linhardt, R. J. Heparin–Protein Interactions. *Angew. Chem. Int. Ed.* **2002**, *41*, 391–412.
7. Whitelock, J. M.; Iozzo, R. V. Heparan Sulfate: A Complex Polymer Charged with Biological Activity. *Chem. Rev.* **2005**, *105*, 2745–2764.
8. Ori, A.; Wilkinson, M. C.; Fernig, D. G. The Heparanome and Regulation of Cell Function: Structures, Functions and Challenges. *Front. Biosci.* **2008**, *13*, 4309–4338.
9. Xu, D.; Esko, J. D. Demystifying Heparan Sulfate–Protein Interactions. *Annu. Rev. Biochem.* **2014**, *83*, 129–157.
10. Mulloy, B.; Forster, M. J. Conformation and Dynamics of Heparin and Heparan Sulfate. *Glycobiology* **2000**, *10*, 1147–1156.
11. Raman, R.; Sasisekharan, V.; Sasisekharan, R. Structural Insights into Biological Roles of Protein-Glycosaminoglycan Interactions. *Chem. Biol.* **2005**, *12*, 267–277.
12. Casu, B.; Choay, J.; Ferro, D. R.; Gatti, G.; Jacquinet, J. C.; Petitou, M.; Provasoli, A.; Ragazzi, M.; Sinaÿ, P.; Torri, G. Controversial Glycosaminoglycan Conformations. *Nature* **1986**, *322*, 215–216.
13. Ferro, D. R.; Provasoli, A.; Ragazzi, M.; Torri, G.; Casu, B.; Gatti, G.; Jacquinet, J. C.; Sinaÿ, P.; Petitou, M.; Choay, J. Evidence for Conformational Equilibrium of the Sulfated L-Iduronate Residue in Heparin and in Synthetic Heparin Mono- and Oligosaccharides—NMR and Force-Field Studies. *J. Am. Chem. Soc.* **1986**, *108*, 6773–6778.
14. Muñoz-García, J. C.; Corzana, F.; de Paz, J. L.; Angulo, J.; Nieto, P. M. Conformations of the Iduronate Ring in Short Heparin Fragments Described by Time-Averaged Distance Restrained Molecular Dynamics. *Glycobiology* **2013**, *23*, 1220–1229.
15. Lever, R.; Mulloy, B.; Page, C. P. *Heparin—A Century of Progress*; Berlin: Springer, 2012.
16. Petitou, M.; van Boeckel, C. A. A. A Synthetic Antithrombin III Binding Pentasaccharide Is now a Drug! What Comes Next? *Angew. Chem. Int. Ed.* **2004**, *43*, 3118–3133.
17. Coombe, D. R.; Kett, W. C. Heparin Mimetics. In: *Heparin—A Century of Progress*; Lever, R., Mulloy, B., Page, C. P., Eds.; Springer: Berlin, 2012; pp 361–383.
18. Ferro, V. Heparan Sulfate Inhibitors and Their Therapeutic Implications in Inflammatory Illnesses. *Expert Opin. Ther. Targets* **2013**, *17*, 965–975.
19. Driguez, P. A.; Potier, P.; Trouilleux, P. Synthetic Oligosaccharides as Active Pharmaceutical Ingredients: Lessons Learned from the Full Synthesis of One Heparin Derivative on a Large Scale. *Nat. Prod. Rep.* **2014**, *31*, 980–989.
20. Poletti, L.; Lay, L. Chemical Contributions to Understanding Heparin Activity: Synthesis of Related Sulfated Oligosaccharides. *Eur. J. Org. Chem.* **2003**, 2999–3024.
21. Karst, N. A.; Linhardt, R. J. Recent Chemical and Enzymatic Approaches to the Synthesis of Glycosaminoglycan Oligosaccharides. *Curr. Med. Chem.* **2003**, *10*, 1993–2031.
22. Codée, J. D. C.; Overkleeft, H. S.; van der Marel, G. A.; van Boeckel, C. A. A. The Synthesis of Well-Defined Heparin and Heparan Sulfate Fragments. *Drug Discov. Today Technol.* **2004**, *1*, 317–326.

23. Noti, C.; Seeberger, P. H. Chemical Approaches to Define the Structure-Activity Relationship of Heparin-Like Glycosaminoglycans. *Chem. Biol.* **2005**, *12*, 731–756.
24. van den Bos, L. J.; Codée, J. D. C.; Litjens, R. E. J. N.; Dinkelaar, J.; Overkleeft, H. S.; van der Marel, G. A. Uronic Acids in Oligosaccharide Synthesis. *Eur. J. Org. Chem.* **2007**, 3963–3976.
25. Dulaney, S. B.; Huang, X. Strategies in Synthesis of Heparin/Heparan Sulfate Oligosaccharides: 2000–Present. *Adv. Carbohydr. Chem. Biochem.* **2012**, *67*, 95–136.
26. Zulueta, M. M. L.; Lin, S.-Y.; Hu, Y.-P.; Hung, S.-C. Synthetic Heparin and Heparan Sulfate Oligosaccharides and Their Protein Interactions. *Curr. Opin. Chem. Biol.* **2013**, *17*, 1023–1029.
27. Zulueta, M. M. L.; Lin, S.-Y.; Hung, S.-C. Chemical Synthesis of Oligosaccharides Based on Heparin and Heparan Sulfate. *Trends Glycosci. Glycotechnol.* **2013**, *25*, 141–158.
28. Grand, E.; Kovensky, J.; Pourceau, G.; Toumieux, S.; Wadouachi, A. Chapter 11 Anionic Oligosaccharides: Synthesis and Applications. *Carbohydrate Chemistry*, Vol. 40; The Royal Society of Chemistry: Cambridge, UK, 2014; pp 195–235.
29. Pellissier, H. Syntheses of L-Iduronyl Synthons. A Review. *Org. Prep. Proced. Int.* **2002**, *34*, 441–465.
30. Hassan, H. H. A. M. Present Status in the Chemistry of Hexuronic Acids Found in Glycosaminoglycans and Their Mimetic Aza-Sugars Analogs. *Mini-Rev. Org. Chem.* **2007**, *4*, 61–74.
31. D'Alonzo, D.; Guaragna, A.; Palumbo, G. Recent Advances in Monosaccharide Synthesis: A Journey into L-Hexose World. *Curr. Org. Chem.* **2009**, *13*, 71–98.
32. Zulueta, M. M. L.; Zhong, Y.-Q.; Hung, S.-C. Synthesis of L-Hexoses and Their Related Biomolecules. *Chem. Commun. (Cambridge)*. **2013**, *49*, 3275–3287.
33. Macher, I.; Dax, K.; Wanek, E.; Weidmann, H. Synthesis of L-Idofuranurono-6,3-Lactone and Its Derivatives *via* Hexodialdodifuranoses. *Carbohydr. Res.* **1980**, *80*, 45–51.
34. Csuk, R.; Hönig, H.; Nimpf, J.; Weidmann, H. A Facile Synthesis of 1,2-*O*-Isopropylidene-β-L-idofuranurono-6,3-lactone. *Tetrahedron Lett.* **1980**, *21*, 2135–2136.
35. Mohamed, S.; Bernhardt, P. V.; Ferro, V. Attempted Synthesis of the Imidazylate of an α-Hydroxylactone Results in Unexpected Chlorination: Synthesis and X-Ray Crystal Structure of 5-Chloro-5-deoxy-1,2-*O*-isopropylidene-β-L-idurono-6,3-lactone. *J. Carbohydr. Chem.* **2014**, *33*, 197–205.
36. Vlahov, I. R.; Linhardt, R. J. Regioselective Synthesis of Derivatives of L-Idopyranuronic Acid—A Key Constituent of Glycosaminoglycans. *Tetrahedron Lett.* **1995**, *36*, 8379–8382.
37. Du, Y. G.; Lin, J. H.; Linhardt, R. J. Regioselective Synthesis of L-Idopyranuronic Acid Derivatives: Intermolecular Aglycon Transfer of Dithioacetal Under Standard Glycosylation Conditions. *J. Carbohydr. Chem.* **1997**, *16*, 1327–1344.
38. Ojeda, R.; de Paz, J. L.; Martín-Lomas, M.; Lassaletta, J. M. A New Route to L-Iduronate Building-Blocks for the Synthesis of Heparin-Like Oligosaccharides. *Synlett* **1999**, ;1316–1318.
39. Jacquinet, J. C.; Petitou, M.; Duchaussoy, P.; Lederman, I.; Choay, J.; Torri, G.; Sinaÿ, P. Synthesis of Heparin Fragments. A Chemical Synthesis of the Trisaccharide *O*-(2-Deoxy-2-sulfamido-3,6-di-*O*-sulfo-α-D-glucopyranosyl)-(1→4)-*O*-(2-*O*-sulfo-α-L-idopyranosyluronic Acid)-(1→4)-2-deoxy-2-sulfamido-6-*O*-sulfo-D-glucopyranose Heptasodium Salt. *Carbohydr. Res.* **1984**, *130*, 221–241.
40. Orgueira, H. A.; Bartolozzi, A.; Schell, P.; Seeberger, P. H. Conformational Locking of the Glycosyl Acceptor for Stereocontrol in the Key Step in the Synthesis of Heparin. *Angew. Chem. Int. Ed.* **2002**, *41*, 2128–2131.

41. Maza, S.; Macchione, G.; Ojeda, R.; López-Prados, J.; Angulo, J.; de Paz, J. L.; Nieto, P. M. Synthesis of Amine-Functionalized Heparin Oligosaccharides for the Investigation of Carbohydrate–Protein Interactions in Microtiter Plates. *Org. Biomol. Chem.* **2012**, *10*, 2146–2163.
42. Maza, S.; Mar Kayser, M.; Macchione, G.; Lopez-Prados, J.; Angulo, J.; de Paz, J. L.; Nieto, P. M. Synthesis of Chondroitin/Dermatan Sulfate-Like Oligosaccharides and Evaluation of Their Protein Affinity by Fluorescence Polarization. *Org. Biomol. Chem.* **2013**, *11*, 3510–3525.
43. Ke, W.; Whitfield, D. M.; Gill, M.; Larocque, S.; Yu, S.-H. A Short Route to L-Iduronic Acid Building Blocks for the Syntheses of Heparin-Like Disaccharides. *Tetrahedron Lett.* **2003**, *44*, 7767–7770.
44. Ke, W.; Whitfield, D. M.; Brisson, J.-R.; Enright, G.; Jarrell, H. C.; Wu, W.-g. Development of Specific Inhibitors for Heparin-Binding Proteins Based on the Cobra Cardiotoxin Structure: An Effective Synthetic Strategy for Rationally Modified Heparin-Like Disaccharides and a Trisaccharide. *Carbohydr. Res.* **2005**, *340*, 355–372.
45. Saito, A.; Wakao, M.; Deguchi, H.; Mawatari, A.; Sobel, M.; Suda, Y. Toward the Assembly of Heparin and Heparan Sulfate Oligosaccharide Libraries: Efficient Synthesis of Uronic Acid and Disaccharide Building Blocks. *Tetrahedron* **2010**, *66*, 3951–3962.
46. Orgueira, H. A.; Bartolozzi, A.; Schell, P.; Litjens, R. E. J. N.; Palmacci, E. R.; Seeberger, P. H. Modular Synthesis of Heparin Oligosaccharides. *Chem. Eur. J.* **2003**, *9*, 140–169.
47. Sinaÿ, P.; Jacquinet, J. C.; Petitou, M.; Duchaussoy, P.; Lederman, I.; Choay, J.; Torri, G. Total Synthesis of a Heparin Pentasaccharide Fragment Having High Affinity for Antithrombin III. *Carbohydr. Res.* **1984**, *132*, C5–C9.
48. Suda, Y.; Bird, K.; Shiyama, T.; Koshida, S.; Marques, D.; Fukase, K.; Sobel, M.; Kusumoto, S. Synthesis and Biological Activity of a Model Disaccharide Containing a Key Unit in Heparin for Binding to Platelets. *Tetrahedron Lett.* **1996**, *37*, 1053–1056.
49. Herczeg, M.; Lázár, L.; Borbás, A.; Lipták, A.; Antus, S. Toward Synthesis of the Isosteric Sulfonate Analogues of the AT-III Binding Domain of Heparin. *Org. Lett.* **2009**, *11*, 2619–2622.
50. Herczeg, M.; Lázár, L.; Mándi, A.; Borbás, A.; Komáromi, I.; Lipták, A.; Antus, S. Synthesis of Disaccharide Fragments of the AT-III Binding Domain of Heparin and Their Sulfonatomethyl Analogues. *Carbohydr. Res.* **2011**, *346*, 1827–1836.
51. Barroca, N.; Jacquinet, J. C. Syntheses of β-D-GalpNAc4SO$_3$-(1→4)-L-IdopA2SO$_3$, a Disaccharide Fragment of Dermatan Sulfate, and of Its Methyl α-L-Glycoside Derivative. *Carbohydr. Res.* **2000**, *329*, 667–679.
52. Schwörer, R.; Zubkova, O. V.; Turnbull, J. E.; Tyler, P. C. Synthesis of a Targeted Library of Heparan Sulfate Hexa- to Dodecasaccharides as Inhibitors of β-Secretase: Potential Therapeutics for Alzheimer's Disease. *Chem. Eur. J.* **2013**, *19*, 6817–6823.
53. van Boeckel, C. A. A.; Beetz, T.; Vos, J. N.; de Jong, A. J. M.; Van Aelst, S. F.; van den Bosch, R. H.; Mertens, J. M. R.; van der Vlugt, F. A. Synthesis of a Pentasaccharide Corresponding to the Antithrombin III Binding Fragment of Heparin. *J. Carbohydr. Chem.* **1985**, *4*, 293–321.
54. Jaurand, G.; Basten, J.; Lederman, I.; van Boeckel, C. A. A.; Petitou, M. Biologically Active Heparin-Like Fragments with a "Non-Glycosamino" Glycan Structure. Part 1: A Pentasaccharide Containing a 3-O-Methyliduronic Acid Unit. *Bioorg. Med. Chem. Lett.* **1992**, *2*, 897–900.
55. Tabeur, C.; Machetto, F.; Mallet, J. M.; Duchaussoy, P.; Petitou, M.; Sinaÿ, P. L-Iduronic Acid Derivatives as Glycosyl Donors. *Carbohydr. Res.* **1996**, *281*, 253–276.
56. Duchaussoy, P.; Jaurand, G.; Driguez, P.-A.; Lederman, I.; Gourvenec, F.; Strassel, J.-M.; Sizun, P.; Petitou, M.; Herbert, J.-M. Design and Synthesis of Heparin Mimetics Able to Inhibit Thrombin. Part 1. Identification of a Hexasaccharide

Sequence Able to Inhibit Thrombin and Suitable for, "Polymerization". *Carbohydr. Res.* **1999**, *317*, 63–84.

57. Prabhu, A.; Venot, A.; Boons, G.-J. New Set of Orthogonal Protecting Groups for the Modular Synthesis of Heparan Sulfate Fragments. *Org. Lett.* **2003**, *5*, 4975–4978.
58. Codée, J. D. C.; Stubba, B.; Schiattarella, M.; Overkleeft, H. S.; van Boeckel, C. A. A.; van Boom, J. H.; van der Marel, G. A. A Modular Strategy Toward the Synthesis of Heparin-Like Oligosaccharides Using Monomeric Building Blocks in a Sequential Glycosylation Strategy. *J. Am. Chem. Soc.* **2005**, *127*, 3767–3773.
59. Polat, T.; Wong, C.-H. Anomeric Reactivity-Based One-Pot Synthesis of Heparin-Like Oligosaccharides. *J. Am. Chem. Soc.* **2007**, *129*, 12795–12800.
60. Tatai, J.; Osztrovszky, G.; Kajtár-Peredy, M.; Fügedi, P. An Efficient Synthesis of L-Idose and L-Iduronic Acid Thioglycosides and Their Use for the Synthesis of Heparin Oligosaccharides. *Carbohydr. Res.* **2008**, *343*, 596–606.
61. Wang, Z.; Xu, Y.; Yang, B.; Tiruchinapally, G.; Sun, B.; Liu, R.; Dulaney, S.; Liu, J.; Huang, X. Preactivation-Based, One-Pot Combinatorial Synthesis of Heparin-Like Hexasaccharides for the Analysis of Heparin–Protein Interactions. *Chem. Eur. J.* **2010**, *16*, 8365–8375.
62. Tiruchinapally, G.; Yin, Z.; El-Dakdouki, M.; Wang, Z.; Huang, X. Divergent Heparin Oligosaccharide Synthesis with Pre-installed Sulfate Esters. *Chem. Eur. J.* **2011**, *17*, 10106–10112.
63. Yang, B.; Yoshida, K.; Yin, Z. J.; Dai, H.; Kavunja, H.; El-Dakdouki, M. H.; Sungsuwan, S.; Dulaney, S. B.; Huang, X. F. Chemical Synthesis of a Heparan Sulfate Glycopeptide: Syndecan-1. *Angew. Chem. Int. Ed.* **2012**, *51*, 10185–10189.
64. Herczeg, M.; Lázár, L.; Bereczky, Z.; Kövér, K. E.; Timári, I.; Kappelmayer, J.; Lipták, A.; Antus, S.; Borbás, A. Synthesis and Anticoagulant Activity of Bio-isosteric Sulfonic-Acid Analogs of the Antithrombin-Binding Pentasaccharide Domain of Heparin. *Chem. Eur. J.* **2012**, *18*, 10643–10652.
65. Roy, S.; El Hadri, A.; Richard, S.; Denis, F.; Holte, K.; Duffner, J.; Yu, F.; Galcheva-Gargova, Z.; Capila, I.; Schultes, B.; Petitou, M.; Kaundinya, G. V. Synthesis and Biological Evaluation of a Unique Heparin Mimetic Hexasaccharide for Structure–Activity Relationship Studies. *J. Med. Chem.* **2014**, *57*, 4511–4520.
66. van Boeckel, C. A. A.; Petitou, M. The Unique Antithrombin III Binding Domain of Heparin: A Lead to New Synthetic Antithrombotics. *Angew. Chem. Int. Ed. Engl.* **1993**, *32*, 1671–1690.
67. Lee, J.-C.; Lu, X.-A.; Kulkarni, S. S.; Wen, Y.-S.; Hung, S.-C. Synthesis of Heparin Oligosaccharides. *J. Am. Chem. Soc.* **2004**, *126*, 476–477.
68. Fügedi, P.; Tatai, J.; Czinege, E. In: *12th European Carbohydrate Symposium,* Grenoble, France, July 6–11, 2003. Abstr. OC012.
69. Fügedi, P.; Tatai, J.; Czinege, E. In: *22nd International Carbohydrate Symposium,* Glasgow, United Kingdom, July 23–27, 2004. Abstr. C29.
70. Hung, S.-C.; Lu, X.-A.; Lee, J.-C.; Chang, M. D.-T.; Fang, S.-l.; Fan, T.-c.; Zulueta, M. M. L.; Zhong, Y.-Q. Synthesis of Heparin Oligosaccharides and Their Interaction with Eosinophil-Derived Neurotoxin. *Org. Biomol. Chem.* **2012**, *10*, 760–772.
71. Xu, P.; Xu, W.; Dai, Y.; Yang, Y.; Yu, B. Efficient Synthesis of a Library of Heparin Tri- and Tetrasaccharides Relevant to the Substrate of Heparanase. *Org. Chem. Front.* **2014**, *1*, 405–414.
72. Zhou, Y.; Lin, F.; Chen, J.; Yu, B. Toward Synthesis of the Regular Sequence of Heparin: Synthesis of Two Tetrasaccharide Precursors. *Carbohydr. Res.* **2006**, *341*, 1619–1629.

73. Arungundram, S.; Al-Mafraji, K.; Asong, J.; Leach, F. E., III; Amster, I. J.; Venot, A.; Turnbull, J. E.; Boons, G.-J. Modular Synthesis of Heparan Sulfate Oligosaccharides for Structure–Activity Relationship Studies. *J. Am. Chem. Soc.* **2009**, *131*, 17394–17405.
74. Chang, C.-H.; Lico, L. S.; Huang, T.-Y.; Lin, S.-Y.; Chang, C.-L.; Arco, S. D.; Hung, S.-C. Synthesis of the Heparin-Based Anticoagulant Drug Fondaparinux. *Angew. Chem. Int. Ed.* **2014**, *53*, 9876–9879.
75. Li, T.; Ye, H.; Cao, X.; Wang, J.; Liu, Y.; Zhou, L.; Liu, Q.; Wang, W.; Shen, J.; Zhao, W.; Wang, P. Total Synthesis of Anticoagulant Pentasaccharide Fondaparinux. *ChemMedChem* **2014**, *9*, 1071–1080.
76. Guedes, N.; Czechura, P.; Echeverria, B.; Ruiz, A.; Michelena, O.; Martin-Lomas, M.; Reichardt, N.-C. Toward the Solid-Phase Synthesis of Heparan Sulfate Oligosaccharides: Evaluation of Iduronic Acid and Idose Building Blocks. *J. Org. Chem.* **2013**, *78*, 6911–6934.
77. Zulueta, M. M. L.; Lin, S.-Y.; Lin, Y.-T.; Huang, C.-J.; Wang, C.-C.; Ku, C.-C.; Shi, Z.; Chyan, C.-L.; Irene, D.; Lim, L.-H.; Tsai, T.-I.; Hu, Y.-P.; Arco, S. D.; Wong, C.-H.; Hung, S.-C. α-Glycosylation by D-Glucosamine-Derived Donors: Synthesis of Heparosan and Heparin Analogues That Interact with Mycobacterial Heparin-Binding Hemagglutinin. *J. Am. Chem. Soc.* **2012**, *134*, 8988–8995.
78. Hu, Y.-P.; Zhong, Y.-Q.; Chen, Z.-G.; Chen, C.-Y.; Shi, Z.; Zulueta, M. M. L.; Ku, C.-C.; Lee, P.-Y.; Wang, C.-C.; Hung, S.-C. Divergent Synthesis of 48 Heparan Sulfate-Based Disaccharides and Probing the Specific Sugar-Fibroblast Growth Factor-1 Interaction. *J. Am. Chem. Soc.* **2012**, *134*, 20722–20727.
79. Lu, L.-D.; Shie, C.-R.; Kulkarni, S. S.; Pan, G.-R.; Lu, X.-A.; Hung, S.-C. Synthesis of 48 Disaccharide Building Blocks for the Assembly of a Heparin and Heparan Sulfate Oligosaccharide Library. *Org. Lett.* **2006**, *8*, 5995–5998.
80. Hu, Y.-P.; Lin, S.-Y.; Huang, C.-Y.; Zulueta, M. M. L.; Liu, J.-Y.; Chang, W.; Hung, S.-C. Synthesis of 3-O-Sulfonated Heparan Sulfate Octasaccharides That Inhibit the Herpes Simplex Virus Type 1 Host–Cell Interaction. *Nat. Chem.* **2011**, *3*, 557–563.
81. Tatai, J.; Fügedi, P. Synthesis of the Putative Minimal FGF Binding Motif Heparan Sulfate Trisaccharides by an Orthogonal Protecting Group Strategy. *Tetrahedron* **2008**, *64*, 9865–9873.
82. Chen, C.; Yu, B. Efficient synthesis of idraparinux, the anticoagulant pentasaccharide. *Bioorg. Med. Chem. Lett.* **2009**, *19*, 3875–3879.
83. Lin, F.; Lian, G.; Zhou, Y. Synthesis of Fondaparinux: Modular Synthesis Investigation for Heparin Synthesis. *Carbohydr. Res.* **2013**, *371*, 32–39.
84. Takahashi, H.; Hitomi, Y.; Iwai, Y.; Ikegami, S. A Novel and Practical Synthesis of L-Hexoses from D-Glycono-1,5-lactones. *J. Am. Chem. Soc.* **2000**, *122*, 2995–3000.
85. Lewis, M. D.; Cha, J. K.; Kishi, Y. Highly Stereoselective Approaches to α- and β-*C*-Glycopyranosides. *J. Am. Chem. Soc.* **1982**, *104*, 4976–4978.
86. Takahashi, H.; Shida, T.; Hitomi, Y.; Iwai, Y.; Miyama, N.; Nishiyama, K.; Sawada, D.; Ikegami, S. Divergent Synthesis of L-Sugars and L-Imino-Sugars from L-Sugars. *Chem. Eur. J.* **2006**, *12*, 5868–5877.
87. Chida, N.; Yamada, E.; Ogawa, S. Synthesis of Methyl (Methyl D-and L-Idopyranosid) Uronates from *myo*-Inositol. *J. Carbohydr. Chem.* **1988**, 7, 555–570.
88. Johansson, M. H.; Samuelson, O. Epimerization and Degradation of 2-O-(4-O-Methyl-α-D-glucopyranosyluronic Acid)-D-Xylitol in Alkaline Medium. *Carbohydr. Res.* **1977**, *54*, 295–299.
89. Baggett, N.; Smithson, A. Synthesis of L-Iduronic Acid Derivatives by Epimerisation of Anancomeric D-Glucuronic Acid Analogues. *Carbohydr. Res.* **1982**, *108*, 59–70.

90. Chiba, T.; Sinaÿ, P. Application of a Radical Reaction to the Synthesis of L-Iduronic Acid Derivatives from D-Glucuronic Acid Analogs. *Carbohydr. Res.* **1986**, *151*, 379–389.
91. Somsák, L.; Ferrier, R. J. Radical-Mediated Brominations at Ring Positions of Carbohydrates. *Adv. Carbohydr. Chem. Biochem.* **1991**, *49*, 37–92.
92. Ferrier, R. J.; Furneaux, R. H. C-5 Bromination of Some Glucopyranuronic Acid Derivatives. *J. Chem. Soc. Perkin Trans.* **1977**, 1996–2000.
93. Chiba, T.; Jacquinet, J.-C.; Sinaÿ, P.; Petitou, M.; Choay, J. Chemical Synthesis of L-Iduronic Acid-Containing Disaccharidic Fragments of Heparin. *Carbohydr. Res.* **1988**, *174*, 253–264.
94. Marra, A.; Dong, X.; Petitou, M.; Sinaÿ, P. Synthesis of Disaccharide Fragments of Dermatan Sulfate. *Carbohydr. Res.* **1989**, *195*, 39–50.
95. Nilsson, M.; Svahn, C.-M.; Westman, J. Synthesis of the Methyl Glycosides of a Tri- and a Tetra-Saccharide Related to Heparin and Heparan Sulphate. *Carbohydr. Res.* **1993**, *246*, 161–172.
96. Medaković, D. An Efficient Synthesis of Methyl 1,2,3,4-Tetra-*O*-acetyl-*β*-L-idopyranuronate. *Carbohydr. Res.* **1994**, *253*, 299–300.
97. Yu, H. N.; Furukawa, J.; Ikeda, T.; Wong, C.-H. Novel Efficient Routes to Heparin Monosaccharides and Disaccharides Achieved via Regio and Stereoselective Glycosidation. *Org. Lett.* **2004**, *6*, 723–726.
98. Voznyi, Y. V.; Kalicheva, I. S.; Galoyan, A. A.; Gusina, N. B. A Convenient Synthesis of Fluorogenic Glycosides of α-L-Iduronic Acid. *Bioorg. Khim.* **1989**, *15*, 1411–1415.
99. Blanchard, S.; Sadilek, M.; Scott, C. R.; Turecek, F.; Gelb, M. H. Tandem Mass Spectrometry for the Direct Assay of Lysosomal Enzymes in Dried Blood Spots: Application to Screening Newborns for Mucopolysaccharidosis I. *Clin. Chem.* **2008**, *54*, 2067–2070.
100. Salamone, S.; Boisbrun, M.; Didierjean, C.; Chapleur, Y. From D-Glucuronic Acid to L-Iduronic Acid Derivatives via a Radical Tandem Decarboxylation-Cyclization. *Carbohydr. Res.* **2014**, *386*, 99–105.
101. Timmer, M. S. M.; Adibekian, A.; Seeberger, P. H. Short De Novo Synthesis of Fully Functionalized Uronic Acid Monosaccharides. *Angew. Chem. Int. Ed.* **2005**, *44*, 7605–7607.
102. Adibekian, A.; Bindschädler, P.; Timmer, M. S. M.; Noti, C.; Schützenmeister, N.; Seeberger, P. H. De Novo Synthesis of Uronic Acid Building Blocks for Assembly of Heparin Oligosaccharides. *Chem. Eur. J.* **2007**, *13*, 4510–4522.
103. Bindschädler, P.; Adibekian, A.; Grünstein, D.; Seeberger, P. H. De Novo Synthesis of Differentially Protected L-Iduronic Acid Glycosylating Agents. *Carbohydr. Res.* **2010**, *345*, 948–955.
104. Hansen, S. U.; Barath, M.; Salameh, B. A. B.; Pritchard, R. G.; Stimpson, W. T.; Gardiner, J. M.; Jayson, G. C. Scalable Synthesis of L-Iduronic Acid Derivatives via Stereocontrolled Cyanohydrin Reaction for Synthesis of Heparin-Related Disaccharides. *Org. Lett.* **2009**, *11*, 4528–4531.
105. Hansen, S. U.; Miller, G. J.; Barath, M.; Broberg, K. R.; Avizienyte, E.; Helliwell, M.; Raftery, J.; Jayson, G. C.; Gardiner, J. M. Synthesis and Scalable Conversion of L-Iduronamides to Heparin-Related Di- and Tetrasaccharides. *J. Org. Chem.* **2012**, *77*, 7823–7843.
106. Hansen, S. U.; Miller, G. J.; Cole, C.; Rushton, G.; Avizienyte, E.; Jayson, G. C.; Gardiner, J. M. Tetrasaccharide Iteration Synthesis of a Heparin-Like Dodecasaccharide and Radiolabelling for In Vivo Tissue Distribution Studies. *Nat. Commun.* **2013**, *4*, 3016/1–3016/9.
107. Hansen, S. U.; Miller, G. J.; Jayson, G. C.; Gardiner, J. M. First Gram-Scale Synthesis of a Heparin-Related Dodecasaccharide. *Org. Lett.* **2013**, *15*, 88–91.

108. Miller, G. J.; Hansen, S. U.; Avizienyte, E.; Rushton, G.; Cole, C.; Jayson, G. C.; Gardiner, J. M. Efficient Chemical Synthesis of Heparin-Like Octa-, Deca- and Dodecasaccharides and Inhibition of FGF2- and $VEGF_{165}$-Mediated Endothelial Cell Functions. *Chem. Sci.* **2013**, *4*, 3218–3222.
109. Hansen, S. U.; Dalton, C. E.; Barath, M.; Kwan, G.; Raftery, J.; Jayson, G. C.; Miller, G. J.; Gardiner, J. M. Synthesis of L-Iduronic Acid Derivatives via [3.2.1] and [2.2.2] L-Iduronic Lactones from Bulk Glucose-Derived Cyanohydrin Hydrolysis: A Reversible Conformationally-Switched Super-Disarmed/Re-Armed Lactone Route to Heparin Disaccharides. *J. Org. Chem.* **2015**, 3777–3789.
110. Lubineau, A.; Gavard, O.; Alais, J.; Bonnaffé, D. New Accesses to L-Iduronyl Synthons. *Tetrahedron Lett.* **2000**, *41*, 307–311.
111. Dilhas, A.; Bonnaffé, D. Efficient Selective Preparation of Methyl 1,2,4-Tri-*O*-acetyl-3-*O*-benzyl-β-L-idopyranuronate from Methyl 3-*O*-Benzyl-L-iduronate. *Carbohydr. Res.* **2003**, *338*, 681–686.
112. Gavard, O.; Hersant, Y.; Alais, J.; Duverger, V.; Dilhas, A.; Bascou, A.; Bonnaffé, D. Efficient Preparation of Three Building Blocks for the Synthesis of Heparan Sulfate Fragments: Towards the Combinatorial Synthesis of Oligosaccharides from Hypervariable Regions. *Eur. J. Org. Chem.* **2003**, 3603–3620.
113. de Paz, J.-L.; Ojeda, R.; Reichardt, N.; Martín-Lomas, M. Some Key Experimental Features of a Modular Synthesis of Heparin-Like Oligosaccharides. *Eur. J. Org. Chem.* **2003**, 3308–3324.
114. Lohman, G. J. S.; Hunt, D. K.; Högermeier, J. A.; Seeberger, P. H. Synthesis of Iduronic Acid Building Blocks for the Modular Assembly of Glycosaminoglycans. *J. Org. Chem.* **2003**, *68*, 7559–7561.
115. Sawant, R. C.; Liao, Y.-J.; Lin, Y.-J.; Badsara, S. S.; Luo, S.-Y. Formal Synthesis of a Disaccharide Repeating Unit (IdoA-GlcN) of Heparin and Heparan Sulfate. *RSC Adv.* **2015**, *5*, 19027–19033.
116. Oh, Y. I.; Sheng, G. J.; Chang, S. K.; Hsieh-Wilson, L. C. Tailored Glycopolymers as Anticoagulant Heparin Mimetics. *Angew. Chem. Int. Ed.* **2013**, *52*, 11796–11799.
117. Sheng, G. J.; Oh, Y. I.; Chang, S. K.; Hsieh-Wilson, L. C. Tunable Heparan Sulfate Mimetics for Modulating Chemokine Activity. *J. Am. Chem. Soc.* **2013**, *135*, 10898–10901.
118. Dondoni, A.; Marra, A.; Massi, A. Carbohydrate Homologation by the Use of 2-(Trimethylsilyl)Thiazole. Preparative Scale Synthesis of Rare Sugars: L-Gulose, L-Idose, and the Disaccharide Subunit of Bleomycin A(2). *J. Org. Chem.* **1997**, *62*, 6261–6267.
119. Rochepeau-Jobron, L.; Jacquinet, J.-C. Diastereoselective Hydroboration of Substituted Exo-Glucals Revisited. A Convenient Route for the Preparation of L-Iduronic Acid Derivatives. *Carbohydr. Res.* **1997**, *303*, 395–406.
120. Alper, P. B.; Hendrix, M.; Sears, P.; Wong, C.-H. Probing the Specificity of Aminoglycoside–Ribosomal RNA Interactions with Designed Synthetic Analogs. *J. Am. Chem. Soc.* **1998**, *120*, 1965–1978.
121. Hinou, H.; Kurosawa, H.; Matsuoka, K.; Terunuma, D.; Kuzuhara, H. Novel Synthesis of L-Iduronic Acid Using Trehalose as the Disaccharidic Starting Material. *Tetrahedron Lett.* **1999**, *40*, 1501–1504.
122. Hinou, H. Disaccharides as Starting Materials for the Synthesis of Valuable Compounds. *Trends Glycosci. Glycotechnol.* **2000**, *12*, 185–190.
123. Kurita, K.; Masuda, N.; Aibe, S.; Murakami, K.; Ishii, S.; Nishimura, S.-I. Synthetic Carbohydrate Polymers Containing Trehalose Residues in the Main Chain: Preparation and Characteristic Properties. *Macromolecules* **1994**, *27*, 7544–7549.
124. Hung, S.-C.; Chen, C.-S. Efficient Synthesis of 1,2,3,4,6-Penta-*O*-acetyl-L-idopyranose. *J. Chin. Chem. Soc.* **2000**, *47*, 1257–1262.

125. Kulkarni, S. S.; Chi, F.-C.; Hung, S.-C. Biologically Potent L-Hexoses and 6-Deoxy-L-hexoses: Their Syntheses and Applications. *J. Chin. Chem. Soc.* **2004**, *51*, 1193–1200.
126. Lu, F.-C.; Lico, L. S.; Hung, S.-C. Synthesis of a Fluorogenic Substrate for α-L-Iduronidase. *ARKIVOC* **2013**, 13–21.
127. Lee, J.-C.; Chang, S.-W.; Liao, C.-C.; Chi, F.-C.; Chen, C.-S.; Wen, Y.-S.; Wang, C.-C.; Kulkarni, S. S.; Puranik, R.; Liu, Y.-H.; Hung, S.-C. From D-Glucose to Biologically Potent L-Hexose Derivatives: Synthesis of α-L-Iduronidase Fluorogenic Detector and the Disaccharide Moieties of Bleomycin A2 and Heparan Sulfate. *Chem. Eur. J.* **2004**, *10*, 399–415.
128. Hung, S.-C.; Thopate, S. R.; Chi, F.-C.; Chang, S.-W.; Lee, J.-C.; Wang, C.-C.; Wen, Y.-S. 1,6-Anhydro-β-L-hexopyranoses as Potent Synthons in the Synthesis of the Disaccharide Units of Bleomycin A2 and Heparin. *J. Am. Chem. Soc.* **2001**, *123*, 3153–3154.
129. Hung, S.-C.; Puranik, R.; Chi, F.-C. Novel Synthesis of 1,2:3,5-Di-*O*-isopropylidene-β-L-idofuranoside and Its Derivatives at C6. *Tetrahedron Lett.* **2000**, *41*, 77–80.
130. Hodosi, G.; Podányi, B.; Kuszmann, J. The Mechanism of the Hydroxyl→Halogen Exchange Reaction in the Presence of Triphenylphosphine, *N*-Bromosuccinimide, and *N,N*-Dimethylformamide: Application of a New Vilsmeier-Type Reagent in Carbohydrate Chemistry. *Carbohydr. Res.* **1992**, *230*, 327–342.
131. Hung, S. C.; Wang, C. C.; Chang, S. W.; Chen, C. S. 1,6-Anhydro-β-L-hexopyranoses as Valuable Building Blocks Toward the Synthesis of L-Gulosamine and L-Altrose Derivatives. *Tetrahedron Lett.* **2001**, *42*, 1321–1324.
132. Bazin, H. G.; Kerns, R. J.; Linhardt, R. J. Regio and Stereoselective Conversion of Δ^4-Uronic Acids to L-Ido- and D-Glucopyranosiduronic Acids. *Tetrahedron Lett.* **1997**, *38*, 923–926.
133. Bazin, H. G.; Wolff, M. W.; Linhardt, R. J. Regio- and Stereoselective Synthesis of β-D-Gluco-, α-L-Ido-, and α-L-Altropyranosiduronic Acids from Δ^4-Uronates. *J. Org. Chem.* **1999**, *64*, 144–152.
134. Cheng, G.; Fan, R.; Hernández-Torres, J. M.; Boulineau, F. P.; Wei, A. *syn*-Additions to 4α-Epoxypyranosides: Synthesis of L-Idopyranosides. *Org. Lett.* **2007**, *9*, 4849–4852.
135. Boulineau, F. P.; Wei, A. Synthesis of L-Sugars from 4-Deoxypentenosides. *Org. Lett.* **2002**, *4*, 2281–2283.
136. Thiem, J.; Ossowski, P. Studies of Hexuronic Acid Ester Glycals and the Synthesis of 2-Deoxy-β-glycoside Precursors. *J. Carbohydr. Chem.* **1984**, *3*, 287–313.
137. Schell, P.; Orgueira, H. A.; Roehrig, S.; Seeberger, P. H. Synthesis and Transformations of D-Glucuronic and L-Iduronic Acid Glycals. *Tetrahedron Lett.* **2001**, *42*, 3811–3814.
138. Adinolfi, M.; Barone, G.; De Lorenzo, F.; Iadonisi, A. Intramolecular Tishchenko Reactions of Protected Hexos-5-uloses: A Novel and Efficient Synthesis of L-Idose and L-Altrose. *Synlett* **1999**, 336–338.
139. Adinolfi, M.; Barone, G.; Iadonisi, A.; Mangoni, L. Diastereoselectivity of the Cyclization of Hexos-5-uloses by Sm_2-Mediated Pinacol Coupling. *Tetrahedron Lett.* **1998**, *39*, 2021–2024.
140. Uenishi, J. i.; Masuda, S.; Wakabayashi, S. Intramolecular Sm^{2+} and Sm^{3+} Promoted Reaction of γ-Oxy-δ-Ketoaldehyde; Stereocontrolled Formation of Pinacol and Lactone. *Tetrahedron Lett.* **1991**, *32*, 5097–5100.
141. Guaragna, A.; Napolitano, C.; D'Alonzo, D.; Pedatella, S.; Palumbo, G. A Versatile Route to L-Hexoses: Synthesis of L-Mannose and L-Altrose. *Org. Lett.* **2006**, *8*, 4863–4866.

142. D'Alonzo, D.; Guaragna, A.; Napolitano, C.; Palumbo, G. Rapid access to 1,6-Anhydro-β-L-hexopyranose Derivatives via Domino Reaction: Synthesis of L-Allose and L-Glucose. *J. Org. Chem.* **2008**, *73*, 5636–5639.
143. Guaragna, A.; D'Alonzo, D.; Paolella, C.; Napolitano, C.; Palumbo, G. Highly Stereoselective De Novo Synthesis of L-Hexoses. *J. Org. Chem.* **2010**, *75*, 3558–3568.
144. Caputo, R.; Guaragna, A.; Palumbo, G.; Pedatella, S. A New and Versatile Allylic Alcohol Anion and Acyl β-Anion Equivalent for Three-Carbon Homologations. *J. Org. Chem.* **1997**, *62*, 9369–9371.
145. Hubschwerlen, C.; Specklin, J. L.; Higelin, J. L-(*S*)-Glyceraldehyde Acetonide [1,3-Dioxolane-4-carboxaldehyde, 2,2-dimethyl-, (*S*)-]. *Org. Synth.* **1995**, *72*, 1–5.
146. Frihed, T. G.; Pedersen, C. M.; Bols, M. Synthesis of All Eight L-Glycopyranosyl Donors Using CH Activation. *Angew. Chem. Int. Ed.* **2014**, *53*, 13889–13893.
147. Frihed, T. G.; Pedersen, C. M.; Bols, M. Synthesis of All Eight Stereoisomeric 6-Deoxy-L-hexopyranosyl Donors—Trends in Using Stereoselective Reductions or Mitsunobu Epimerizations. *Eur. J. Org. Chem.* **2014**, 7924–7939.

Glycosylation of Cellulases: Engineering Better Enzymes for Biofuels

Eric R. Greene*[,1], Michael E. Himmel†, Gregg T. Beckham‡, Zhongping Tan*

*Department of Chemistry and Biochemistry and BioFrontiers Institute, University of Colorado, Boulder, Colorado, USA

†Biosciences Center, National Renewable Energy Laboratory, Golden, Colorado, USA

‡National Bioenergy Center, National Renewable Energy Laboratory, Golden, Colorado, USA

[1]Current address: Department of Molecular and Cellular Biology, University of California at Berkeley, Berkeley, California, 94720, USA.

Contents

1. Introduction 65
2. Glycosylation of Cellulose-Degrading Enzymes 66
   2.1 Introduction 66
   2.2 Glycan Structures Found on *Tr*Cel7A and Additional Secreted Cellulases 68
   2.3 Implications of N-Glycosylation of the *Tr*Cel7A Catalytic Domain 72
   2.4 Implications of O-Glycosylation of *Tr*Cel7A Linker Domain 76
   2.5 Implications of O-Glycosylation of the CBM 79
3. Recombinant Expression of Fungal Cellulases 82
   3.1 Expression of *T. reesei* Cellulases in *Saccharomyces cerevisiae* 82
   3.2 Glycoengineered Strains of and Heterologous Protein Expression in *Pichia pastoris* 84
   3.3 Glycosylation and Engineering of Expressed Proteins in *Aspergillus* Species 86
4. Modifications by Glycan-Trimming Enzymes 88
   4.1 Introduction 88
   4.2 Secreted Glycan-Active α-Mannosidases from *T. reesei* 89
   4.3 Secreted α-Mannosidases from Additional Fungi 91
   4.4 *endo*-β-*N*-Acetylglucosaminidases Secreted by Fungi 92
5. Summary and Future Perspectives 96
Acknowledgments 96
Appendix 1. Molecular Dynamics Simulation of a Linker Interacting with Crystalline Cellulose 97
References 97

Advances in Carbohydrate Chemistry and Biochemistry, Volume 72
ISSN 0065-2318
http://dx.doi.org/10.1016/bs.accb.2015.08.001

ABBREVIATIONS

Ala L-alanine
Arg L-arginine
Asn L-asparagine
CAZy carbohydrate-active enzyme
CAZymes carbohydrate-active enzymes
CBM carbohydrate-binding module
CD catalytic domain
CNX/CRT calnexin/calreticulin
ENGase *endo*-*N*-acetylglucosaminidase
ER endoplasmic reticulum
ER-MnsI Class I ER mannosidase
FucT α-1,3-fucosyltransferase
GalT galactose-1-phosphate uridylyltransferase
GH glycoside hydrolase
GI glucosidase I
GII glucosidase II
Glc D-glucose
GlcNAc 2-(acetylamino)-2-deoxy-D-glucose
Gln L-glutamine
Gly glycine
GnT-I *N*-acetylglucosamine transferase I
GnT-II *N*-acetylglucosaminyl transferase II
GnT-IV *N*-acetylglucosaminetransferase IV
Man D-mannopyranose
Mnn1 α-1,3-mannosyltransferase
Mnn4 mannosylphosphatetransferase
Mns-I Class I Golgi mannosidase
Mns-II Class II Golgi mannosidase
ND not determined
NMR nuclear magnetic resonance
Och1 α-1,6-mannosyltransferase
OST oligosaccharidetransferase complex
P phosphate
PNGase F peptide-*N*-glycosidase F
Pro L-proline
Ser L-serine
SiaT α-2,8-sialyltransferase I
Thr L-threonine
Trp L-tryptophan
UGGT UDP-glucose glycoprotein:glucosyltransferase
XylT β-1,3-xylosyltransferase
β3GalT β-1,3-galactosyltransferase

1. INTRODUCTION

Cellulose is a key structural component of plant cell walls and is the most abundant source of renewable carbon on Earth. The biological conversion of the stored potential energy in cellulose to biologically derived fuels (biofuels) has gained much attention over the past few decades as the drive to shift human energy dependence from fossil fuels to renewable sources continues. The biological depolymerization of cellulose found in lignocellulosic biomass is primarily achieved from the action of synergistic cellulase enzymes. Cellulases are capable of breaking down insoluble crystalline cellulose into soluble sugars that can then be fed to ethanologens to produce bioethanol or other engineered microorganisms to produce other fuel precursors.[1,2] Currently, the saccharification of cellulose in recalcitrant lignocellulosic biomass by cellulases remains costly, thus hindering the commercial bioethanol production process.[3,4]

One reason that the enzymatic saccharification step of bioalcohol production is rate limiting is due to the inherent challenges of degrading complex mixtures of polysaccharides and phenolic polymers present in plant cell walls. The three most prevalent lignocellulosic components are cellulose, hemicellulose, and lignin. Numerous methods are available to process biomass to isolate its crude components, although complete purification of cellulose from lignocellulosic biomass is too costly.[5] Lignin is a heterogeneous aromatic polymer found in higher plants that consists of assembled phenolic esters likely synthesized via radical coupling reactions. Large lignin complexes are hydrophobic and can carry a charge depending upon the pretreatment method applied to the lignocellulosic feedstock. Hemicellulose includes a diversity of polysaccharides that act to tether cellulose microfibrils together in plant cell walls. Cellulose is a polymer of β-1,4-linked D-glucose monomers. Cellulose fibers aggregate in aqueous solution. These aggregated fibers can form water-insoluble crystalline structures (microfibrils), and these structures provide plants cell walls with necessary stability. The intrinsic complexity of cellulosic substrates makes saccharification difficult and costly at the process scale.[6]

Moreover, cellulases are inherently slow enzymes that must overcome the obstacles inherent to a highly heterogeneous substrate, thus compounding existing cost challenges in commercializing bioalcohol production. One strategy to overcome these obstacles is to engineer and evolve cellulases for improved activity.[3,4,6–9] However, to efficiently engineer enzymes for improved activity, detailed knowledge of cellulase properties is required. This knowledge includes understanding reaction mechanisms, structural motifs that contribute to stability and activity, and the posttranslational modifications that can alter these functions. One of the most prevalent, but least studied posttranslational modifications to fungal cellulases, is glycosylation.[7,10–12]

Glycosylation is the biochemical addition of saccharide moieties to proteins and falls into two broad classes in eukaryotes: O-linked and N-linked glycosylation. N-Linked glycosylation pathways have been well studied in eukaryotic cells and involve a conserved mechanism for the production of $GlcMan_9GlcNAc_2$-dolichol phosphate that is transferred to asparagine residues in nascent polypeptides with the conserved Asn-X-Ser/Thr consensus sequence (where X can be any amino acid that is not proline) in the endoplasmic reticulum (ER) (Fig. 1A). Various enzymes in the ER and Golgi further modify these glycans creating diverse structures that are specific for various domains of life (Fig. 1).[10,13] O-Linked glycosylation is performed in the ER or Golgi and involves the addition of typically one to three saccharide units to either serine or threonine residues.[11] O-Glycosylation does not occur on an amino acid consensus sequence like N-glycosylation and is therefore more difficult to predict from genomic data.[11,14]

Glycosylation is found in ~50% of eukaryotic proteins and is typically required for cellulase secretion in filamentous fungi.[10,12,13,15,16] Despite the important role for glycosylation in cellulase secretion, relatively little is known about the biochemical consequences of protein glycosylation of fungal cellulases. Moreover, an exhaustive survey of structure–function relationships between glycans and any single cellulase has not yet been reported, likely stemming from the difficulties inherent in studying the glycosylation process. Despite these challenges, many studies have elucidated various roles for glycans in cellulases including but not limited to the following: proteolytic stability, catalytic activity modulation, thermal stability, solubility promotion, pH stability, and conformational stabilization.[7,17–26] Given the myriad of crucial protein properties affected by glycosylation, controlling and optimizing this process in fungal cellulases is crucial for engineering the optimal activity and stability required to achieve efficient saccharification of biomass for biofuels production.[4,7,16]

2. GLYCOSYLATION OF CELLULOSE-DEGRADING ENZYMES

2.1 Introduction

Trichoderma reesei is well known to produce high titers of cellulases, whose synergistic action can saccharify lignocellulosic biomass into soluble and

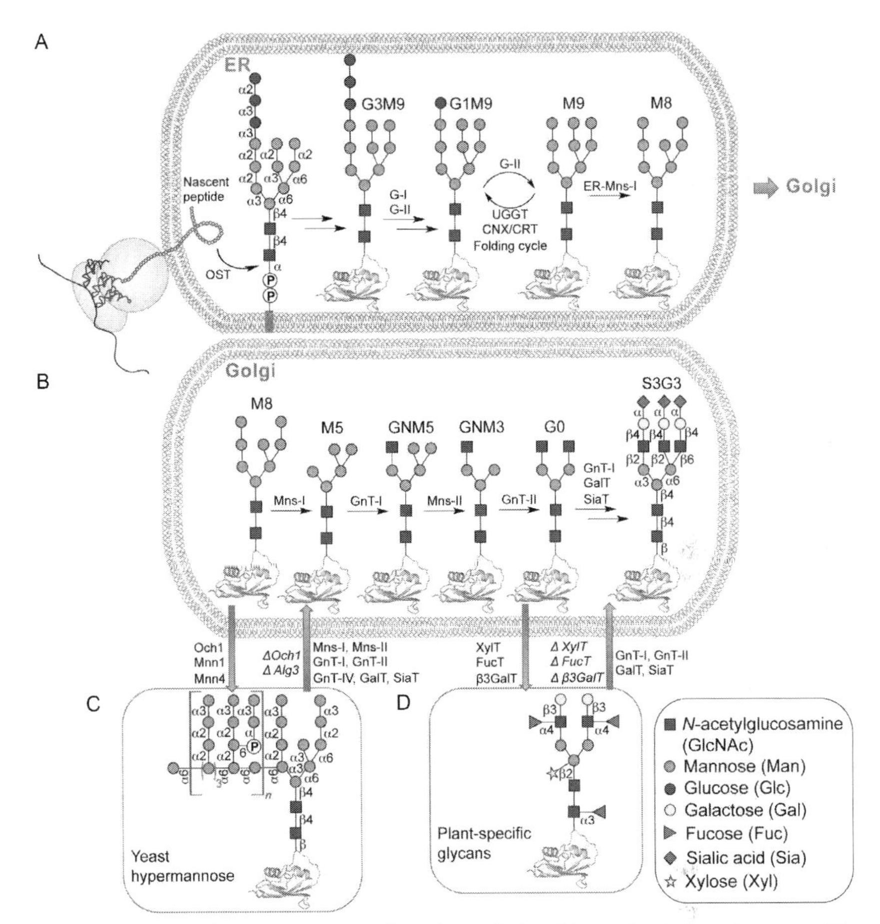

Figure 1 Graphical representation of N-glycosylation biosynthesis in eukaryotes. The dolichol phosphate intermediate, nascent growing polypeptide docked on the ER membrane and *N*-glycan biosynthesis in the ER (A) and Golgi apparatus (B) are depicted. Examples of hypermannose structures found in yeast and complex-type *N*-glycans present in plants are also given in (C) and (D), respectively. Circled letter P represents a phosphate moiety.

fermentable sugars.[8,27] These cellulases fall into four broad categories based on catalytic action on cellulose: exoglucanases, endoglucanases, β-glucosidases, and lytic polysaccharide monooxygenases. The most abundant cellulase secreted by *T. reesei* is a Family 7 glycoside hydrolase termed Cel7A (CBH I or *Tr*Cel7A), which is a cellobiohydrolase that is specific for the reducing end of cellulose. *Tr*Cel7A accounts for 50–60% of all secreted proteins from *T. reesei*.[28,29] Structurally *Tr*Cel7A is a multimodular enzyme containing a catalytic domain (CD) and a carbohydrate-binding module (CBM) attached via a heavily O-glycosylated linker. The catalytic domain contains *N*-linked

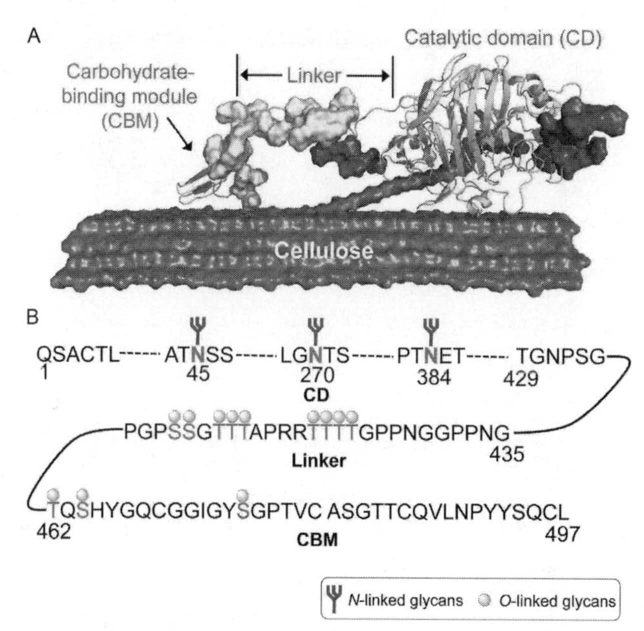

Figure 2 Molecular snapshot of a catalytically engaged *Tr*Cel7A on crystalline cellulose with domains labeled (A). Cyan (light gray in the print version) represents the polypeptide chain. Blue (dark gray in the print version) represents the *N*-linked glycans. Yellow (light gray in the print version) represents the *O*-linked glycans. Green (dark gray in the print version) represents crystalline cellulose. Amino acid sequence scheme of *Tr*Cel7A with locations of glycosylation sites labeled in red (gray in the print version) (B).

glycans and the CBM also contains *O*-linked glycans[24,25,30–32] (Fig. 2). Large discrepancies exist in the *Tr*Cel7A glycan patterns and structures when secreted by different *T. reesei* strains and also when the fungi are grown with different media and even at different pH.[33,34]

2.2 Glycan Structures Found on *Tr*Cel7A and Additional Secreted Cellulases

Filamentous fungi produce high mannose-type *N*-glycans and highly diverse *O*-glycans composed primarily of mannose with instances of glucose and charged moieties, such as sulfate and phosphate groups (Fig. 3 and 4).[10,11,31,32,35] High-mannose, phosphorylated *N*-linked glycans are known to exist in secreted *Tr*Cel7A.[29,36] Various groups have demonstrated that the *N*-linked glycans on the catalytic domain of *Tr*Cel7A secreted by RUT-C30 exist as large charged glycans, such as P_2Man_{11}-$GlcNAc_2$ and $P_1Man_9GlcNAc_2$, in which the phosphate group was shown to link two mannose residues through a phosphodiester linkage to

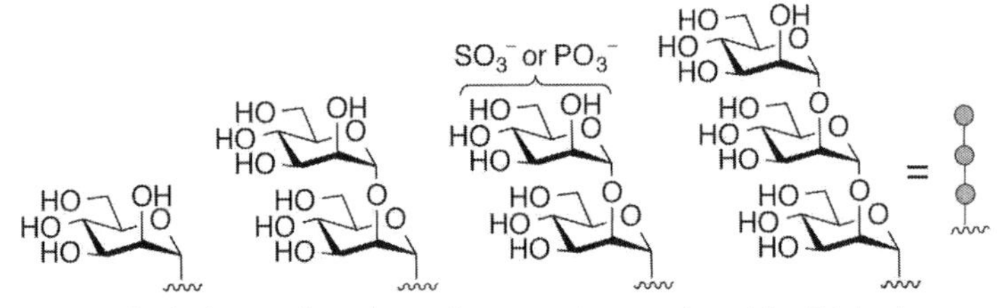

Figure 3 *O*-Linked glycans found on glycoproteins produced by *Trichoderma* species.

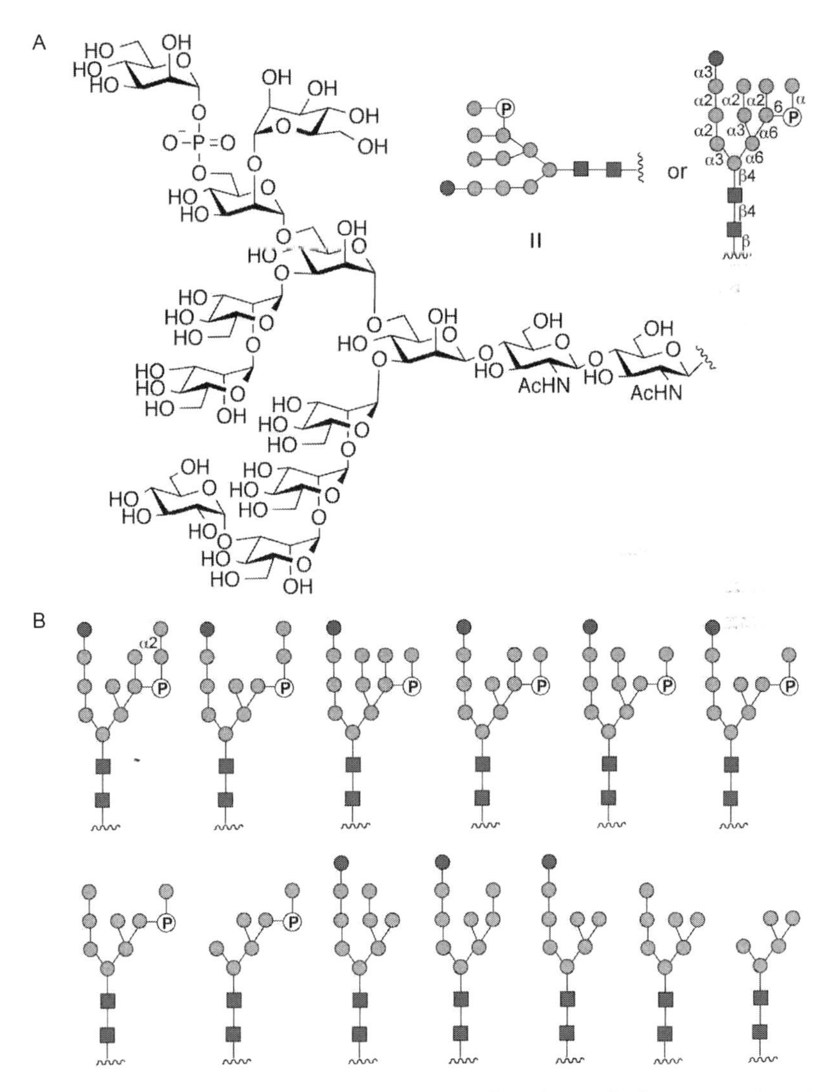

Figure 4 High-mannose-type *N*-glycan structures found in *Trichoderma reesei* RUT-C30 strain. (A) The chemical structure of a representative glycan. (B) Cartoon depiction of different high mannose charged and uncharged *N*-glycan moieties.

uncharged glycans, such as $GlcMan_9GlcNAc_2$.[29,33,34,36–38] Additional studies of different strains of *T. reesei* reported the smallest potential *N*-linked glycan GlcNAc[20,30] on *Tr*Cel7A or the mammalian high-mannose-type $Man_5GlcNAc_2$ *N*-glycans.[20,39] These structures are readily found in *T. reesei* strains QM9414, QM6a, or VTT-D-80133 (Fig. 5).[33,34,39]

One of the reasons for the dramatic difference in *N*-glycan structure between these *T. reesei* strains is due to a frameshift mutation in the glucosidase II alpha subunit gene involved in *N*-glycan maturation (Fig. 1).[40] This mutation results in improper *N*-glycan processing and gives rise to the larger, unique *N*-glycans found in the RUT-C30 strain. Also important is the discovery of a unique glycan-trimming enzyme, Endo T, secreted by *T. reesei* that is responsible for creating the unique GlcNAc residue found at N-linked glycosylation sites in other strains of *T. reesei*.[41,42] However, this enzyme is not always present in secretomic mixtures and therefore the presence of a single GlcNAc *N*-glycan is not as common, as will be discussed at greater length in Section 4.[43,44]

These cropped glycan structures have also been noted in other cellulases produced by *T. reesei*. The endoglucanase, Cel7B, contains five potential N-linked glycosylation sites, two of which were shown to be natively occupied by glycans. Cel7B secreted by *T. reesei* QM 9414 can also contain a single GlcNAc residue at the N-linked glycosylation sites,[45] yet under different culture conditions will preferentially produce $Man_5GlcNAc_2$.[46] The same N-linked glycosylation pattern was also found in the crystal structure for Cel61B, a lytic polysaccharide monooxygenase (LPMO) that oxidatively cleaves glycosidic bonds in cellulose via a metal-dependent mechanism.[47] Lastly, a β-glucosidase that was homologously expressed in *T. reesei* had three of the four N-linked glycosylation sites occupied by the single GlcNAc

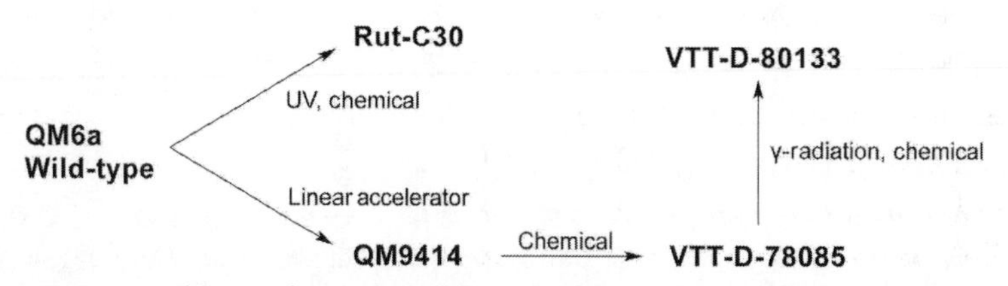

Figure 5 Ancestral hierarchy and mutagenesis source of *Trichoderma reesei* wild-type strain QM6a to form cellulase hyperproducing strains. UV represents ultraviolet radiation treatment of cells. Chemical represents treatment of cells with various mutagenic compounds. Linear accelerator represents treatment of cells with particles generated by a linear accelerator. γ-Radiation represents treatment of cells with γ-rays. *Ancestral information was obtained and adapted from Stals* et al.[33,34]

residue, whereas the fourth residue was reported to be of the high-mannose type.[48] Furthermore, it was noted that a small population of *N*-glycans of Cel7B, Cel6A, and Cel5A contained a single GlcNAc residue, indicating the presence of Endo T in the secretome of RUT-C30.[35] These data further demonstrate that RUT-C30 stains are capable of producing significantly different N-glycosylation patterns compared to its ancestor strain QM6a. This highly heterogeneous representation of glycan structure is to be expected, given the different growth conditions in each of the above studies that likely give raise to different expression patterns of Endo T and carbon catabolism.

The *O*-linked glycan structures of filamentous fungi also display a wide structural variation similar to the vast diversity of *N*-glycans. O-Linked glycosylation of the *Tr*Cel7A linker and CBM occurs on Ser/Thr residues and can contain single *O*-linked mannose residues or poly-mannosylated and charged glycans.[31,32] To our knowledge, no study has yet determined the exact structure of endogenous linkages in the *Tr*Cel7A *O*-linked glycans; however, *Trichoderma* species are capable of producing linear α-1,2-mannosidic chains up to three mannose residues long, phosphate/sulfate-Man-α-1,2-Man, and branched glycans which include α-1,6-linkages to either glucose or mannose (Fig. 3).[11] Due to the high abundance of O-linked glycosylation sites and the lack a consensus sequence for *O*-glycan placement, the heterogeneity of the *O*-glycans of *Tr*Cel7A has been difficult to thoroughly analyze.[14,49] Because the precise *O*-glycan structures have yet to be confirmed in *Tr*Cel7A, the effects that different carbohydrate moieties have on the functions of this enzyme remain largely unknown. Although our knowledge of the effects of glycosylation on the properties of *Tr*Cel7A is incomplete, the glycosylation of *Tr*Cel7A has been the most studied of all cellulases and may serve as an effective model for future studies of other GH7 enzymes and even general cellulase glycosylation effects.

Most early studies of the glycosylation effects of enzymes adopted the "all or nothing" approach, whereby the enzyme activities and properties were compared between fully glycosylated and deglycosylated proteins. Protein deglycosylation is typically achieved by endoglycosidases (PNGase F or Endo H), or expression in *Escherichia coli*, or treatment of fungi with tunicamycin (a glycosylation pathway inhibitor). However, it was found that *Tr*Cel7A forms large inclusion bodies in *E. coli*, likely due to the 10 disulfide bonds found in the CD[50] required for proper folding of the catalytic domain and/or the hydrophobic nature of the enzyme that requires glycosylation for proper solubility,[17] making this approach unavailable for *Tr*Cel7A. However, further studies into the effects of N-glycosylation of

*Tr*Cel7A could be conducted through mutagenesis of the four known N-X-S/T motifs in the catalytic domain and expression of *Tr*Cel7A homologously[22] in other fungal hosts, such as *Pichia pastoris*, *Saccharomyces cerevisiae*, and *Aspergillus niger* var. *awamori*.[18,20,21,26,51–57] Furthermore, advances in computer simulation have permitted a more detailed understanding of the O-glycosylation of the linker domain and CBM.[19,23,24,58] Lastly, the advancements in new chemical technologies enabled the synthesis of glycoforms of the CBM allowing for elucidation of glycosylation functionality for this domain, corroborating computational simulation.[25] Taken together, these new approaches have greatly aided in elucidating the full implications of glycosylation of *Tr*Cel7A.

2.3 Implications of N-Glycosylation of the *Tr*Cel7A Catalytic Domain

The catalytic domain of *Tr*Cel7A is responsible for the cleavage of cellulose into cellobiose and contains four N-X-S/T motifs, three of which, N45, N270, and N384, are known to be natively glycosylated.[21,30,32,59] The locations of these glycosylation sites are shown in Fig. 2. N45 is near the entrance of the catalytic tunnel, N270 is near the exit of catalytic tunnel, and N384 is in a flexible loop region.[18,20,22,30,32,50] As mentioned previously, the *N*-linked glycans on the CD display a large variance in structure as a function of growth conditions. Glycan-trimming enzymes are likely responsible for these widely different *N*-glycan structures.[33,34,41,42] Given that the locations of the *N*-glycans at N45 and N270 are immediately near the entrance and exit of the catalytic tunnel of *Tr*Cel7A, respectively, as well as the crucial W40 residue,[60,61] it is plausible that the markedly different glycan structures reported play a role in the catalytic activity of *Tr*Cel7A. Although there are many valid experimental approaches to quantify CBH activity, these methods are rarely directly comparable. Despite this limitation, there exists data that favor the hypothesis that glycosylation can both aid or hinder the catalytic activity of *Tr*Cel7A dependent on glycan structure and location.[18,20–22]

Initially, hyperglycosylation of *Tr*Cel7A when expressed heterologously in fungal hosts was found to be detrimental to hydrolytic activity when directly compared to a wild-type glycovariant.[18,20,21,26,51] When *Tr*Cel7A was expressed in *Aspergillus niger* var. *awamori*, it was shown to have a large increase in mass from *N*-linked glycans while retaining relatively similar O-linked glycosylation mass, but with a higher average relative distribution when compared to natively glycosylated *Tr*Cel7A.[20] Marked decreases in hydrolytic rate and extent were found on both highly crystalline bacterial cellulose and amorphous cellulose.[20] The catalytic activity of the recombinant *Tr*Cel7A was partially recovered upon treatment with PNGase F, an enzyme that wholly removes high-mannose-type *N*-linked glycans at the Asn-GlcNAc linkage.[20] However, *Tr*Cel7A is unstable without *N*-glycans,[17] which is a likely source of noted experimental error.[20] Analysis of the bound concentration of *Tr*Cel7A (recombinant and wild-type) found that much more recombinant *Tr*Cel7A was bound to cellulose, suggesting that the hydrolytic drop was likely due to nonproductive adsorption or nonproductive active-site binding of recombinant *Tr*Cel7A to the cellulose surface.[20] Indeed, changes in the glycosylation pattern of the CBM affect adsorption to cellulose.[25,62–64] However, the general conclusion from these studies suggests that larger glycans seem to inhibit cellulose adsorption rather than dramatically increasing cellulose affinity. Alternatively, the larger *N*-glycans of the recombinant *Tr*Cel7A may affect the k_{on}/k_{off} rates for *Tr*Cel7A when catalytically engaged onto a cellulose fiber.

The mechanistic steps of crystalline cellulose hydrolysis by *Tr*Cel7A are complicated and comprise the CBM-mediated adsorption to cellulose with a $K_d = \sim 1$–$10\ \mu M$,[23,25,65–73] CD detection of a free reducing end of a cellulose fiber, catalytic engagement, and threading of the free reducing end into the CD with a $K_d = \sim 0.1$–$1\ \mu M$,[74–80] transglycosylation reaction, elimination reaction, cellobiose expulsion from the CD, processive motion forward by one cellobiose unit,[81–83] and eventually dissociation of the catalytically bound enzyme with cellulose.[84–90] The catalytic steps of *Tr*Cel7A catalysis can be consolidated to a k_{on} rate involving just catalytic engagement of a free cellulose fiber, a k_{cat} rate involving primarily the catalytic cleavage of cellobiose from cellulose, and finally a k_{off} rate involving the dissociation of the bound *Tr*Cel7A–cellulose fiber complex.[89] The *Tr*Cel7A k_{on}/k_{off} rates in this simplified model have both been independently implicated to be the potential rate-limiting step for cellulose conversion in homogeneous systems containing only *Tr*Cel7A and crystalline cellulose,[77,80,89–92] while in mixed systems of cellulases, the k_{cat} rate has been suggested to be rate limiting.[93] However, the exact mechanism for these delays is not currently understood.

The adsorption of the CBM-linker region to cellulose is known to be relatively weak and reversible.[23–25,66–73,94,95] However, evidence suggests that the binding of the free cellulose chain to the CD is less reversible, as *Tr*Cel7A has been observed to "stall" on various substrates.[77,90–93,95–99] This stalling behavior is thought to be caused by irregularities of the cellulose encountered by the catalytically engaged *Tr*Cel7A, such as noncellulosic components of lignocellulosic biomass substrates[100–114] or amorphous cellulose components.[89,90,93,98,115] This is why many research efforts have been aimed at diversifying the enzymes present in hydrolytic enzyme cocktails to include endoglucanases and LPMOs that help to overcome the apparent barriers presented to cellobiohydrolases.[116] However, it is also possible that the "stalling" behavior and low k_{off} rates observed for *Tr*Cel7A could also be an artifact from *N*-glycan interaction with the cellulose surface. Indeed, it has been demonstrated that hyperglycosylation of *Tr*Cel7A expressed in *A. niger* not only lowered the catalytic efficiency of the enzyme, but also increased the total amount of enzyme bound to the substrate for both crystalline and amorphous cellulose.[20] Because there is a much higher affinity of *Tr*Cel7A for a free cellulose chain in the CD than the CBM-linker adsorption to the cellulose surface, it is more likely that the larger *N*-glycans present on the CD in the case of the recombinant enzyme affect the "stalling" behavior or the k_{off} rate of the enzyme. Indeed, N45 is proximal to the free cellulose entrance to the catalytic tunnel and the crucial W40 residue that has been implicated in threading cellulose into the active-site tunnel of the CD.[60,61] It is possible that a large glycan here may hinder expulsion of a free cellulose chain from the CD or also aid in stalling the enzyme at amorphous regions, considering that molecular dynamics simulations have suggested that the architecture and flexibility of the active-site loops can influence the processivity of cellobiohydrolases.[117–119] Subsequently, it has also been postulated that large *N*-glycans on the CD may inhibit the initial threading mechanism of *Tr*Cel7A.

When large glycans at N384 in an *A. niger* recombinant enzyme were removed via alanine mutation, there was a boost in activity of the mutant recombinant enzyme of 70% over the enzyme with the large *N*-glycans present, although both enzymes were less active than enzyme from the wild-type host.[18] It was suggested by the authors that the larger *N*-glycan at N384 could hinder the catalytic efficiency of *Tr*Cel7A by potentially acting as an unnecessary spacer between the CD and cellulose surface, affecting the ability of *Tr*Cel7A to thread a free cellulose chain into the active-site tunnel.[18] Indeed, the binding of *Tr*Cel7A to a soluble substrate was not influenced by the presence or absence of native glycans, further implicating *Tr*Cel7A *N*-glycans in activity on insoluble crystalline cellulose substrates.[22] However, these postulations are speculative, and additional studies that show definitively that these large differences in activity are due to changes in *N*-glycan structure have yet to be completed. What is more certain are the pronounced effects of N-glycosylation on the folding and thermal stability of *Tr*Cel7A.

Thermal stability is a highly desirable trait of cellulases as stable cellulases permit faster and more facile bioprocessing at higher temperatures.[120] Glycosylation typically confers enhanced thermal stability, which is indeed true for *Tr*Cel7A CD. Adney and coworkers determined that the native enzyme and hyperglycosylated variant *Tr*Cel7A expressed in *A. niger* var. *awanori* had the same thermal stability when measured via DSC.[17] However, when each glycosylated Asn was systematically mutated to Ala, it was found that only the N384A mutation caused a significant melting temperature change (i.e., 2 °C).[18] These data were corroborated by another recent study that systematically mutated each glycosylated Asn to Gln and expressed the recombinant enzymes endogenously in *T. reesei*. The authors also found the largest drop in melting temperature for the N384Q mutation on the same order found by Adney and coworkers.[18,22] Moreover, it was found in both studies that there were activity losses for each glycosylation mutant.[18,22] Beyond thermal stability conferred by N-glycosylation of the *Tr*Cel7A CD, there is also evidence that glycosylation plays a significant role in *Tr*Cel7A folding.

N-Glycosylation is known to stabilize proteins when located within loop regions and near aromatic amino acids.[121–123] Evidence from far-UV circular dichroism suggests that the *N*-glycan at position 384 thermally stabilizes the secondary structure in the loop region of the CD at high temperatures.[22] Evidence suggests that phenylalanine residues that are adjacent or up to three positions away from the conserved Asn-X-S/T motif could be stabilized through interaction with the first GlcNAc residue of *N*-glycans.[121–123] Indeed, a phenylalanine at position 273, near the protein surface and within a reverse turn in the *Tr*Cel7A CD, is likely to be stabilized by the *N*-glycan at N270.[56,121,122] Although mutation of this position does not significantly affect the thermal stability of the CD, it is likely that the N270 glycans help to stabilize the terminal end of the active-site tunnel as has been suggested.[18,20,22] Indeed, active-site stabilization is a common postulation that explains the glycosylation requirement for full enzyme activity,[124] and select mutations to these loops regions have also been implicated in activity enhancement of the enzyme.[118–125] Additionally, all three glycosylation sites are found within loop regions, and these may be prerequisite for stabilization of these regions of the CD and likely play a role in correct folding of these regions. Lending credence to this theory was the finding that the N-glycosylation of the *Tr*Cel7A CD aided in chaperone-independent protein folding, as the nonglycosylated mutant was still found to be secreted at the same rate as the fully glycosylated wild-type enzyme and did not upregulate the production of chaperone genes in *T. reesei*, but did activate the unfolded protein response (UPR) through a separate mechanism.[22]

2.4 Implications of O-Glycosylation of *Tr*Cel7A Linker Domain

The linker domain of *Tr*Cel7A serves a variety of functions for the whole enzyme, including its role as a spacer between the CBM and CD, an extension of the CBM adsorption capacity to cellulose, and enhanced thermal stability. [126,127] The linker domains of *Tr*Cel7A and the majority of multimodular fungal cellulases are rich in Ser, Thr, Gly, and Pro amino acids and depleted in aromatic and most aliphatic amino acids (except Gly). The Ser/Thr residues are generally hyper-O-glycosylated prior to secretion in fungi. Linker regions also typically lack Asn-X-Ser/Thr motifs and have not yet been reported to contain any *N*-linked glycans. Additionally, linkers typically lack aromatic and hydrophobic residues, thus preventing hydrophobic protein folding.[58,128] Although the exact functions of glycosylation of this domain are not fully understood, various studies have indicated that glycans on fungal linkers aid in their functionality.

One of the proposed actions of glycans on cellulases is to act as proteolytic protectors. Linker regions are intrinsically disordered, but are usually found to adopt an extended conformation such that the peptide backbone is highly susceptible to proteolytic attack. Filamentous fungi endogenously secrete various proteases to act in concert with glycoside hydrolases to degrade both plant cell walls and plant proteins as a nutrient source.[44,129–133] The O-glycosylation of linkers has been shown to increase the proteolytic stability of multimodular enzymes.[70,134,135] The heavy glycosylation on the exposed linear polypeptide backbone of the linker region serves to sterically hinder proteolysis. These observations were also corroborated by studies of the O-glycosylation of the *Tr*Cel7A CBM, where it was noted that there exists a glycan density-dependent effect on CBM stability to proteolytic hydrolysis. However, the steric hindrance of proteases encountering glycoprotein substrates is a well-documented phenomenon,[25,136,137] and other actions of *O*-linked glycans on linkers have also been proposed.

Both computational simulation studies and circular dichroism evidence have demonstrated that the linker of *Tr*Cel7A lacks any defined secondary structure.[58] Furthermore, the flexibility of the linker has thus far hindered the crystal structural determination of the full-length enzyme. Schmuck *et al.* were able to resolve the structure to between 0.5 and 1 nm and showed that the enzyme adopts the commonly represented tadpole structure.[138]

Recently, *Tr*Cel7A protein crystals obtained by Meilleur and coworkers were subjected to small-angle X-ray diffraction, and the structure of the CD was readily available, but the density of the CBM-linker domains could not be resolved due to heterogeneity of this structure in the crystal, which is likely due to the high flexibility exhibited by the linker domain.[139] There is simulation and experimental evidence to show that the *Tr*Cel7A linker and other similar linker domains can adopt both a relaxed and an extended conformation. These discrete structures are hypothesized to propagate allosteric changes between the two domains.[126–128]

Both small-angle neutron and X-ray scattering experiments and molecular dynamics simulations have helped elucidate the structural and functional roles of the *Tr*Cel7A linker and other similar linker domains. Small-angle neutron scattering showed that the natively glycosylated linker of *Tr*Cel7A adopts a more extended conformation upon a change in pH from 7.0 to the optimal enzymatic pH of 5.0.[140] This conformational effect is likely due to the linker domains' relatively high isoelectric point conferred by the presence of two Arg residues in the sequence of the linker domain.[140,141] Because the optimal pH for *Tr*Cel7A is 5.0, this suggests that the extended conformation of the linker is preferred for its function. This same extended conformation has also been observed in the homologous Cel7A of *Trichoderma harzianum* revealed by small-angle X-ray scattering.[142] Furthermore, the authors note that it is likely that the mobile glycans on the linker played a role in extending the conformation by sterically limiting the available conformations of the linker.[142] Moreover, this phenomenon was also documented in the *Humicola insolens* Cel45, another enzyme of similar architecture to Cel7A where glycans promoted an extended conformation of the linker domain in solutions.[143] Molecular dynamics simulations of the *Tr*Cel7A linker confirmed the high level of disorder in this domain and also corroborated that the presence of glycans promoted this same extended conformation of the linker.[19,58,126] This extended conformation also seems to be important in bacterial cellulase linkers, and an increase in the Pro content of these linkers is responsible for the preservation of the extended confirmation where glycosylation cannot serve the same purpose.[58,144] But how and why are the extended conformation and the relaxed conformations of disordered linkers seemingly conserved?

Beckham and coworkers studied the sequences of fungal and bacterial linkers in cellulases and found that although the sequences are hardly conserved, the amino acid composition of the linkers was very similar in all sequences studied.[58] These results were corroborated in another study detailing the amino acid contents of α-amylase linkers that showed that these disordered linkers contained much higher relative levels of Gly, Pro, Ser, and Thr residues in their structures.[145] It was also found that there is a dependence on enzyme functionality with linker length, suggesting that linkers are optimized based on enzymatic activity.[58,135] Furthermore, Beckham and coworkers found that in fungal linkers, the abundance of Ser and Thr residues scaled linearly with linker length, but Pro residues did not scale with length.[58] This further implicates the glycosylation of fungal linkers as the mechanism of linker extension as these dense Ser/Thr regions are also predicted and shown to be hyper-O-glycosylated.[14] Additionally, the existing "inch-worm" molecular motion model implicates the linker domain length and stiffness as crucial for proper movement on crystalline cellulose.[127] Due to the crucial function of glycans in stabilizing extended linker length, it is also possible that the glycans may also affect the stiffness of the linker and therefore play another role in Cel7A function. Moreover, due to the high prevalence of hyper-O-glycosylation of linker regions in other fungal enzymes, it is also highly likely that the functional roles of the *Tr*Cel7A linker glycosylation are further conserved in these enzymes.[14]

Recently, it was shown through a combination of computational simulation and chemical synthesis that glycosylated CBM linkers are able to interact with crystalline cellulose.[23] This study showed that binding affinity of the native glycosylated CBM linker from *T. reesei* QM9414 was an order of magnitude higher than the unglycosylated CBM following predictions initially obtained from molecular dynamics simulation. (See Appendix I for molecular dynamics simulations of this process.) Furthermore, it was postulated by computational simulation that the *O*-glycans found on the CBM domain[32] could interact with cellulose (Fig. 2). Therefore, it is reasonable to presume that the extent of O-glycosylation of the linker may contribute to this affinity enhancement. However, although the enhanced affinity to cellulose is generally accepted as beneficial to enzyme functionality overall,[76,146] there is also evidence that the linker can participate in nonproductive binding to lignin through electrostatic[141] or hydrophobic interactions.[109,110] It was found that by manipulating the isoelectric point of the linker domain through the mutation of an Arg to a Ser, that *Tr*Cel7A bound less to lignin in lignocellulosic biomass and retained a higher specific activity in the presence of lignin.[141] It is likely that the Arg-to-Ser mutation also enhanced the glycosylation of the linker, and we recently found that glycosylation can modulate the affinity of the *Tr*Cel7A CBM to lignin (unpublished results). Therefore, glycoengineering of the linker domain may be a beneficial method for enhancing the overall properties of *Tr*Cel7A and other multimodular enzymes for biofuel production.

2.5 Implications of O-Glycosylation of the CBM

The Family 1 CBM of *Tr*Cel7A is a Type-A CBM that is primarily responsible for adsorption of the enzyme to crystalline cellulose.[64] The NMR structure of the CBM was reported in 1989, wherein it was shown that the 36-amino acid peptide folds into a cysteine-knot structure with two antiparallel β-sheets and displays three tyrosine residues on a flat face.[147] The three tyrosine residues were shown to preferentially bind to the hydrophobic face of crystalline cellulose, likely through hydrophobic interactions.[65–67,94] Until recently, glycosylation was not explicitly realized to be present in Family 1 CBMs. However, careful analysis of mass spectrometric data showed that two amino acids belonging to the CBM were in fact O-glycosylated, with one and as many as three mannose residues with unresolved linkages.[24,32] Moreover, it was found that among Family 1 CBMs, there is also a highly conserved Ser/Thr residue at position 14 (Figs. 2 and 6) that is also near the binding interface of the CBM and is likely to also be glycosylated, though no experimental evidence has showed this yet.[24,32]

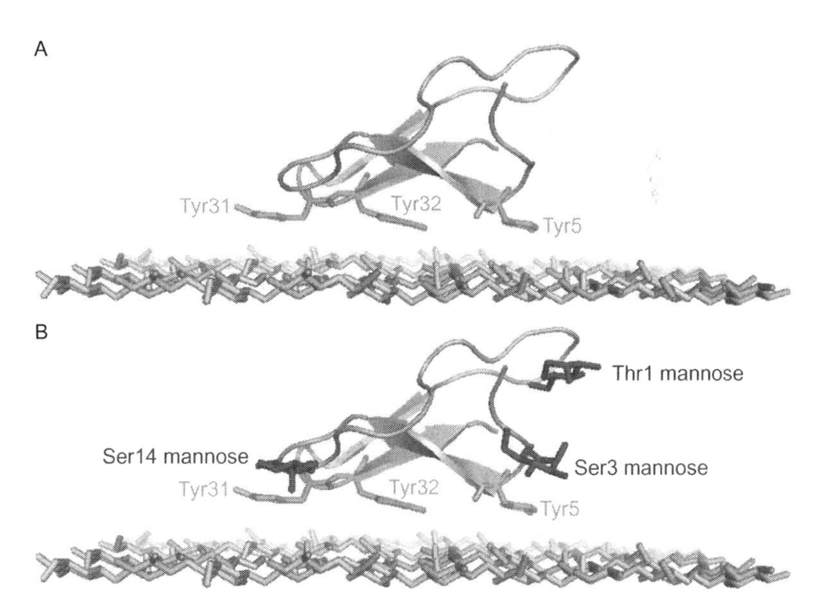

Figure 6 Visualization of O-glycosylation of the *Tr*Cel7A CBM bound to crystalline cellulose. (A) The nonglycosylated CBM structure.[147] (B) The optimally glycosylated variant of the *Tr*Cel7A CBM as determined by Tan and coworkers.[25]

In an effort to assess the impact that these glycans may have on the adsorption activity of the *Tr*Cel7A CBM, molecular dynamics studies were carried out and showed that the affinity of the CBM for crystalline cellulose was markedly increased with the attachment of a single *O*-linked mannose residue to each of the three putative glycosylation sites at Thr1, Ser3, and Ser14 (Fig. 6). It was also noted that the binding enhancement was diminished if the glycan attached had a larger α-1,2-linked dimannose motif attached.[24] Furthermore, it was observed that different glycosylation sites conferred different abilities to enhance the affinity of the CBM for cellulose, where Thr1 was found to only minimally contribute to binding affinity, but Ser3 and Ser14 were found to dramatically modulate crystalline cellulose affinity.[24]

In a follow-up study, a library of 20 *Tr*Cel7A CBM glycoforms were chemically synthesized and characterized for their binding affinity for crystalline cellulose, and many results from the molecular dynamics studies were corroborated.[25] This study upheld results from the molecular dynamics simulation and found that the largest binding affinity enhancement of the CBM for crystalline cellulose was obtained through the covalent attachment of a monomannose motif to each of the three putative glycosylation sites.[25] Furthermore, it was found that certain glycosylation sites were more important for binding enhancement properties where Ser3 was most important, followed by Ser14, and finally by Thr1.[25] This result signified that glycosylation-based cellulose affinity enhancement is both site and pattern specific for the Family 1 CBM. Additionally, alternatives to glycan attachment, such as α-1,6-mannose linkages, also confer similar beneficial binding characteristics as α-1,2-mannose linkages, suggesting that overall glycan size is a more important factor for Family 1 CBM binding affinity enhancement than exact glycan structure (unpublished results). Furthermore, it was shown that attachment of glycans of increasing size eliminated the adsorption enhancements observed, likely through a steric hindrance mechanism and similar to other results.[25]

Data obtained for a Family 2a CBM also showed that the attachment of large glycans to a CBM was detrimental toward CBM binding affinity to cellulose.[62,63] The Family 2a CBM from *Cellulomonas fimi* also belongs to the Type-A group of CBMs along with the Family 1 CBM from *Tr*Cel7A and is specific for crystalline cellulose. The Family 2a CBM also displays a flat hydrophobic binding face.[64] However, attempts to glycoengineer the CBM to have better binding properties through the addition of glycosylation sites were met with limited success. Both large *N*-glycans and native *O*-glycans added to the CBM via heterologous expression in *P. pastoris* and through genetic addition of Asn-X-S/T motifs caused large decreases in the affinity of the CBM for crystalline cellulose.[62,63] Beyond modulating the affinity the CBM for crystalline cellulose, preliminary evidence suggests that glycosylation patterns also modulate binding specificity to various lignocellulosic biomass components (unpublished results).

As mentioned before, lignocellulosic biomass consists of many different components, such as cellulose, hemicellulosic components, pectins, and lignin.[5,6] Moreover, various chemical and mechanical pretreatment techniques further alter the composition and chemical nature of lignocellulosic biomass substrates, which concomitantly changes how efficiently different mixtures of cellulases saccharify these substrates.[5,148–150] Key to this degradation is targeting enzymes to their proper substrate using their linked CBM, a finding which has recently gained much attention.[151] With the preliminary data demonstrating that glycosylation patterns also vary the substrate specificity of some CBMs, the door is now open to apply a rationally based glyc oengineering strategy for CBMs to modulate their specificity for preferred cell-wall components. Moreover, it may also be possible to hinder CBM adsorption to nonpreferential cell-wall components, such as lignin. Some cellulases have been shown to nonproductively adsorb to lignin, which hinders not only their activity but also their ability to be recycled.[100–114] Some evidence suggests that the CBM of these cellulases is responsible for the nonproductive adsorption, given the hydrophobic nature of both lignin and Type-A CBMs.[152] However, preliminary data suggest that the adsorption of the *Tr*Cel7A CBM to lignin can be reduced via engineered glycosylation. Taken together, we conclude that there is a necessity for Family 1 CBM glycosylation in *Tr*Cel7A and that smaller glycans are preferred for maximal performance, although the form and history of the substrate remains important.

Beyond contributions to activity of the CBM, glycosylation also contributes to the thermal and proteolytic stability of the CBM. Recently, it was found that the CBM-linker domain of *Tr*Cel7A is the thermal stabilizing domain that contributes to the overall thermal stability of *Tr*Cel7A.[153] The library of 20 CBM glycoforms was also tested for thermal stability where it was found that O-glycosylation greatly increased the thermal transition temperature of this domain by up to 16 °C.[25] Additionally, it was also found that position Ser3 was the important glycosylation site within this domain responsible for the most thermal stabilizing enhancement.[25] Not only is the position 14 Ser/Thr highly conserved through Family 1 CBMs, but also the Ser/Thr at position 3 is equally highly conserved, suggesting that glycosylation at this position was evolved to stabilize and protect this module. Additionally, it was demonstrated by Tan and coworkers that attachment of *O*-glycans to the CBM conferred robust proteolytic stability against thermolysin degradation.[25] It was also shown that this property followed glycans in a size-dependent manner with the glycoforms with the largest glycans showing the highest half-life to thermolysin degradation and glycoforms, with smaller glycans possessing lower half-lives.[25] This result is suggestive of a steric hindrance mechanism that has commonly been proposed for this glycosylation phenomenon.[136,137] Taken together, glycosylation is a quintessential posttranslational modification to the *Tr*Cel7A CBM conferring multiple beneficial properties when properly sized glycans are present at certain glycosylation sites.

Glycosylation is essential for *Tr*Cel7A activity and stability. Though the most optimal glycosylation pattern has remained elusive, future studies will elucidate ideal cellulase glycans. Protein engineering efforts have uncovered numerous beneficial mutations to cellulases, especially in *Tr*Cel7A, that stabilize cellulases on the same magnitude as optimal glycans.[25,120,154,155] However, these mutational studies have not taken into account these general trends regarding glycosylation patterns. With optimal glycosylation patterns, these enhanced mutants will likely perform even better, paving the way to more efficient lignocellulosic bioprocessing. These studies are aided by the development of more sophisticated constitutive expression systems that help stabilize both expression and glycosylation patterns.[156]

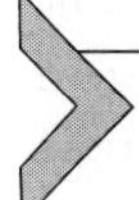

3. RECOMBINANT EXPRESSION OF FUNGAL CELLULASES

3.1 Expression of *T. reesei* Cellulases in *Saccharomyces cerevisiae*

There are many advantages to utilizing *Saccharomyces cerevisiae* to express cellulases. These include the relative ease of growing *S. cerevisiae*, the similarity of yeast glycosylation machinery to several industrially relevant fungi, mature genetic technology for facile incorporation and removal of genes, and the potential to condense the processing of bioethanol from lignocellulosic biomaterials into a single step.[157,158] However, many problems prevent efficient secretion of catalytically competent cellulases in *S. cerevisiae* including improper folding, N- and O-glycosylation, and low secretion yields. Despite these challenges, many groups have reported successful expression and secretion of cellulases in *S. cerevisiae*.

One of the first attempts to express *Tr*Cel7A in *S. cerevisiae* was partially successful.[52] The enzyme was expressed, and some enzyme could be recovered from the periplasmic space; however, some *Tr*Cel7A formed intracellular inclusion bodies, and other *Tr*Cel7A molecules were extensively hyperglycosylated, and the activity of *Tr*Cel7A was drastically diminished.[52] Later, it came as no surprise that expression of cellulases also induced the unfolded protein response in yeast.[53] Though the authors were also able to show more efficient secretion of hyperglycosylated Cel6A, secretion yields were still low, and the recombinant enzyme also displayed reduced activity.[52] Many years later, another attempt was made at efficient expression and secretion of cellobiohydrolases in *S. cerevisiae.*[159,160] The authors were able to express and more readily secrete four different cellobiohydrolases from GH6 and GH7,[159] an endoglucanase, and a β-glucosidase in *S. cerevisiae.*[160] Although these enzymes were all hyperglycosylated, the authors reported that the cellobiohydrolases displayed only a reduction in activity on crystalline cellulose and that the recombinant enzymes retained activity on amorphous cellulose. The successful growth of yeast on amorphous cellulose was the first demonstration of a one-step bioethanol production.[159,160] In contrast, endoglucanases do not appear to show a decrease in activity due to hyperglycosylation or hypoglycosylation, as activity when expressed in either *S. cerevisiae* or *E. coli* does not appear to vary significantly.[160–163] Despite these encouraging results, the activity of cellobiohydrolases expressed in *S. cerevisiae* remained low.

The expression of more active cellobiohydrolases by *S. cerevisiae* was aided by various structure-guided mutagenesis studies that could be performed readily in *S. cerevisiae.*[55,56,125,164–166] These studies were able to generate mutated forms of cellobiohydrolases in *S. cerevisiae* with enhanced activity and stability compared to wild-type recombinant cellobiohydrolases. Furthermore, other groups were able to target the protein secretion machinery of *S. cerevisiae* to promote secretion of properly folded cellulases with more optimal glycosylation patterns. Due to the presence of many disulfide bonds in *Tr*Cel7A, it was found that upregulating the expression of protein disulfide isomerase in *S. cerevisiae* aided in the secretion of properly folded cellobiohydrolases.[125,164,167] Additionally, manipulation of glycosylation machinery in *S. cerevisiae* was shown to aid in the proper enzymatic function of recombinant fungal cellulosomes (multienzyme cellulase complexes).[120,168–170] When applied to recombinant fungal cellobiohydrolases, it was found that disruption of *S. cerevisiae* genes involved in glycosylation promoted secretion of highly active cellobiohydrolases that were not hyperglycosylated.[167] Additionally, the endogenous GH47 α-1,2-mannosidase from *T. reesei*, responsible for trimming glycans (see Section 4), was also stably expressed in the engineered *S. cerevisiae*.[170] With the additional glycan trimming provided by the α-1,2-mannosidase, *Tr*Cel7A was secreted more efficiently and retained most of its endogenous catalytic function on cellulose.[170] Taken together, through new recombinant *S. cerevisiae* technology, the expression and secretion of fully functional fungal cellulases are becoming readily attainable.

3.2 Glycoengineered Strains of and Heterologous Protein Expression in *Pichia pastoris*

Another fungal yeast, *P. pastoris*, is a common host for the expression and secretion of glycoproteins due to its ability to secrete high yields of protein cheaply and quickly. Furthermore, due to its similarly low biohazard level with *S. cerevisiae*,[171,172] this host has been used for the expression and secretion of various cellulases, including cellulases with relatively well-characterized glycans.[26,51,173–177] From these studies, it is inferred that *P. pastoris* is a generally better host than *S. cerevisiae* for the secretion of multimodular processive cellulases primarily due to its capability to produce high mannose-type *N*-glycans with a lesser extent of hypermannosylation than that of the *N*-glycans of *S. cerevisiae* (Fig. 7).[10]

The primary processive cellulase with characterized glycosylation, *Tr*Cel7A, has been recombinantly expressed in *P. pastoris* and efficiently secreted.[26,51] It was found that the secreted recombinant Cel7A was also hyper-N-glycosylated when compared to the enzymes native counterpart secreted in *T. reesei*.[26,51] The hyperglycosylation of Cel7A did not appear to drastically change the pH optimum, the thermal optimum, or its activity on soluble substrates of the recombinant enzyme; however, there was a drastic reduction in activity on insoluble substrates, namely, crystalline cellulose.[26,51] Furthermore, through a detailed analysis of the secondary structure of Cel7A through circular dichroism, it was found that there were perturbations in the secondary structure of the *P. pastoris* recombinant enzyme compared to that of the native enzyme.[51] Upon further inspection with N-glycosylation site mutants of Cel7A expressed in *P. pastoris*, it was found the N64 was glycosylated in the recombinant enzyme.[21] Even through N64 contains the consensus sequence for N-glycosylation, glycans here have never been found when produced in different strains of *T. reesei*.[29–34,37,38] It was found that mutation of N64 to serine when expressed in *P. pastoris* resulted in a gain of activity along with the subsequently lower molecular weight of the protein.[21] Despite the apparent

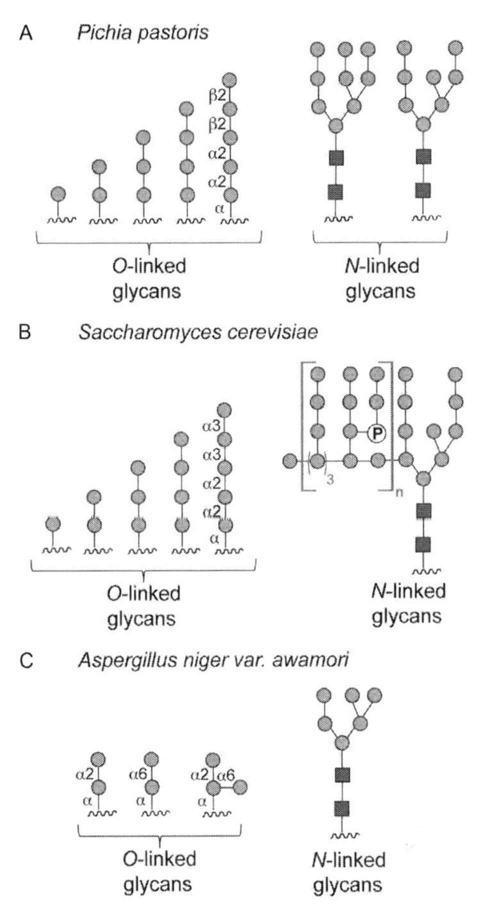

Figure 7 Representative glycan structures found in filamentous fungi referenced herein. (A) *O*- and *N*-linked glycans found in *Pichia pastoris*. (B) *O*- and *N*-linked glycans found in *Saccharomyces cerevisiae*. (C) *O*- and *N*-linked glycans found in *Aspergillus niger* var. *awamori*.

challenges to express cellobiohydrolases in *P. pastoris*, nonprocessive *T. reesei* endoglucanase expression and secretion has been largely successful.[173–177] These enzymes retain full activity, pH stability, and thermal stability in *P. pastoris*. Furthermore, given the single module nature of many of these enzymes, some groups have also reported relatively successful expression of endoglucanases in the glycosylation-deficient microbe *E. coli*.[162,163] This finding signifies that glycosylation may not be as important in globular enzymes in terms of activity as it is for multimodular enzymes.

It is known that the N-glycosylation patterns of *P. pastoris* are different from those in *T. reesei* (Figs. 4 and 7).[10] Beyond the structural differences in glycans between the two organisms, they also appear to have different glycosylation machinery that accounts for the difference noted by Zhuang and coworkers that different N-glycosylation sites are occupied in native versus recombinant enzymes.[21] Therefore, further work is needed in order to engineer *P. pastoris* strains to secrete proteins with limited glycosylation site addition and limited glycan molecular weight to better reflect the glycosylation of native recombinant proteins. Luckily, large efforts to humanize the glycosylation of secreted proteins from *P. pastoris* have already been completed, and the technology for glycoengineering in *P. pastoris* is relatively well defined.

Given the extraordinary capability of *P. pastoris* to secrete high yields of glycoproteins, it has received much attention from the biomedical community as a potential host of biotherapeutic glycoproteins.[178–183] Therefore, many efforts were aimed at humanizing the glycosylation pattern of *P. pastoris*. One of the first such studies involved cloning and expressing the *T. reesei msd1* gene encoding the α-1,2-mannosidase gene into *P. pastoris* to obtain mammalian high-mannose-type *N*-glycans in other recombinant proteins.[184] Similar to other studies of this enzyme, the authors found that the mannosidase could cleave ~85% of the mannose from reporter N-glycosylated recombinant proteins achieving mammalian high-mannose-type *N*-glycans in *P. pastoris*.[184] Many more studies also aimed to produce other human *N*-glycans were conducted such as glycans with terminal galactose residues,[185] complex-type *N*-glycans,[186] and terminally sialylated glycans.[187] Newer genomic engineering technology has also enabled researchers to target many different genes involved in glycosylation machinery of *P. pastoris* simultaneously or sequentially allowing for the targeting of genes in either the ER or the Golgi.[188] These same technologies could readily be applied to *P. pastoris* to produce enzymes with similar glycosylation patterns to other cellulose-producing fungi. Indeed, the *P. pastoris* strain developed by Contreras and coworkers would likely produce processive cellulases with higher recombinant catalytic efficiency due to the relative impairment of hypermannosylation in this strain.[184] However, more work needs to be done to optimize secretion of glycosylation competent cellulases in *P. pastoris*.

3.3 Glycosylation and Engineering of Expressed Proteins in *Aspergillus* Species

Another set of fungal hosts for recombinant cellulase production that are also capable of endogenous cellulase production are from the *Aspergillus* genus. These fungi are also able to secrete high yields of glycoproteins, and technology has been developed to allow for facile genetic manipulation of these species of fungi. Furthermore, the glycosylation patterns and structures of species within this genus have also been relatively well characterized, and some glycoengineering studies have been attempted.[10,189] This genus typically produces high-mannose-type *N*-glycans similar to those of other filamentous fungi, but notably, these fungi also produce some glycan structures that are similar to those produced in *T. reesei*, making heterologous expression between these fungi promising from a glycoengineering standpoint (Fig. 7). It has been demonstrated in *A. oryzae* that *N*-glycans typically contain five to eight mannose residues connected to the conserved chitobiose core,[10,190,191] but that other structures like $PMan_{11}GlcNAc_2$ can also exist.[192] Genetic addition of a Class I α-mannosidase into *A. oryzae* could readily produce $Man_5GlcNAc_2$ *N*-glycan structures,[190] similar to the glycan structures of the *T. reesei* parental strains.[33,34] Furthermore, larger *N*-glycans have been mentioned in *A. niger*, such as $GlcMan_9GlcNAc_2$ structures[193,194] and even smaller *N*-glycan structures, such as $Man_{3-4}GlcNAc_2$.[195,196] All of these glycans are highly similar to those found in various strains of *T. reesei*. It was also found that the pH of the growth solution could affect the glycosylation pattern of secreted proteins from *A. niger*,[193] potentially though similar mechanisms mentioned for *T. reesei*.

Many of the cellulases from *T. reesei* with relatively well-defined functions for glycans have also been expressed in various species of *Aspergillus*. Cel7A along with another processive cellulase, Cel6A, and three endoglucanases from *T. reesei* were expressed in *A. oryzae* and found to be hyperglycosylated when compared to the native enzymes in *T. reesei*.[197] When hydrolyzing crystalline cellulose, the two cellobiohydrolases exhibited a significantly decreased rate of hydrolysis compared to that of natively glycosylated cellulases.[197] Moreover, the endoglucanases expressed in *A. oryzae* did not display a significant activity drop, which is in agreement with most studies of the activity differences of endoglucanases subjected to changes in glycosylation pattern. This first study was complemented by other studies previously mentioned. Adney and coworkers also showed lower activity on crystalline cellulose of *Tr*Cel7A expressed in *Aspergillus niger* var. *awamori* and cited comparatively hyperglycosylation of the recombinant enzyme as the reason.[18,20] To our knowledge, the presence of glycosylation at N64 in the recombinant *Tr*Cel7A expressed in *Aspergillus* was neither confirmed nor denied in any of the aforementioned studies.

These studies sum up common themes in heterologous expression of cellulases in fungal hosts. Namely that:

1. Glycosylation must be finely tuned in processive cellulases for peak performance.
2. Optimal glycan structures and glycosylation patterns for processive cellulases are yet unknown.
3. Glycosylation in nonprocessive cellulases is required, but has not yet been shown to drastically impact catalytic activity in general.

This conclusion indicates that when choosing expression hosts, careful consideration regarding the glycosylation machinery of the host is paramount for successful secretion of a fully active cellulase. Moreover, care is required when genetically modifying the glycosylation machinery of expression hosts, as in many cases, deletions of certain glycosylation genes result in abnormal cell-wall function. This phenomenon occurs in *S. cerevisiae*,[169,198] in *Trichoderma* sp.,[16,199–201] in *Aspergillus* sp.,[189,202–204] and in *P. pastoris*.[188,205] Given the intrinsically difficult nature of manipulating and deleting endogenous glycosylation machinery genes in fungal hosts, an alternate and attractive strategy is to genetically add or engineer other glycan-active (trimming) enzymes, glycosyltransferases, or glycoside hydrolases, to promote certain glycosylation patterns in recombinant enzymes.

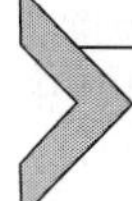

4. MODIFICATIONS BY GLYCAN-TRIMMING ENZYMES

4.1 Introduction

Fungal cellulases have been shown to produce a myriad of *N*- and *O*-linked glycans that change as a function of strain type, growth medium, and pH.[33,34] Furthermore, it has been shown that fungi also secrete various glycan-active glycoside hydrolases that are capable of enzymatic deglycosylation within fungal secretomes leading to even higher degrees of glycan heterogeneity. These glycoside hydrolases belong to Families GH18B5, GH38, GH47, GH85, GH89, and GH92 and can be explored readily at the curated CAZy database.[206] These enzymes have been demonstrated to either act in an exo-fashion or an endo-fashion on protein glycans. With the presence of glycan-active glycoside hydrolases in secretomic mixtures, the heterogeneous nature of fungal cellulase glycosylation is compounded. In the following section, glycan-active enzymes will be described, and the implications of these enzymes in mixtures will be discussed with emphasis on *T. reesei* as the model fungal cellulase producer.

4.2 Secreted Glycan-Active α-Mannosidases from *T. reesei*

T. reesei is a major producer of commercial cellulases for enzymatic saccharification of lignocellulosic biomass. Glycosylation is ubiquitously found in secreted proteins of *T. reesei* and affects protein stability and activity. Accordingly, many more studies have been conducted to elucidate the roles of *T. reesei* cellulase glycosylation and glycan-active glycoside hydrolases in secretomic mixtures than in other filamentous fungi, making this organism an essential model organism for future studies into cellulase glycosylation.[33,34,37,38,41,42,207,208]

The first two glycan-active glycoside hydrolases to be discovered in the secretome of *T. reesei* was found by Neustroev and coworkers.[207] The authors discovered and partially characterized two distinct α-mannosidases with activity on both *N*- and *O*-linked glycans; however, only one enzyme could be successfully purified for further characterization. The authors found that this α-mannosidase could cleave α-1,2-, α-1,3-, and α-1,6-mannosidic bonds with kinetic preference for α-1,2- and α-1,3-mannosidic linkages (Fig. 8). Furthermore, this enzyme was only found active postsecretionally, with no activity being noted in the cytosol of *T. reesei*. This study also showed that up to 30% of all mannose present in secreted glycoproteins could be released by a larger dosage of this enzyme to the mixture over 48 h, which promoted the production of the $Man_5GlcNAc_2$ *N*-glycan structure in neutral *N*-linked glycans. Unfortunately, no sequence information was presented in the original study to positively correlate the identity of

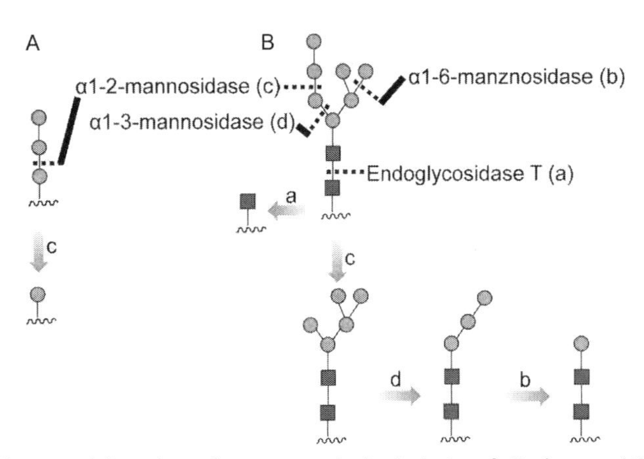

Figure 8 Structural locations for enzymatic hydrolysis of *O*-glycans (A) and high mannose-type *N*-glycans (B) by glycan-active glycosyl hydrolases from *T. reesei*.

this α-mannosidase as belonging to GH47 or GH92 to correlate activity and quantity to newer quantitative proteomic data.

Two years later, another α-mannosidase was discovered in *T. reesei* QM9414 with specificity for α-1,2-mannosidic linkages;[208] however, this enzyme was unable to cleave the unique phosphomannosidic linkages found in *T. reesei* *N*-glycans.[38] This enzyme belongs to the GH47 family, and its expression was only found under cellulose growth conditions, indicating that this glycan trimming may be under similar transcriptional control as major cellulases (Table 2). The structure of the *T. reesei* α-1,2-mannosidase was solved in 2001 and shows sequential and structural similarity to eukaryotic Class I ER mannosidase proteins.[209] The authors also showed that the *T. reesei* α-1,2-mannosidase has a larger active-site conformation than the active site of the homologous *Saccharomyces cerevisiae* enzyme, which allows for the cleavage of four mannose residues instead of the typical single mannose cleavage observed in other Class I ER mannosidase enzymes.[209] This suggests that this enzyme may have evolutionarily gained a secretion signal as an accessory enzyme for mannan degradation, given its mannan activity and ability to hydrolyze multiple mannose residues (Table 1). Alternatively, this α-1,2-mannosidase may have gained a secretion signal to aid in deglycosylating secreted enzymes, given the potential beneficial properties conferred to Cel7A by smaller glycans over larger glycans[17,18,20–22,25,26] and the fact that exogenous mannose residues are more readily incorporated into additional *N*-glycan synthesis than mannose converted from glucose.[210]

Although it was noted that no mannosidase activity was found in the growth media of the *T. reesei* RUT-C30 growth media,[208] this enzyme is readily found in the secretome of RUT-C30.[43,44] Two scenarios explain this discrepancy: first, the GH47 α-mannosidase activity is pH dependent (Table 1), and the pH range of the broth may have been outside this range.[33,34,208] Second, it is possible that the α-mannosidase gene was not

Table 1 Comparison of Characterized *T. reesei* Secreted Glycan-Active Mannosidases

Mannosidase	GH Family	pH Optimum	Glycosylated	Mannan Active	References
α-1,2/3/6-Mannosidase	38 or 47 or 92	6.5	Yes	Yes α-1,6-Linkages only	207
α-1,2-Mannosidase	47	5.0	Yes	Yes	38,208

transcriptionally induced potentially due to inducing agents lacking in the growth media, such as cellulose metabolites.[211] Indeed, it appears that *T. reesei* upregulates the transcription and secretion of α-1,2-mannosidase in the presence of cellulose in the growth broth.[43,211] Interestingly, in commercial enzyme mixtures, this α-mannosidase is missing entirely along with many proteases.[44] It was also found by Adney and coworkers that the *N*-linked glycans of *Tr*Cel7A in commercial mixtures were relatively small, consisting primarily of only a single GlcNAc residue or a di-GlcNAc moiety,[20] which is quite distinct from the glycoforms typically found for the cellulase hypersecreting strain RUT-C30 that does include secretion of the GH47 α-mannosidase.[44]

4.3 Secreted α-Mannosidases from Additional Fungi

Other industrially relevant fungi are also known to secrete glycan-active α-mannosidases that show similar activities to those identified in *T. reesei.* These enzymes follow a similar paradigm as the *T. reesei* α-mannosidases and are also mannan active in addition to being glycan active. One of the first extracellular mannosidases to be discovered was found to be active on α-1,3/6-mannosidic linkages in both yeast mannans and also on glycoproteins.[212] This enzyme was found in the culture media from a *Cellulomonas* species and was also determined to be active at more alkaline pH than the mannosidases from *T. reesei* with a pH active range between 6.5 and 8.[212] Enzymes with similar, but more specific, substrate preference were also discovered in *Aspergillus phoenicis.*[213] The authors characterized two α-mannosidases, one with specificity for α-1,6-mannosidic linkages and the other for α-1,2-mannosidic linkages. Interestingly, although the α-1,3-mannosidase was mannan active, the α-1,6-mannosidase was not. Furthermore, the α-1,6-mannosidase was found to be highly thermostable with optimal activity at 60 °C and also displayed optimal activity in an acidic environment with pH ranges between pH 4.0 and 4.5.[213] Many other α-mannosidases have also been detected in secretomic mixtures from other species of *Trichoderma*, such as *T. harzianum*,[214,215] and other organisms, such as *N. crassa*,[216] *Aspergillus fumigatus*,[217,218] *Penicillium decumbens*,[219] *Amanita bisporigera*,[44] *Laccaria bicolor*,[44] *Coprinopsis cinerea*,[44] *Galerina marginata*,[44] and *Phanerochaete chrsosporium.*[220] Although there is a high prevalence of these enzymes in secretomic mixtures, the glycan activity of the enzymes quantified is still putative.

4.4 *endo*-β-*N*-Acetylglucosaminidases Secreted by Fungi

Beyond the high variability in the high-mannose-type *N*-glycans found in *T. reesei* that can be partially explained by differential expression or presence of glucosidases and mannosidases, there existed the additional question regarding the presence of the single GlcNAc glycan in secreted *T. reesei* proteins found in various studies. This was likely a consequence of a secreted *endo*-β-*N*-acetylglucosaminidase (ENGase)—a group of enzymes capable of cleaving the chitobiose core of *N*-glycans as has been previously demonstrated in fungi.[16,30] In 2010, it was uncovered that this unique glycan structure was a consequence of a novel secreted glycoside hydrolase (Endo T) from Family 18 group B5 chitinases that also functioned as a ENGase, and it was the first secreted ENGase to be discovered in fungi (Fig. 8).[42] The discovery of Endo T prompted the reannotation of this particular CAZy group (GH18B5) to include subgroup B5 for ENGases (GH18B5a and GH18B5b), where the latter is classified based on intracellular localization and the former on extracellular secretion.[41,201] This same study also found ENGase *N*-glycan activity within seven other *Trichoderma* species and noted that this activity was also present in a commercial enzymatic preparation. Indeed, the same year another secreted ENGase was purified and characterized from the fungus *Flammulina velutipes* named Endo FV.[221] Additionally, a GH18B5 enzyme (Endo Tv) was purified form a commercial chitinase mixture of enzymes derived from *Trichoderma viride* that was characterized to be high-mannose *N*-glycan active.[222] The structure of Endo T was solved in 2012, and interestingly, although the fungal ENGase shares low sequence homology with the bacterial ENGases, their respective folds and catalytic residues are nearly identical.[41] The fungal ENGase also shares similar specificity and activity to the bacterial ENGases.[41,42,221] From the collected sequence data on the characterized fungal ENGases, it was found that the sequences of the fungal ENGases more closely matched chitinases than bacterial ENGases, suggesting that this class of fungal enzymes likely evolved from chitinases or vice versa.[222] GH18B5 activity was also observed in *Neurospora crassa* cultures as confirmed by GH18 deletions.[223] Beyond extracellular glycoprotein processing, another homologous intracellular ENGase in *T. reesei* is thought to be involved in ER-associated glycoprotein degradation pathways.[41,201] There is also a homolog to this intracellular ENGase in *T. viride*.[224] Glycoside hydrolase Family 18 is not the only CAZy family to contain glycan-active enzymes; GH85 also contains *N*-glycan-active enzymes from bacteria and eukaryotes. Though much less work has been done on fungal GH85 enzymes, two have been identified and have been shown to be *N*-glycan active and able to trim high-mannose-type *N*-glycans to the single GlcNAc residue.[225,226] Although these enzymes do not have the same prevalence in extracellular media as the GH18B5 enzymes, they still play a role in intracellular processing of *N*-glycans and may be useful targets for glycoengineering fungi to produce particular *N*-glycans. Indeed, both fungal GH85 enzymes mentioned above have been exploited for their transglycosylation activity to incorporate synthetic oligosaccharides into Asn-GlcNAc scaffolds.[227–230]

Many efforts have been directed to elucidating the exact molecules responsible for inducing cellulase upregulation in *T. reesei*, with the goal of facilitating more efficiently secreted relevant enzymes for industrial purposes.[211] Using the latest advances in proteomic semiquantitative mass spectrometry, quantitative PCR, and whole RNA microarrays, researchers have gained insight into the regulation of cellulase genes and cellulase secretion. These studies have also amassed significant data about the preferential expression secretion of glycan-active glycoside hydrolases based on inducing agent, and these data uphold the hypothesis of Claeyssens and coworkers that different inducing agents change not only the types and quantities of cellulases but also glycan-trimming enzymes.[33,34] A recent study found that in the case of *T. reesei*, the correlation between transcriptomic upregulation and enhanced protein presence in the secretome was 70.17%, meaning that a review of transcriptomic and secretomic studies is warranted to elucidate the potential *T. reesei* inducing agents for glycan-active glycoside hydrolases.[211] Data from the most recent and relevant studies of the expression and secretion of cellulases and glycan-active glycoside hydrolases are presented in Table 2.

Recently, semiquantitative mass spectrometry of secreted proteins from *T. reesei* detected many glycan-active glycoside hydrolases and was also showed the effects of different growth media on the secretion levels of these enzymes.[43] The characterized α-1,2-mannosidase[208] was secreted at an appreciable level in *T. reesei* QM6a, but was hardly produced in RUT-C30.[43] This suggests that the mutations accumulated in the RUT-C30 strain may have also affected the ability of this strain to secrete various glycoside hydrolases. Furthermore, the data presented show a partial induction of this enzyme in response to *T. reesei* growth on cellulose over starch, suggesting that cellulose metabolites may contribute to inducing expression and secretion of this enzyme.[43] Indeed, the same α-1,2-mannosidase was also revealed in the secretome of RUT-C30 grown on pretreated corn stover.[44] Various

Table 2 Secretomic and Transcriptomic Evidence of Glycan-Modifying Enzymes by *T. reesei*

T. reesei Strain	Carbon Source	α-1,2-Mannosidase *msd1*	Endo T	Transcriptional or Secretomic Study	References
RUT-C30	Corn stover	Detected	ND	Secretomic	44
QM6a RUT-C30	Cellulose, starch, glucose	Upregulated on cellulose detected on others	ND	Secretomic	43
QM9414	Wheat straw, lactose, glucose	Wheat straw upregulated	ND	Transcriptomic	231
QM9414	Lactose	Partially regulated by LAE1	Not regulated by LAE1	Transcriptomic	232
QM9414	Lactose	ND	Not regulated by LAE1 Growth rate dependent	Transcriptomic	233
QM6a	Wheat straw, lactose, glucose	ND	Upregulated on lactose and wheat straw	Transcriptomic	234
QM9414	Cellulose, lactose	Upregulated more by cellulose	Upregulated more by cellulose	Transcriptomic	235
QM9414	Glucose	ND	Not regulated by LAE1	Transcriptomic	236
QM9414	Cellulose, sophorose, glucose	Cellulose upregulated	ND	Secretomic and transcriptomic	211

other α-mannosidases were also identified in the study; however, it is not known if these are glycan active despite the fact that other glycan-active mannosidases have been characterized in *T. reesei* secretomes.[207] Interestingly, though Endo T activity has been described in the secretome of RUT-C30,[31,35] no presence was found in either the QM6a or the RUT-C30 strain,[43] even though the transcription of Endo T has been noted in RUT-C30.[211,231] However, another similar study of the *T. reesei* QM9414 secretome readily detected Endo T in proteomic mixtures and also showed transcriptomic changes in the Endo T gene based on substrate.[211]

This latter study showed that Endo T was only secreted by *T. reesei* QM9414 in the presence of crystalline cellulose.[211] The authors also identified the transcription factor LAE1 as being primarily responsible for cellulose-specific gene induction of various cellulases and hemicellulases including Endo T.[211] However, when *T. reesei* QM9414 is grown with lactose or glucose as the primary carbon source, there appears to be no dependence of LAE1 on the expression of Endo T.[232–234] This suggests that there is yet another cellulose-metabolite-specific transcription factor and alternate expression systems responsible for the induction of Endo T expression. Indeed, many other transcription factors were also identified, and some seem to be cellulose-metabolite specific.[211] Interestingly, although the expression of Endo T does not appear to be linked to LAE1, the expression of the primary extracellular α-1,2-mannosidase does appear to be linked to the LAE1 transcription factor as the expression of this enzyme is downregulated in a *T. reesei*-null LAE1 mutant.[233] Despite how useful these data are for uncovering the subtleties of cellulase and glycan-active glycoside hydrolase expression and secretion, without an exhaustive and focused study toward the expression of glycan-active enzymes, the induction mechanism of these enzymes will remain unknown.

Ongoing efforts are directed at elucidating the optimal glycosylation patterns of cellulases, especially *Tr*Cel7A, in order to inform rationally based glycoengineering efforts. It is possible that if the induction mechanisms of CAZymes are revealed, growth media can be optimized for diverse enzyme production in *T. reesei* and for promoting proper glycan-active glycoside hydrolase secretion as a method for glycoengineering secreted cellulases for enhanced activity. Furthermore, it would be possible to genetically engineer *T. reesei* strains to include only select glycan-active glycoside hydrolases in various cellulase induction groups, such that certain glycan patterns can be promoted in cellulases where glycosylation pattern is of paramount importance. Genetic engineering of *T. reesei* to produce human glycosylation patterns has already been successfully implemented,[199,235,236] so further engineering to promote alternate optimal glycosylation patterns is readily attainable when ideal glycosylation pattern targets are resolved.

5. SUMMARY AND FUTURE PERSPECTIVES

Glycosylation is a highly prevalent posttranslational modification to fungal cellulases that is required for optimal enzymatic function. We know that glycans on cellulases are crucial for maintaining thermal, proteolytic, and pH stability as well as playing crucial roles in processive cellulase catalytic activity on crystalline cellulose. However, we lack crucial insight into the fine structure–function relationships of glycans with their cognate cellulases. Most glycan activities are site and structure specific, necessitating a detailed knowledge of both glycan site and structure to support future studies. However, given the general trend that minimal glycans are required for optimal activity, future cellulase engineering efforts should synergistically incorporate both beneficial mutations and optimal glycosylation patterns. Heterologous expression of cellulases must be performed carefully with respect to compatibility of host glycosylation patterns to endogenous glycosylation patterns of recombinant cellulases, as differences in these structures often lead to reduced activity and stability. Expression host engineering is a promising strategy to produce cellulases with optimized glycosylation patterns and, therefore, optimal or enhanced properties. The best strategy for host–strain glycoengineering in fungi is to knock-in glycan-active enzymes that can either add different sugar moieties to glycans or trim back excessive glycosylation. Many glycan-active glycoside hydrolases have been identified in fungi, and there are reasonably many more currently unknown. Through combined study of specific glycan structure–function relationships in cellulases and application of optimal glycosylation patterns to commercial cellulases, the efficient saccharification of lignocellulosic biomass can become an attainable reality.

ACKNOWLEDGMENTS

The authors thank the University of Colorado Boulder, the Butcher seed grant, and the National Science Foundation (CHE-1454925) for financial support. Work at NREL was supported by the U.S. Department of Energy under Contract No. DE-AC36-08GO28308. Funding for this work was also provided by the DOE Office of Energy Efficiency and Renewable Energy, Bioenergy Technologies Office. Additionally, we thank all of those who have contributed to the advancement of the science of cellulase glycosylation.

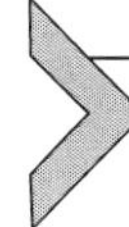

APPENDIX 1. MOLECULAR DYNAMICS SIMULATION OF A LINKER INTERACTING WITH CRYSTALLINE CELLULOSE

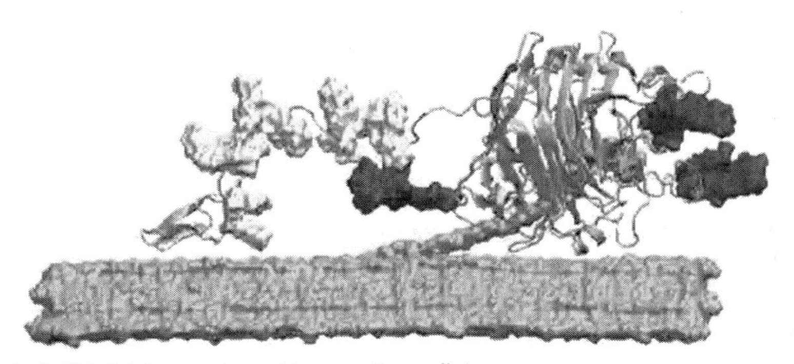

Movie 1 *Tr*Cel7A interacting with crystalline cellulose.

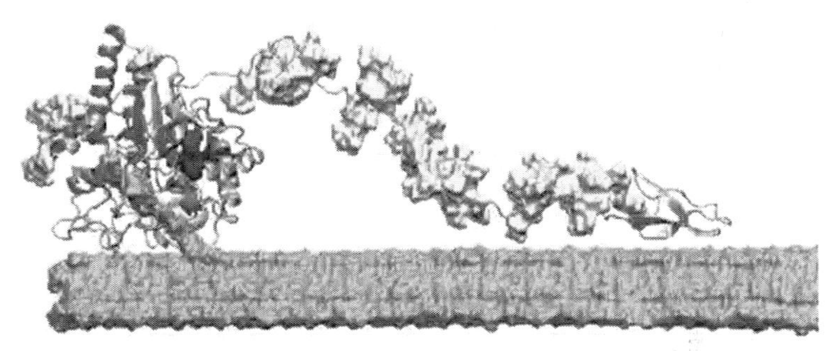

Movie 2 *Tr*Cel6A interacting with crystalline cellulose.

REFERENCES

1. Bond-Watts, B. B.; Bellerose, R. J.; Chang, M. C. Enzyme Mechanism as a Kinetic Control Element for Designing Synthetic Biofuel Pathways. *Nat. Chem. Biol.* **2011**, *7*, 222–227.
2. Wen, M.; Bond-Watts, B. B.; Chang, M. C. Production of Advanced Biofuels in Engineered E. coli. *Curr. Opin. Chem. Biol.* **2013**, *17*, 472–479.
3. Wilson, D. B. Cellulases and Biofuels. *Curr. Opin. Biotechnol.* **2009**, *20*, 295–299.
4. Banerjee, G.; Scott-Craig, J. S.; Walton, J. D. Improving Enzymes for Biomass Conversion: A Basic Research Perspective. *BioEnergy Res.* **2010**, *3*, 82–92.
5. Scheper, T. *Advances in Biochemical Engineering/Biotechnology*; Vol. 108. Springer-Verlag: Berlin, Heidelberg, 2007.
6. Himmel, M. E.; Ding, S. Y.; Johnson, D. K.; Adney, W. S.; Nimlos, M. R.; Brady, J. W.; Foust, T. D. Biomass Recalcitrance: Engineering Plants and Enzymes for Biofuels Production. *Science* **2007**, *315*, 804–807.

7. Beckham, G. T.; Dai, Z.; Matthews, J. F.; Momany, M.; Payne, C. M.; Adney, W. S.; Baker, S. E.; Himmel, M. E. Harnessing Glycosylation to Improve Cellulase Activity. *Curr. Opin. Biotechnol.* **2012**, *23*, 338–345.
8. Seiboth, B.; Ivanova, C.; Seidl-Seiboth, V. *Trichoderma reesei*: A Fungal Enzyme Producer for Cellulosic Biofuels. In: *Biofuel Production—Recent Developments and Prospects*; Bernardes, M. A. D. S. Ed.; InTech: Rijeka, Croatia, 2011.
9. Serpa, V. I.; Polikarpov, I. *Enzymes in Bioenergy. Routes to Cellulosic Ethanol*, Springer Science, New York: New York, 2011, pp. 97–113.
10. Deshpande, N.; Wilkins, M. R.; Packer, N.; Nevalainen, H. Protein Glycosylation Pathways in Filamentous Fungi. *Glycobiology* **2008**, *18*, 626–637.
11. Goto, M. Protein O-Glycosylation in Fungi: Diverse Structures and Multiple Functions. *Biosci. Biotechnol. Biochem.* **2007**, *71*, 1415–1427.
12. Kubicek, C. P.; Panda, T.; Schreferl-Kunar, G.; Gruber, F.; Messner, R. O-Linked but not N-Linked Glycosylation Is Necessary for the Secretion of Endoglucanases I and II by *Trichoderma reesei*. *Can. J. Microbiol.* **1987**, *33*, 698–703.
13. Dejgaard, S.; Nicolay, J.; Taheri, M.; Thomas, D. Y.; Bergeron, J. M. The ER Glycoprotein Qualitiy Control System. *Curr. Issues Mol. Biol.* **2004**, *6*, 29–42.
14. Gonzalez, M.; Brito, N.; Gonzalez, C. High Abundance of Serine/Threonine-Rich Regions Predicted to be Hyper-O-Glycosylated in the Secretory Proteins Coded by Eight Fungal Genomes. *BMC Microbiol.* **2012**, *12*, 213.
15. Hart, G. W.; Copeland, R. J. Glycomics Hits the Big Time. *Cell* **2010**, *143*, 672–676.
16. Kruszewska, J. S.; Perlinska-Lenart, U.; Gorka-Niec, W.; Orlowski, J.; Zembek, P.; Palamarczyk, G. Alterations in Protein Secretion Caused by Metabolic Engineering of Glycosylation Pathways in Fungi. *Acta Biochim. Pol.* **2008**, *55*, 447–456.
17. Adney, W. S.; Chou, Y.-C.; Decker, S. R.; Ding, S.-Y.; Baker, J. O.; Kunkel, G.; Vinzant, T. B.; Himmel, M. E. Heterologous Expression of *Trichoderma reesei* 1,4-β-D-Glucan Cellobiohydrolase (Cel 7A). *ACS Symp. Ser.* **2003**, *855*, 403–437.
18. Adney, W. S.; Jeoh, T.; Beckham, G. T.; Chou, Y.-C.; Baker, J. O.; Michener, W.; Brunecky, R.; Himmel, M. E. Probing the Role of *N*-Linked Glycans in the Stability and Activity of Fungal Cellobiohydrolases by Mutational Analysis. *Cellulose* **2009**, *16*, 699–709.
19. Beckham, G. T.; Bomble, Y. J.; Matthews, J. F.; Taylor, C. B.; Resch, M. G.; Yarbrough, J. M.; Decker, S. R.; Bu, L.; Zhao, X.; McCabe, C.; Wohlert, J.; Bergenstrahle, M.; Brady, J. W.; Adney, W. S.; Himmel, M. E.; Crowley, M. F. The O-Glycosylated Linker from the Trichoderma reesei Family 7 Cellulase Is a Flexible, Disordered Protein. *Biophys. J.* **2010**, *99*, 3773–3781.
20. Jeoh, T.; Michener, W.; Himmel, M. E.; Decker, S. R.; Adney, W. S. Implications of Cellobiohydrolase Glycosylation for Use in Biomass Conversion. *Biotechnol. Biofuels* **2008**, *1*, 10.
21. Wu, G.; Wei, L.; Liu, W.; Lin, J.; Wang, L.; Qu, Y.; Zhuang, G. Asn64-Glycosylation Affects *Hypocrea jecorina* (Syn. *Trichoderma reesei*) Cellobiohydrolase Cel7A Activity Expressed in *Pichia pastoris*. *World J. Microbiol. Biotechnol.* **2010**, *26*, 323–328.
22. Qi, F.; Zhang, W.; Zhang, F.; Chen, G.; Liu, W. Deciphering the Effect of the Different N-Glycosylation Sites on the Secretion, Activity, and Stability of Cellobiohydrolase I from *Trichoderma reesei*. *Appl. Environ. Microbiol.* **2014**, *80*, 3962–3971.
23. Payne, C. M.; Resch, M. G.; Chen, L.; Crowley, M.; Himmel, M. E.; Taylor, L. E., II; Sandgren, M.; Stahlberg, J.; Stals, I.; Tan, Z.; Beckham, G. T. Glycosylated Linkers in Multimodular Lignocellulose-Degrading Enzymes Dynamically Bind to Cellulose. *Proc. Natl. Acad. Sci. U.S.A.* **2013**, *110*, 14646–14651.
24. Taylor, C. B.; Talib, M. F.; McCabe, C.; Bu, L.; Adney, W. S.; Himmel, M. E.; Crowley, M. F.; Beckham, G. T. Computational Investigation of Glycosylation Effects on a Family 1 Carbohydrate-Binding Module. *J. Biol. Chem.* **2012**, *287*, 3147–3155.

25. Chen, L.; Drake, M. R.; Resch, M. G.; Greene, E. R.; Himmel, M. E.; Chaffey, P. K.; Beckham, G. T.; Tan, Z. Specificity of O-Glycosylation in Enhancing the Stability and Cellulose Binding Affinity of Family 1 Carbohydrate-Binding Modules. *Proc. Natl. Acad. Sci. U.S.A.* **2014**, *111*, 7612–7617.
26. Boer, H.; Teeri, T.; Koivula, A. Characterization of *Trichoderma reesei* Cellbiohydrolase Cel7A Secreted from *Pichia pastoris* Using Two Different Promoters. *Biotechnol. Bioeng.* **2000**, *69*, 486–494.
27. Schmoll, M.; Kubicek, C. P. Regulation of Trichoderma Cellulase Formation: Lessons in Molecular Biology from an Industrial Fungus. *Acta Microbiol. Immunol.* **2003**, *50*, 125–145.
28. Lynd, L. R.; Weimer, P. J.; van Zyl, W. H.; Pretorius, I. S. Microbial Cellulose Utilization: Fundamentals and Biotechnology. *Microbiol. Mol. Biol. Rev.* **2002**, *66*, 506–577.
29. Maras, M.; De Bruyn, A.; Schraml, J.; Herdewijn, P.; Claeyssens, M.; Fiers, W.; Contreras, R. Structural Characterization of *N*-Linked Oligosaccharides from Cellbiohydrolase I Secreted by the Filamentous Fungus *Trichoderma reesei* RUTC 30. *Eur. J. Biochem.* **1997**, *245*, 617–625.
30. Klarskov, K.; Piens, K.; Stahlberg, J.; Hoj, P. B.; Van Beeumen, J.; Claeyssens, M. Cellobiohydrolase I from *Trichoderma reesei*: Identification of an Active-Site Nucleophile and Additional Information on Sequence Including the Glycosylation Pattern of the Core Protein. *Carbohydr. Res.* **1997**, *304*, 143–154.
31. Hui, J. P. M.; Lanthier, P.; White, T. C.; McHugh, S. G.; Yaguchi, M.; Roy, R.; Thibault, P. Characterization of Cellbiohydrolase I (Cel7A) Glycoforms from Extracts of *Trichoderma reesei* Using Capillary Isoelectric Focusing and Electrospray Mass Spectrometry. *J. Chromatogr. B* **2001**, *752*, 349–368.
32. Harrison, M. J.; Nouwens, A. S.; Jardine, D. R.; Zachara, N. E.; Gooley, A. A.; Nevalainen, H.; Packer, N. Modified Glycosylation of Cellobiohydrolase I from a High Cellulase-Producing Mutant Strain of *Trichoderma reesei. Eur. J. Biochem.* **1998**, *256*, 119–127.
33. Stals, I.; Sandra, K.; Devreese, B.; Van Beeumen, J.; Claeyssens, M. Factors Influencing Glycosylation of Trichoderma reesei Cellulases. II: N-Glycosylation of Cel7A Core Protein Isolated from Different Strains. *Glycobiology* **2004**, *14*, 725–737.
34. Stals, I.; Sandra, K.; Geysens, S.; Contreras, R.; Van Beeumen, J.; Claeyssens, M. Factors Influencing Glycosylation of *Trichoderma reesei* Cellulases. I: Postsecretorial Changes of the O- and N-Glycosylation Pattern of Cel7A. *Glycobiology* **2004**, *14*, 713–724.
35. Hui, J. P. M.; White, T. C.; Thibault, P. Identification of Glycan Structure and Glycosylation Sites in Cellobiohydrolase II and Endoglucanases I and II from *Trichoderma reesei. Glycobiology* **2002**, *12*, 837–849.
36. De Bruyn, A.; Maras, M.; Schraml, J.; Herdewijn, P.; Contreras, R. NMR Evidence for a Novel Asparagine-Linked Oligosaccharide on Cellobiohydrolase I from *Trichoderma reesei* RUTC 30. *FEBS Lett.* **1997**, *405*, 111–113.
37. Sandra, K.; Devreese, B.; Van Beeumen, J.; Stals, I.; Claeyssens, M. The Q-Trap Mass Spectrometer, a Novel Tool in the Study of Protein Glycosylation. *J. Am. Soc. Mass Spectrom.* **2004**, *15*, 413–423.
38. Sandra, K.; Van Beeumen, J.; Stals, I.; Sandra, P.; Claeyssens, M.; Devreese, B. Characterization of Cellbiohydrolase I *N*-Glycans and Differentiation of Their Phosphorylated Isomers by Capillary Electrophoresis-O-Trap Mass Spectrometry. *Anal. Chem.* **2004**, *76*, 5878–5886.
39. Salovuori, I.; Makarow, M.; Rauvala, H.; Knowles, J. K. C.; Kaariainen, L. Low Molecular Weight High-Mannose Type Glycans in a Secreted Protein of the Filamentous Fungus *Trichoderma reesei. Nat. Biotechnol.* **1987**, *5*, 152–156.
40. Geysens, S.; Pakula, T.; Uusitalo, J.; Dewerte, I.; Penttila, M.; Contreras, R. Cloning and Characterization of the Glucosidase II Alpha Subunit Gene of *Trichoderma*

reesei: A Frameshift Mutation Results in the Aberrant Glycosylation Profile of the Hypercellulolytic Strain Rut-C30. *Appl. Environ. Microbiol.* **2005**, *71*, 2910–2924.

41. Stals, I.; Karkehabadi, S.; Kim, S.; Ward, M.; Van Landschoot, A.; Devreese, B.; Sandgren, M. High Resolution Crystal Structure of the *endo-N*-Acetyl-β-D-Glucosaminidase Responsible for the Deglycosylation of *Hypocrea jecorina* Cellulases. *PLoS One* **2012**, 7, e40854.
42. Stals, I.; Samyn, B.; Sergeant, K.; White, T.; Hoorelbeke, K.; Coorevits, A.; Devreese, B.; Claeyssens, M.; Piens, K. Identification of a Gene Coding for a Deglycosylating Enzyme in *Hypocrea jecorina*. *FEMS Microbiol. Lett.* **2010**, *303*, 9–17.
43. Adav, S. S.; Chao, L. T.; Sze, S. K. Protein Abundance in Multiplexed Samples (PAMUS) for Quantitation of *Trichoderma reesei* Secretome. *J. Proteomics* **2013**, *83*, 180–196.
44. Nagendran, S.; Hallen-Adams, H. E.; Paper, J. M.; Aslam, N.; Walton, J. D. Reduced Genomic Potential for Secreted Plant Cell-Wall-Degrading Enzymes in the Ectomycorrhizal Fungus *Amanita bisporigera*, Based on the Secretome of *Trichoderma reesei*. *Fungal Genet. Biol.* **2009**, *46*, 427–435.
45. Garcia, R.; Cremata, J. A.; Quintero, O.; Montesino, R.; Benkestock, K.; Stahlberg, J. Characterization of Protein Glycoforms with *N*-Linked Neutral and Phosphorylated Oligosaccharides: Studies on the Glycosylation of Endoglucanase I (Cel7B) from *Trichoderma reesei*. *Biotechnol. Appl. Biochem.* **2001**, *33*, 141–152.
46. Eriksson, T.; Stals, I.; Collen, A.; Tjerneld, F.; Claeyssens, M.; Stalbrand, H.; Brumer, H. Heterogeneity of Homologously Expressed *Hypocrea jecorina* (*Trichoderma reesei*) Cel7B Catalytic Module. *Eur. J. Biochem.* **2004**, *271*, 1266–1276.
47. Karkehabadi, S.; Hansson, H.; Kim, S.; Piens, K.; Mitchinson, C.; Sandgren, M. The First Structure of a Glycoside Hydrolase Family 61 Member, Cel61B from *Hypocrea jecorina*, at 1.6 Å Resolution. *J. Mol. Biol.* **2008**, *383*, 144–154.
48. Wei, W.; Chen, L.; Zou, G.; Wang, Q.; Yan, X.; Zhang, J.; Wang, C.; Zhou, Z. N-Glycosylation Affects the Proper Folding, Enzymatic Characteristics and Production of a Fungal β-Glucosidase. *Biotechnol. Bioeng.* **2013**, *110*, 3075–3084.
49. Christiansen, M. N.; Kolarich, D.; Nevalainen, H.; Packer, N.; Jensen, P. H. Challenges of Determining O-Glycospeptide Heterogeneity: A Fungal Glucanase Model System. *Anal. Chem.* **2010**, *82*, 3500–3509.
50. Divne, C.; Stahlberg, J.; Reinikainen, T.; Ruohonen, L.; Pettersson, G.; Knowles, J. K. C.; Teeri, T.; Jones, T. A. The Three-Dimensional Crystal Structure of the Catalytic Core of Cellobiohydrolase I from *Trichoderma reesei*. *Science* **1994**, *265*, 524–528.
51. Godbole, S.; Decker, S. R.; Nieves, R. A.; Adney, W. S.; Vinzant, T. B.; Baker, J. O.; Thomas, S. R.; Himmel, M. E. Cloning and Expression of *Trichoderma reesei* Cellobiohydrolase I in *Pichia pastoris*. *Biotechnol. Prog.* **1999**, *15*, 828–833.
52. Penttila, M.; Andre, L.; Lehtovaara, P.; Bailey, M.; Teeri, T.; Knowles, J. K. C. Efficient Secretion of two Fungal Cellobiohydrolases by *Saccharomyces cerevisiae*. *Gene* **1988**, *63*, 103–112.
53. Ilmen, M.; den Haan, R.; Brevnova, E.; McBride, J.; Wiswall, E.; Froehlich, A.; Koivula, A.; Voutilainen, S. P.; Siika-Aho, M.; la Grange, D. C.; Thorngren, N.; Ahlgren, S.; Mellon, M.; Deleault, K.; Rajgarhia, V.; van Zyl, W. H.; Penttila, M. High Level Secretion of Cellobiohydrolases by *Saccharomyces cerevisiae*. *Biotechnol. Biofuels* **2011**, *4*, 30.
54. Voutilainen, S. P.; Boer, H.; Linder, M. B.; Puranen, T.; Rouvinen, J.; Vehmaanperä, J.; Koivula, A. Heterologous Expression of *Melanocarpus albomyces* Cellobiohydrolase Cel7B, and Random Mutagenesis to Improve Its Thermostability. *Enzyme Microb. Technol.* **2007**, *41*, 234–243.

55. Voutilainen, S. P.; Murray, P. G.; Tuohy, M. G.; Koivula, A. Expression of *Talaromyces emersonii* Cellobiohydrolase Cel7A in *Saccharomyces cerevisiae* and Rational Mutagenesis to Improve Its Thermostability and Activity. *Protein Eng. Des. Sel.* **2010**, *23*, 69–79.
56. Voutilainen, S. P.; Nurmi-Rantala, S.; Penttila, M.; Koivula, A. Engineering Chimeric Thermostable GH7 Cellobiohydrolases in *Saccharomyces cerevisiae*. *Appl. Microbiol. Biotechnol.* **2014**, *98*, 2991–3001.
57. Voutilainen, S. P.; Puranen, T.; Siika-Aho, M.; Lappalainen, A.; Alapuranen, M.; Kallio, J.; Hooman, S.; Viikari, L.; Vehmaanpera, J.; Koivula, A. Cloning, Expression, and Characterization of Novel Thermostable Family 7 Cellobiohydrolases. *Biotechnol. Bioeng.* **2008**, *101*, 515–528.
58. Sammond, D. W.; Payne, C. M.; Brunecky, R.; Himmel, M. E.; Crowley, M. F.; Beckham, G. T. Cellulase Linkers Are Optimized Based on Domain Type and Function: Insights from Sequence Analysis, Biophysical Measurements, and Molecular Simulation. *PLoS One* **2012**, 7, e48615.
59. Martinez, D.; Berka, R. M.; Henrissat, B.; Saloheimo, M.; Arvas, M.; Baker, S. E.; Chapman, J.; Chertkov, O.; Coutinho, P. M.; Cullen, D.; Danchin, E. G.; Grigoriev, I. V.; Harris, P.; Jackson, M.; Kubicek, C. P.; Han, C. S.; Ho, I.; Larrondo, L. F.; de Leon, A. L.; Magnuson, J. K.; Merino, S.; Misra, M.; Nelson, B.; Putnam, N.; Robbertse, B.; Salamov, A. A.; Schmoll, M.; Terry, A.; Thayer, N.; Westerholm-Parvinen, A.; Schoch, C. L.; Yao, J.; Barabote, R.; Nelson, M. A.; Detter, C.; Bruce, D.; Kuske, C. R.; Xie, G.; Richardson, P.; Rokhsar, D. S.; Lucas, S. M.; Rubin, E. M.; Dunn-Coleman, N.; Ward, M.; Brettin, T. S. Genome Sequencing and Analysis of the Biomass-Degrading Fungus *Trichoderma reesei* (syn. *Hypocrea jecorina*). *Nat. Biotechnol.* **2008**, *26*, 553–560.
60. Ghattyvenkatakrishna, P. K.; Alekozai, E. M.; Beckham, G. T.; Schulz, R.; Crowley, M. F.; Uberbacher, E. C.; Cheng, X. Initial Recognition of a Cellodextrin Chain in the Cellulose-Binding Tunnel May Affect Cellobiohydrolase Directional Specificity. *Biophys. J.* **2013**, *104*, 904–912.
61. Nakamura, A.; Tsukada, T.; Auer, S.; Furuta, T.; Wada, M.; Koivula, A.; Igarashi, K.; Samejima, M. The Tryptophan Residue at the Active Site Tunnel Entrance of *Trichoderma reesei* Cellobiohydrolase Cel7A Is Important for Initiation of Degradation of Crystalline Cellulose. *J. Biol. Chem.* **2013**, *288*, 13503–13510.
62. Boraston, A. B.; Sandercock, L. E.; Warren, R. A. J.; Kilburn, D. G. O-Glycosylation of a Recombinant Carbohydrate-Binding Module Mutant Secreted by *Pichia pastoris*. *J. Mol. Microbiol. Biotechnol.* **2003**, *5*, 29–36.
63. Boraston, A. B.; Warren, R. A. J.; Kilburn, D. G. Glycosylation by *Pichia pastoris* Decreases the Affinity of a Family 2a Carbohydrate-Binding Module from *Cellulomonas fimi*: A Functional and Mutational Analysis. *Biochem. J.* **2001**, *358*, 423–430.
64. Boraston, A. B.; Bolam, D. N.; Gilbert, H. J.; Davies, G. J. Carbohydrate-Binding Modules: Fine-Tuning Polysaccharide Recognition. *Biochem. J.* **2004**, *382*, 769–781.
65. Lehtio, J.; Sugiyama, J.; Gustavsson, M.; Fransson, L.; Linder, M.; Teeri, T. T. The Binding Specificity and Affinity Determinants of Family 1 and Family 3 Cellulose Binding Modules. *Proc. Natl. Acad. Sci. U.S.A.* **2003**, *100*, 484–489.
66. Linder, M.; Lindeberg, G.; Reinikainen, T.; Teeri, T.; Pettersson, G. The Difference in Affinity Between Two Fungal Cellulose-Binding Domains Is Dominated by a Single Amino Acid Substitution. *FEBS Lett.* **1995**, *372*, 96–98.
67. Linder, M.; Mattinen, M.-L.; Kontteli, M.; Lindeberg, G.; Stahlberg, J.; Drakenberg, T.; Reinikainen, T.; Pettersson, G.; Annila, A. Identification of Functionally Important Amino Acids in the Cellulose-Binding Domain of *Trichoderma reesei* Cellobiohydrolase I. *Protein Sci.* **1995**, *4*, 1056–1064.

68. Pinto, R.; Amaral, A. L.; Carvalho, J.; Ferreira, E. C.; Mota, M.; Gama, M. Development of a Method Using Image Analysis for the Measurement of Cellulose-Binding Domains Adsorbed onto Cellulose Fibers. *Biotechnol. Prog.* **2007**, *23*, 1492–1497.
69. Pinto, R.; Amaral, A. L.; Ferreira, E. C.; Mota, M.; Vilanova, M.; Ruel, K.; Gama, M. Quantification of the CBD-FITC Conjugates Surface Coating on Cellulose Fibres. *BMC Biotechnol.* **2008**, *8*, 1.
70. Pinto, R.; Carvalho, J.; Mota, M.; Gama, M. Large-Scale Production of Cellulose-Binding Domains. Adsorption Studies Using CBD-FITC Conjugates. *Cellulose* **2006**, *13*, 557–569.
71. Pinto, R.; Moreira, S.; Mota, M.; Gama, M. Studies on the Cellulose-Binding Domains Adsorption to Cellulose. *Langmuir* **2004**, *20*, 1409–1413.
72. Sugimoto, N.; Igarashi, K.; Samejima, M. Cellulose Affinity Purification of Fusion Proteins Tagged with Fungal Family 1 Cellulose-Binding Domain. *Protein Expr. Purif.* **2012**, *82*, 290–296.
73. Sugimoto, N.; Igarashi, K.; Wada, M.; Samejima, M. Adsorption Characteristics of Fungal Family 1 Cellulose-Binding Domain from *Trichoderma reesei* Cellobiohydrolase I on Crystalline Cellulose: Negative Cooperative Adsorption via a Steric Exclusion Effect. *Langmuir* **2012**, *28*, 14323–14329.
74. Medve, J.; Stahlberg, J.; Tjerneld, F. Isotherms for Adsorption of Cellobiohydrolase I and II from *Trichoderma reesei* on Microcrystalline Cellulose. *Appl. Biochem. Biotechnol.* **1997**, *66*, 39–56.
75. Palonen, H.; Tenkanen, M.; Linder, M. Dynamic Interaction of *Trichoderma reesei* Cellobiohydrolases Cel6A and Cel7A and Cellulose at Equilibrium and During Hydrolysis. *Appl. Environ. Microbiol.* **1999**, *65*, 5229–5233.
76. Takashima, S.; Ohno, M.; Hidaka, M.; Nakamura, A.; Masaki, H.; Uozumi, T. Correlation Between Cellulose Binding and Activity of Cellulose-Binding Domain Mutants of *Humicola grisea* Cellobiohydrolase 1. *FEBS Lett.* **2007**, *581*, 5891–5896.
77. Igarashi, K.; Uchihashi, T.; Koivula, A.; Wada, M.; Kimura, S.; Okamoto, T.; Penttila, M.; Ando, T.; Samejima, M. Traffic Jams Reduce Hydrolytic Efficiency of Cellulase on Cellulose Surface. *Science* **2011**, *333*, 1279–1282.
78. Gao, D.; Chundawat, S. P.; Uppugundla, N.; Balan, V.; Dale, B. E. Binding Characteristics of *Trichoderma reesei* Cellulases on Untreated, Ammonia Fiber Expansion (AFEX), and Dilute-Acid Pretreated Lignocellulosic Biomass. *Biotechnol. Bioeng.* **2011**, *108*, 1788–1800.
79. Gao, J.; Thomas, D. A.; Sohn, C. H.; Beauchamp, J. L. Biomimetic Reagents for the Selective Free Radical and Acid–Base Chemistry of Glycans: Application to Glycan Structure Determination by Mass Spectrometry. *J. Am. Chem. Soc.* **2013**, *135*, 10684–10692.
80. Cruys-Bagger, N.; Tatsumi, H.; Borch, K.; Westh, P. A Graphene Screen-Printed Carbon Electrode for Real-Time Measurements of Unoccupied Active Sites in a Cellulase. *Anal. Biochem.* **2014**, *447*, 162–168.
81. Payne, C. M.; Jiang, W.; Shirts, M. R.; Himmel, M. E.; Crowley, M.; Beckham, G. T. Glycoside Hydrolase Processivity Is Directly Related to Oligosaccharide Binding Free Energy. *J. Am. Chem. Soc.* **2013**, *135*, 18831–18839.
82. Knott, B. C.; Crowley, M.; Himmel, M. E.; Stahlberg, J.; Beckham, G. T. Carbohydrate–Protein Interactions That Drive Processive Polysaccharide Translocation in Enzymes Revealed from a Computation Study of Cellobiohydrolase Processivity. *J. Am. Chem. Soc.* **2014**, *136*, 8810–8819.
83. Knott, B. C.; Momeni, M. H.; Crowley, M.; Mackenzie, L. F.; Gotz, A. W.; Sandgren, M.; Withers, S. G.; Stahlberg, J.; Beckham, G. T. The Mechanism of Cellulose Hydrolysis by a Two-Step, Retaining Cellobiohydrolase Elucidate by

Structural and Transition Path Sampling Studies. *J. Am. Chem. Soc.* **2014**, *136*, 321–329.
84. Bu, L.; Beckham, G. T.; Shirts, M. R.; Nimlos, M. R.; Adney, W. S.; Himmel, M. E.; Crowley, M. F. Probing Carbohydrate Product Expulsion from a Processive Cellulase with Multiple Absolute Binding Free Energy Methods. *J. Biol. Chem.* **2011**, *286*, 18161–18169.
85. Bu, L.; Nimlos, M. R.; Shirts, M. R.; Stahlberg, J.; Himmel, M. E.; Crowley, M. F.; Beckham, G. T. Product Binding Varies Dramatically Between Processive and Nonprocessive Cellulase Enzymes. *J. Biol. Chem.* **2012**, *287*, 24807–24813.
86. Taylor, C. B.; Payne, C. M.; Himmel, M. E.; Crowley, M. F.; McCabe, C.; Beckham, G. T. Binding Site Dynamics and Aromatic-Carbohydrate Interactions in Processive and Non-Processive Family 7 Glycoside Hydrolases. *J. Phys. Chem. B* **2013**, *117*, 4924–4933.
87. Beckham, G. T.; Stahlberg, J.; Knott, B. C.; Himmel, M. E.; Crowley, M. F.; Sandgren, M.; Sorlie, M.; Payne, C. M. Towards a Molecular-Level Theory of Carbohydrate Processivity in Glycoside Hydrolases. *Curr. Opin. Biotechnol.* **2014**, *27*, 96–106.
88. Yan, S.; Li, T.; Yao, L. Mutational Effects on the Catalytic Mechanism of Cellobiohydrolase I from *Trichoderma reesei*. *J. Phys. Chem. B* **2011**, *115*, 4982–4989.
89. Cruys-Bagger, N.; Elmerdahl, J.; Praestgaard, E.; Borch, K.; Westh, P. A Steady-State Theory for Processive Cellulases. *FEBS J.* **2013**, *280*, 3952–3961.
90. Cruys-Bagger, N.; Tatsumi, H.; Ren, G. R.; Borch, K.; Westh, P. Transient Kinetics and Rate-Limiting Steps for the Processive Cellobiohydrolase Cel7A: Effects of Substrate Structure and Carbohydrate Binding Domain. *Biochemistry* **2013**, *52*, 8938–8948.
91. Igarashi, K.; Koivula, A.; Wada, M.; Kimura, S.; Penttila, M.; Samejima, M. High Speed Atomic Force Microscopy Visualizes Processive Movement of *Trichoderma reesei* Cellobiohydrolase I on Crystalline Cellulose. *J. Biol. Chem.* **2009**, *284*, 36186–36190.
92. Igarashi, K.; Uchihashi, T.; Koivula, A.; Wada, M.; Kimura, S.; Penttila, M.; Ando, T.; Samejima, M. Visualization of Cellobiohydrolase I from *Trichoderma reesei* Moving on Crystalline Cellulose Using High-Speed Atomic Force Microscopy. *Methods Enzymol.* **2012**, *510*, 169–182.
93. Jalak, J.; Kurasin, M.; Teugjas, H.; Valjamae, P. endo-exo Synergism in Cellulose Hydrolysis Revisited. *J. Biol. Chem.* **2012**, *287*, 28802–28815.
94. Linder, M.; Teeri, T. The Cellulose-Binding Domain of the Major Cellobiohydrolase of *Trichoderma reesei* Exhibits True Reversibility and a High Exchange Rate on Crystalline Cellulose. *Proc. Natl. Acad. Sci. U.S.A.* **1996**, *93*, 12251–12255.
95. Jung, J.; Sethi, A.; Gaiotto, T.; Han, J. J.; Jeoh, T.; Gnanakaran, S.; Goodwin, P. M. Binding and Movement of Individual Cel7A Cellobiohydrolases on Crystalline Cellulose Surfaces Revealed by Single-Molecule Fluorescence Imaging. *J. Biol. Chem.* **2013**, *288*, 24164–24172.
96. Fox, J. M.; Levine, S. E.; Clark, D. S.; Blanch, H. W. Initial- and Processive-Cut Products Reveal Cellobiohydrolase Rate Limitations and the Role of Companion Enzymes. *Biochemistry* **2012**, *51*, 442–452.
97. Ganner, T.; Bubner, P.; Eibinger, M.; Mayrhofer, C.; Plank, H.; Nidetzky, B. Dissecting and Reconstructing Synergism: In Situ Visualization of Cooperativity Among Cellulases. *J. Biol. Chem.* **2012**, *287*, 43215–43222.
98. Jalak, J.; Valjamae, P. Mechanism of Initial Rapid Rate Retardation in Cellobiohydrolase Catalyzed Cellulose Hydrolysis. *Biotechnol. Bioeng.* **2010**, *106*, 871–883.
99. Kurasin, M.; Valjamae, P. Processivity of Cellobiohydrolases is Limited by the Substrate. *J. Biol. Chem.* **2011**, *286*, 169–177.

100. Barsberg, S.; Selig, M. J.; Felby, C. Impact of Lignins Isolated from Pretreated Lignocelluloses on Enzymatic Cellulose Saccharification. *Biotechnol. Lett.* **2013**, *35*, 189–195.
101. Kumar, L.; Arantes, V.; Chandra, R.; Saddler, J. The Lignin Present in Steam Pretreated Softwood Binds Enzymes and Limits Cellulose Accessibility. *Bioresour. Technol.* **2012**, *103*, 201–208.
102. Kumar, R.; Hu, F.; Sannigrahi, P.; Jung, S.; Ragauskas, A. J.; Wyman, C. E. Carbohydrate Derived-Pseudo-Lignin Can Retard Cellulose Biological Conversion. *Biotechnol. Bioeng.* **2013**, *110*, 737–753.
103. Martin-Sampedro, R.; Rahikainen, J. L.; Johansson, L. S.; Marjamaa, K.; Laine, J.; Kruus, K.; Rojas, O. J. Preferential Adsorption and Activity of Monocomponent Cellulases on Lignocellulose Thin Films with Varying Lignin Content. *Biomacromolecules* **2013**, *14*, 1231–1239.
104. Nakagame, S.; Chandra, R. P.; Kadla, J. F.; Saddler, J. N. Enhancing the Enzymatic Hydrolysis of Lignocellulosic Biomass by Increasing the Carboxylic Acid Content of the Associated Lignin. *Biotechnol. Bioeng.* **2011**, *108*, 538–548.
105. Nakagame, S.; Chandra, R. P.; Kadla, J. F.; Saddler, J. N. The Isolation, Characterization and Effect of Lignin Isolated from Steam Pretreated Douglas-fir on the Enzymatic Hydrolysis of Cellulose. *Bioresour. Technol.* **2011**, *102*, 4507–4517.
106. Nakagame, S.; Chandra, R. P.; Saddler, J. N. The Effect of Isolated Lignins, Obtained from a Range of Pretreated Lignocellulosic Substrates, on Enzymatic Hydrolysis. *Biotechnol. Bioeng.* **2010**, *105*, 871–879.
107. Nonaka, H.; Kobayashi, A.; Funaoka, M. Behavior of Lignin-Binding Cellulase in the Presence of Fresh Cellulosic Substrate. *Bioresour. Technol.* **2013**, *135*, 53–57.
108. Nonaka, H.; Kobayashi, A.; Funaoka, M. Lignin Isolated from Steam-Exploded Eucalyptus Wood Chips by Phase Separation and Its Affinity to *Trichoderma reesei* Cellulase. *Bioresour. Technol.* **2013**, *140*, 431–434.
109. Palonen, H.; Tjerneld, F.; Zacchi, G.; Tenkanen, M. Adsorption of *Trichoderma reesei* CBH I and EG II and Their Catalytic Domains on Steam Pretreated Softwood and Isolated Lignin. *J. Biotechnol.* **2004**, *107*, 65–72.
110. Pareek, N.; Gillgren, T.; Jonsson, L. J. Adsorption of Proteins Involved in Hydrolysis of Lignocellulose on Lignins and Hemicelluloses. *Bioresour. Technol.* **2013**, *148*, 70–77.
111. Rahikainen, J.; Mikander, S.; Marjamaa, K.; Tamminen, T.; Lappas, A.; Viikari, L.; Kruus, K. Inhibition of Enzymatic Hydrolysis by Residual Lignins from Softwood—Study of Enzyme Binding and Inactivation on Lignin-Rich Surface. *Biotechnol. Bioeng.* **2011**, *108*, 2823–2834.
112. Rahikainen, J. L.; Evans, J. D.; Mikander, S.; Kalliola, A.; Puranen, T.; Tamminen, T.; Marjamaa, K.; Kruus, K. Cellulase-Lignin Interactions—The Role of Carbohydrate-Binding Module and pH in Non-Productive Binding. *Enzyme Microb. Technol.* **2013**, *53*, 315–321.
113. Rahikainen, J. L.; Martin-Sampedro, R.; Heikkinen, H.; Rovio, S.; Marjamaa, K.; Tamminen, T.; Rojas, O. J.; Kruus, K. Inhibitory Effect of Lignin During Cellulose Bioconversion: The Effect of Lignin Chemistry on Non-Productive Enzyme Adsorption. *Bioresour. Technol.* **2013**, *133*, 270–278.
114. Rodrigues, A. C.; Leitao, A. F.; Moreira, S.; Felby, C.; Gama, M. Recycling of Cellulases in Lignocellulosic Hydrolysates Using Alkaline Elution. *Bioresour. Technol.* **2012**, *110*, 526–533.
115. Shang, B. Z.; Chu, J.-W. Kinetic Modeling at Single-Molecule Resolution Elucidates the Mechanisms of Cellulase Synergy. *ACS Catal.* **2014**, *4*, 2216–2225.
116. Hu, J.; Arantes, V.; Pribowo, A.; Saddler, J. The Synergistic Action of Accessory Enzymes Enhances the Hydrolytic Potential of a "Cellulase Mixture" but Is Highly Substrate Specfic. *Biotechnol. Biofuels* **2013**, *6*, 112.

117. Payne, C. M.; Jiang, W.; Shirts, M. R.; Himmel, M. E.; Crowley, M. F.; Beckham, G. T. Glycoside Hydrolase Processivity Is Directly Related to Oligosaccharide Binding Free Energy. *J. Am. Chem. Soc.* **2013**, *135*, 18831–18839.
118. Boer, H.; Koivula, A. The Relationship Between Thermal Stability and pH Optimum Studied with Wild-Type and Mutant *Trichoderma reesei* Cellobiohydrolase Cel7A. *Eur. J. Biochem.* **2003**, *270*, 841–848.
119. von Ossowski, I.; Ståhlberg, J.; Koivula, A.; Piens, K.; Becker, D.; Boer, H.; Harle, R.; Harris, M.; Divne, C.; Mahdi, S.; Zhao, Y.; Driguez, H.; Claeyssens, M.; Sinnott, M. L.; Teeri, T. T. Engineering the exo-Loop of *Trichoderma reesei* Cellobiohydrolase, Cel7A. A Comparison with *Phanerochaete chrysosporium* Cel7D. *J. Mol. Biol.* **2003**, *333*, 817–829.
120. Trudeau, D. L.; Lee, T. M.; Arnold, F. H. Engineered Thermostable Fungal Cellulases Exhibit Efficient Synergistic Cellulose Hydrolysis at Elevated Temperatures. *Biotechnol. Bioeng.* **2014**, *111*, 2390–2397.
121. Culyba, E. K.; Price, J. L.; Hanson, S. R.; Dhar, A.; Wong, C. H.; Gruebele, M.; Powers, E. T.; Kelly, J. W. Protein Native-State Stabilization by Placing Aromatic Side Chains in N-Glycosylated Reverse Turns. *Science* **2011**, *331*, 571–575.
122. Price, J. L.; Powers, D. L.; Powers, E. T.; Kelly, J. W. Glycosylation of the Enhanced Aromatic Sequon Is Similarly Stabilizing in Three Distinct Reverse Turn Contexts. *Proc. Natl. Acad. Sci. U.S.A.* **2011**, *108*, 14127 14132.
123. Price, J. L.; Shental-Bechor, D.; Dhar, A.; Turner, M. J.; Powers, E. T.; Gruebele, M.; Levy, Y.; Kelly, J. W. Context-Dependent Effects of Asparagine Glycosylation on Pin WW Folding Kinetics and Thermodynamics. *J. Am. Chem. Soc.* **2010**, *132*, 15359–15367.
124. Skropeta, D. The Effect of Individual *N*-Glycans on Enzyme Activity. *Bioorg. Med. Chem.* **2009**, *17*, 2645–2653.
125. Dana, C. M.; Saija, P.; Kal, S. M.; Bryan, M. B.; Blanch, H. W.; Clark, D. S. Biased Clique Shuffling Reveals Stabilizing Mutations in Cellulase Cel7A. *Biotechnol. Bioeng.* **2012**, *109*, 2710–2719.
126. Zhao, X.; Rignall, T. R.; McCabe, C.; Adney, W. S.; Himmel, M. E. Molecular Simulation Evidence for Processive Motion of *Trichoderma reesei* Cel7A During Cellulose Depolymerization. *Chem. Phys. Lett.* **2008**, *460*, 284–288.
127. Ting, C. L.; Makarov, D. E.; Wang, Z.-G. A Kinetic Model for the Enzymatic Action of Cellulase. *J. Phys. Chem. B* **2009**, *113*, 4970–4977.
128. Ma, B.; Tsai, C. J.; Haliloglu, T.; Nussinov, R. Dynamic Allostery: Linkers Are not Merely Flexible. *Structure* **2011**, *19*, 907–917.
129. Dienes, D.; Börjesson, J.; Hägglund, P.; Tjerneld, F.; Lidén, G.; Réczey, K.; Stålbrand, H. Identification of a Trypsin-Like Serine Protease from *Trichoderma reesei* QM9414. *Enzyme Microb. Technol.* **2007**, *40*, 1087–1094.
130. Eneyskaya, E. V.; Kulminskaya, A. A.; Savel'ev, A. N.; Savel'ev, N. V.; Shabalin, K. A.; Neustroev, K. N. Acid Protease from *Trichoderma reesei*: Limited Proteolysis of Fungal Carbohydrates. *Appl. Microbiol. Biotechnol.* **1999**, *52*, 226–231.
131. Haab, D.; Hagspiel, K.; Szakmary, K.; Kubicek, C. P. Formation of the Extracellular Proteases from *Trichoderma reesei* QM 9414 Involved in Cellulase Degradation. *J. Biotechnol.* **1990**, *16*, 187–198.
132. Yike, I. Fungal Proteases and Their Pathophysiological Effects. *Mycopathologia* **2011**, *171*, 299–323.
133. Hagspiel, K.; Haab, D.; Kubicek, C. P. Protease Activity and Proteolytic Modification of Cellulase from a *Trichoderma reesei* QM 9414 Selectant. *Appl. Microbiol. Biotechnol.* **1989**, *32*, 61–67.
134. Langsford, M. L.; Gilkes, N. R.; Singh, B.; Moser, B.; Miller, R. C., Jr.; Warren, R. A. J.; Kilburn, D. G. Glycosylation of Bacterial Cellulases Prevents Proteolytic Cleavage Between Functional Domains. *FEBS Lett.* **1987**, *225*, 163–167.

135. Srisodsuk, M.; Reinikainen, T.; Penttila, M.; Teeri, T. Role of the Interdomain Linker Peptide of *Trichoderma reesei* Cellobiohydrolase I in Its Interaction with Crystalline Cellulose. *J. Biol. Chem.* **1993**, *268*, 20756–20761.
136. Bernard, B. A.; Newton, S. A.; Olden, K. Effect of Size and Location of the Oligosaccharide Chain on Protease Degradation of Bovine Pancreatic Ribonuclease. *J. Biol. Chem.* **1983**, *258*, 12198–12202.
137. West, C. M. Current Ideas on the Significance of Protein Glycosylation. *Mol. Cell. Biochem.* **1986**, *72*, 3–20.
138. Schmuck, M.; Pilz, I.; Hayn, M.; Esterbauer, H. Investigation of Cellobiohydrolase from *Trichoderma reesei* by Small Angle X-Ray Scatter. *Biotechnol. Lett.* **1986**, *8*, 397–402.
139. Bodenheimer, A. M.; Cuneo, M. J.; Swartz, P. D.; He, J.; O'Neill, H. M.; Myles, D. A.; Evans, B. R.; Meilleur, F. Crystallization and Preliminary X-Ray Diffraction Analysis of *Hypocrea jecorina* Cel7A in Two New Crystal Forms. *Acta Crystallogr. F Struct. Biol. Commun.* **2014**, *70*, 773–776.
140. Pingali, S. V.; O'Neill, H. M.; McGaughey, J.; Urban, V. S.; Rempe, C. S.; Petridis, L.; Smith, J. C.; Evans, B. R.; Heller, W. T. Small Angle Neutron Scattering Reveals pH-Dependent Conformational Changes in *Trichoderma reesei* Cellobiohydrolase I: Implications for Enzymatic Activity. *J. Biol. Chem.* **2011**, *286*, 32801–32809.
141. Scott, B. R.; St-Pierre, P.; Lavigne, J.; Masri, N.; White, T. C.; Tomashek, J. J. Construction of Lignin-Resistant *Trichoderma reesei* Cellulase Variants with Modified Linker Peptides for Use in Cellulose Hydrolysis. *PCT Int. Appl.* **2010**. WO 2010096931 A1 20100902.
142. Lima, L. H. F.; Serpa, V. I.; Rosseto, F. R.; Sartori, G. R.; Oliveira Neto, M.; Martínez, L.; Polikarpov, I. Small-Angle X-ray Scattering and Structural Modeling of Full-Length: Cellobiohydrolase I from *Trichoderma harzianum*. *Cellulose* **2013**, *20*, 1573–1585.
143. Receveur, V.; Czjzek, M.; Schulein, M.; Panine, P.; Henrissat, B. Dimension, Shape, and Conformational Flexibility of a Two Domain Fungal Cellulase in Solution Probed by Small Angle X-Ray Scattering. *J. Biol. Chem.* **2002**, *277*, 40887–40892.
144. Poon, D. K.; Withers, S. G.; McIntosh, L. P. Direct Demonstration of the Flexibility of the Glycosylated Proline-Threonine Linker in the *Cellulomonas fimi* Xylanase Cex Through NMR Spectroscopic Analysis. *J. Biol. Chem.* **2007**, *282*, 2091–2100.
145. Feller, G.; Dehareng, D.; Lage, J. L. How to Remain Nonfolded and Pliable: The Linkers in Modular α-Amylases as a Case Study. *FEBS J.* **2011**, *278*, 2333–2340.
146. Srisodsuk, M.; Lehtio, J.; Linder, M.; Margolles-Clark, E.; Reinikainen, T.; Teeri, T. *Trichoderma reesei* Cellobiohydrolase I with an Endoglucanase Cellulose-Binding Domain: Action on Bacterial Microcrystalline Cellulose. *J. Biotechnol.* **1997**, *57*, 49–57.
147. Kraulis, P. J.; Clore, G. M.; Nilges, M.; Jones, T. A.; Pettersson, G.; Knowles, J. K. C.; Gronenborn, A. M. Determination of the Three-Dimensional Solution Structure of the C-Terminal Domain of Cellobiohydrolase I from *Trichoderma reesei*. A Study Using Nuclear Magnetic Resonance and Hybrid Distance Geometry-Dynamical Simulated Annealing. *Biochemistry* **1989**, *28*, 7241–7257.
148. Arantes, V.; Saddler, J. N. Access to Cellulose Limits the Efficiency of Enzymatic Hydrolysis: The Role of Amorphogenesis. *Biotechnol. Biofuels* **2010**, *3*, 4.
149. Chundawat, S. P.; Beckham, G. T.; Himmel, M. E.; Dale, B. E. Deconstruction of Lignocellulosic Biomass to Fuels and Chemicals. *Annu. Rev. Chem. Biomol. Eng.* **2011**, *2*, 121–145.
150. Hall, M.; Bansal, P.; Lee, J. H.; Realff, M. J.; Bommarius, A. S. Cellulose Crystallinity—A Key Predictor of the Enzymatic Hydrolysis Rate. *FEBS J.* **2010**, *277*, 1571–1582.

151. Fox, J. M.; Jess, P.; Jambusaria, R. B.; Moo, G. M.; Liphardt, J.; Clark, D. S.; Blanch, H. W. A Single-Molecule Analysis Reveals Morphological Targets for Cellulase Synergy. *Nat. Chem. Biol.* **2013**, *9*, 356–361.
152. Le Costaouec, T.; Pakarinen, A.; Varnai, A.; Puranen, T.; Viikari, L. The Role of Carbohydrate Binding Module (CBM) at High Substrate Consistency: Comparison of *Trichoderma reesei* and *Thermoascus aurantiacus* Cel7A (CBHI) and Cel5A (EGII). *Bioresour. Technol.* **2013**, *143*, 196–203.
153. Hall, M.; Rubin, J.; Behrens, S. H.; Bommarius, A. S. The Cellulose-Binding Domain of Cellobiohydrolase Cel7A from *Trichoderma reesei* Is Also a Thermostabilizing Domain. *J. Biotechnol.* **2011**, *155*, 370–376.
154. Bayram Akcapinar, G.; Venturini, A.; Martelli, P. L.; Casadio, R.; Sezerman, U. O. Modulating the Thermostability of Endoglucanase I from *Trichoderma reesei* Using Computational Approaches. *Protein Eng., Des. Sel.* **2015**, *28*, 127–135.
155. Zhang, S.; Wang, Y.; Song, X.; Hong, J.; Zhang, Y.; Yao, L. Improving *Trichoderma reesei* Cel7B Thermostability by Targeting the Weak Spots. *J. Chem. Inf. Model.* **2014**, *54*, 2826–2833.
156. Linger, J. G.; Taylor, L. E., 2nd.; Baker, J. O.; Vander Wall, T.; Hobdey, S. E.; Podkaminer, K.; Himmel, M. E.; Decker, S. R. A Constitutive Expression System for Glycosyl Hydrolase Family 7 Cellobiohydrolases in *Hypocrea jecorina. Biotechnol. Biofuels* **2015**, *8*, 45.
157. den Haan, R.; Kroukamp, H.; Mert, M.; Bloom, M.; Görgens, J. F.; van Zyl, W. H. Engineering *Saccharomyces cerevisiae* for Next Generation Ethanol Production. *J. Chem. Technol. Biotechnol.* **2013**, *88*, 983–991.
158. den Haan, R.; Kroukamp, H.; van Zyl, J.-H. D.; van Zyl, W. H. Cellobiohydrolase Secretion by Yeast: Current State and Prospects for Improvement. *Process Biochem.* **2013**, *48*, 1–12.
159. Den Haan, R.; McBride, J. E.; Grange, D. C. L.; Lynd, L. R.; Van Zyl, W. H. Functional Expression of Cellobiohydrolases in *Saccharomyces cerevisiae* Towards One-Step Conversion of Cellulose to Ethanol. *Enzyme Microb. Technol.* **2007**, *40*, 1291–1299.
160. Den Haan, R.; Rose, S. H.; Lynd, L. R.; van Zyl, W. H. Hydrolysis and Fermentation of Amorphous Cellulose by Recombinant *Saccharomyces cerevisiae. Metab. Eng.* **2007**, *9*, 87–94.
161. Qin, Y.; Wei, X.; Liu, X.; Wang, T.; Qu, Y. Purification and Characterization of Recombinant Endoglucanase of *Trichoderma reesei* Expressed in *Saccharomyces cerevisiae* with Higher Glycosylation and Stability. *Protein Expr. Purif.* **2008**, *58*, 162–167.
162. Kwon, I.; Ekino, K.; Goto, M.; Furukawa, K. Heterologous Expression and Characterization of Endoglucanase I (EGI) from *Trichoderma viride* HK-75. *Biosci. Biotechnol. Biochem.* **1999**, *63*, 1714–1720.
163. Nakazawa, H.; Okada, K.; Kobayashi, R.; Kubota, T.; Onodera, T.; Ochiai, N.; Omata, N.; Ogasawara, W.; Okada, H.; Morikawa, Y. Characterization of the Catalytic Domains of *Trichoderma reesei* Endoglucanase I, II, and III, Expressed in *Escherichia coli. Appl. Microbiol. Biotechnol.* **2008**, *81*, 681–689.
164. Dana, C. M.; Dotson-Fagerstrom, A.; Roche, C. M.; Kal, S. M.; Chokhawala, H. A.; Blanch, H. W.; Clark, D. S. The Importance of Pyroglutamate in Cellulase Cel7A. *Biotechnol. Bioeng.* **2014**, *111*, 842–847.
165. Heinzelman, P.; Komor, R.; Kanaan, A.; Romero, P.; Yu, X.; Mohler, S.; Snow, C.; Arnold, F. Efficient Screening of Fungal Cellobiohydrolase Class I Enzymes for Thermostabilizing Sequence Blocks by SCHEMA Structure-Guided Recombination. *Protein Eng. Des. Sel.* **2010**, *23*, 871–880.
166. Heinzelman, P.; Snow, C. D.; Wu, I.; Nguyen, C.; Villalobos, A.; Govindarajan, S.; Minshull, J.; Arnold, F. H. A Family of Thermostable Fungal Cellulases Created by Structure-Guided Recombination. *Proc. Natl. Acad. Sci. U.S.A.* **2009**, *106*, 5610–5615.

167. Xu, L.; Shen, Y.; Hou, J.; Peng, B.; Tang, H.; Bao, X. Secretory Pathway Engineering Enhances Secretion of Cellobiohydrolase I from *Trichoderma reesei* in *Saccharomyces cerevisiae*. *J. Biosci. Bioeng.* **2014**, *117*, 45–52.
168. Suzuki, H.; Imaeda, T.; Kitagawa, T.; Kohda, K. Deglycosylation of Cellulosomal Enzyme Enhances Cellulosome Assembly in *Saccharomyces cerevisiae*. *J. Biotechnol.* **2012**, *157*, 64–70.
169. Wang, T. Y.; Huang, C. J.; Chen, H. L.; Ho, P. C.; Ke, H. M.; Cho, H. Y.; Ruan, S. K.; Hung, K. Y.; Wang, I. L.; Cai, Y. W.; Sung, H. M.; Li, W. H.; Shih, M. C. Systematic Screening of Glycosylation- and Trafficking-Associated Gene Knockouts in *Saccharomyces cerevisiae* Identifies Mutants with Improved Heterologous Exocellulase Activity and Host Secretion. *BMC Biotechnol.* **2013**, *13*, 71.
170. Xu, L.; Shen, Y.; Hou, J.; Tang, H.; Wang, C.; Bao, X. Promotion of Extracellular Activity of Cellobiohydrolase I from *Trichoderma reesei* by Protein Glycosylation Engineering in *Saccharomyces cerevisiae*. *Curr. Synth. Syst. Biol.* **2014**, *2*, 100111. http://dx.doi.org/10.4172/2332-0737.1000111.
171. Cereghino, G. P. L.; Cereghino, J. L.; Ilgen, C.; Cregg, J. M. Production of Recombinant Proteins in Fermenter Cultures of the Yeast *Pichia pastoris*. *Curr. Opin. Biotechnol.* **2002**, *13*, 329–332.
172. Daly, R.; Hearn, M. T. Expression of Heterologous Proteins in *Pichia pastoris*: A Useful Experimental Tool in Protein Engineering and Production. *J. Mol. Recognit.* **2005**, *18*, 119–138.
173. Akcapinar, G. B.; Gul, O.; Sezerman, U. Effect of Codon Optimization on the Expression of *Trichoderma reesei* Endoglucanase 1 in *Pichia pastoris*. *Biotechnol. Prog.* **2011**, *27*, 1257–1263.
174. Bayram Akcapinar, G.; Gul, O.; Sezerman, U. O. From In Silico to In Vitro: Modelling and Production of *Trichoderma reesei* Endoglucanase 1 and Its Mutant in *Pichia pastoris*. *J. Biotechnol.* **2012**, *159*, 61–68.
175. Boonvitthya, N.; Bozonnet, S.; Burapatana, V.; O'Donohue, M. J.; Chulalaksananukul, W. Comparison of the Heterologous Expression of *Trichoderma reesei* Endoglucanase II and Cellobiohydrolase II in the Yeasts *Pichia pastoris* and *Yarrowia lipolytica*. *Mol. Biotechnol.* **2013**, *54*, 158–169.
176. Mitrovic, A.; Flicker, K.; Steinkellner, G.; Gruber, K.; Reisinger, C.; Schirrmacher, G.; Camattari, A.; Glieder, A. Thermostability Improvement of Endoglucanase Cel7B from *Hypocrea pseudokoningii*. *J. Mol. Catal. B Enzym.* **2014**, *103*, 16–23.
177. Zahri, S.; Zamani, M. R.; Motallebi, M.; Sadeghi, M. Cloning and Characterization of *cbh*II Gene from *Trichoderma parceramosum* and Its Expression in *Pichia pastoris*. *Iran. J. Biotechnol.* **2005**, *3*, 204–215.
178. Bretthauer, R. K. Genetic Engineering of *Pichia pastoris* to Humanize N-Glycosylation of Proteins. *Trends Biotechnol.* **2003**, *21*, 459–462.
179. Damasceno, L. M.; Huang, C. J.; Batt, C. A. Protein Secretion in *Pichia pastoris* and Advances in Protein Production. *Appl. Microbiol. Biotechnol.* **2012**, *93*, 31–39.
180. Hamilton, S. R.; Gerngross, T. U. Glycosylation Engineering in Yeast: The Advent of Fully Humanized Yeast. *Curr. Opin. Biotechnol.* **2007**, *18*, 387–392.
181. Li, H.; Sethuraman, N.; Stadheim, T. A.; Zha, D.; Prinz, B.; Ballew, N.; Bobrowicz, P.; Choi, B. K.; Cook, W. J.; Cukan, M.; Houston-Cummings, N. R.; Davidson, R.; Gong, B.; Hamilton, S. R.; Hoopes, J. P.; Jiang, Y.; Kim, N.; Mansfield, R.; Nett, J. H.; Rios, S.; Strawbridge, R.; Wildt, S.; Gerngross, T. U. Optimization of Humanized IgGs in Glycoengineered *Pichia pastoris*. *Nat. Biotechnol.* **2006**, *24*, 210–215.
182. Vogl, T.; Hartner, F. S.; Glieder, A. New Opportunities by Synthetic Biology for Biopharmaceutical Production in *Pichia pastoris*. *Curr. Opin. Biotechnol.* **2013**, *24*, 1094–1101.

183. De Pourcq, K.; De Schutter, K.; Callewaert, N. Engineering of Glycosylation in Yeast and Other Fungi: Current State and Perspectives. *Appl. Microbiol. Biotechnol.* **2010**, *87*, 1617–1631.
184. Callewaert, N.; Laroy, W.; Cadirgi, H.; Geysens, S.; Saelens, X.; Min Jou, W.; Contreras, R. Use of HDEL-Tagged *Trichoderma reesei* Mannosyl Oligosaccharide 1,2-α-D-Mannosidase for *N*-Glycan Engineering in *Pichia pastoris*. *FEBS Lett.* **2001**, *503*, 173–178.
185. Bobrowicz, P.; Davidson, R. C.; Li, H.; Potgieter, T. I.; Nett, J. H.; Hamilton, S. R.; Stadheim, T. A.; Miele, R. G.; Bobrowicz, B.; Mitchell, T.; Rausch, S.; Renfer, E.; Wildt, S. Engineering of an Artificial Glycosylation Pathway Blocked in Core Oligosaccharide Assembly in the Yeast *Pichia pastoris*: Production of Complex Humanized Glycoproteins with Terminal Galactose. *Glycobiology* **2004**, *14*, 757–766.
186. Hamilton, S. R.; Bobrowicz, P.; Bobrowicz, B.; Davidson, R. C.; Li, H.; Mitchell, T.; Nett, J. H.; Rausch, S.; Stadheim, T. A.; Wischnewski, H.; Wildt, S.; Gerngross, T. U. Production of Complex Human Glycoproteins in Yeast. *Science* **2003**, *301*, 1244–1246.
187. Hamilton, S. R.; Davidson, R. C.; Sethuraman, N.; Nett, J. H.; Jiang, Y.; Rios, S.; Bobrowicz, P.; Stadheim, T. A.; Li, H.; Choi, B. K.; Hopkins, D.; Wischnewski, H.; Roser, J.; Mitchell, T.; Strawbridge, R. R.; Hoopes, J.; Wildt, S.; Gerngross, T. U. Humanization of Yeast to Produce Complex Terminally Sialylated Glycoproteins. *Science* **2006**, *313*, 1441–1443.
188. Nett, J. H.; Stadheim, T. A.; Li, H.; Bobrowicz, P.; Hamilton, S. R.; Davidson, R. C.; Choi, B. K.; Mitchell, T.; Bobrowicz, B.; Rittenhour, A.; Wildt, S.; Gerngross, T. U. A Combinatorial Genetic Library Approach to Target Heterologous Glycosylation Enzymes to the Endoplasmic Reticulum or the Golgi Apparatus of *Pichia pastoris*. *Yeast* **2011**, *28*, 237–252.
189. Geysens, S.; Whyteside, G.; Archer, D. B. Genomics of Protein Folding in the Endoplasmic Reticulum, Secretion Stress and Glycosylation in the Aspergilli. *Fungal Genet. Biol.* **2009**, *46*, S121–S140.
190. Akao, T.; Yamaguchi, M.; Yahara, A.; Yoshiuchi, K.; Fujita, H.; Yamada, O.; Akita, O.; Ohmachi, T.; Asada, Y.; Yoshida, T. Cloning and Expression of 1,2-α-Mannosidase Gene (*fmanIB*) from Filamentous Fungus *Aspergillus oryzae*: *In Vivo* Visualization of the FmanIBp-GFP Fusion Protein. *Biosci. Biotechnol. Biochem.* **2006**, *70*, 471–479.
191. Eriksen, S. H.; Jensen, B.; Olsen, J. Effect of N-Linked Glycosylation on Secretion, Activity, and Stability of α-Amylase from *Aspergillus oryzae*. *Curr. Microbiol.* **1998**, *37*, 117–122.
192. Nakao, Y.; Kozutsumi, Y.; Funakoshi, I.; Kawasaki, T.; Yamashina, I.; Mutsaers, J. H. G. M.; Van Halbeek, H.; Vliegenthart, J. F. G. Structures of Oligosaccharides on β-Galactosidase from *Aspergillus oryzae*. *J. Biochem.* **1987**, *102*, 171–179.
193. Wallis, G. L. F.; Swift, R. J.; Atterbury, R.; Trappe, S.; Rinas, U.; Hemming, F. W.; Wiebe, M. G.; Trinci, A. P. J.; Peberdy, J. F. The Effect of pH on Glucoamylase Production, Glycosylation and Chemostat Evolution of *Aspergillus niger*. *Biochim. Biophys. Acta* **2001**, *1527*, 112–122.
194. Wallis, G. L. F.; Swift, R. J.; Hemming, F. W.; Trinci, A. P. J.; Peberdy, J. F. Glucoamylase Overexpression and Secretion in *Aspergillus niger*: Analysis of Glycosylation. *Biochim. Biophys. Acta* **1999**, *1472*, 576–586.
195. Woosley, B.; Xie, M.; Wells, L.; Orlando, R.; Garrison, D.; King, D.; Bergmann, C. Comprehensive Glycan Analysis of Recombinant *Aspergillus niger endo*-Polygalacturonase C. *Anal. Biochem.* **2006**, *354*, 43–53.
196. Woosley, B. D.; Kim, Y. H.; Kumar Kolli, V. S.; Wells, L.; King, D.; Poe, R.; Orlando, R.; Bergmann, C. Glycan Analysis of Recombinant *Aspergillus niger endo*-Polygalacturonase A. *Carbohydr. Res.* **2006**, *341*, 2370–2378.

197. Takashima, S.; Iikura, H.; Akira, N.; Hidaka, M.; Masaki, H.; Uozumi, T. Overproduction of Recombinant *Trichoderma reesei* Cellulases by *Aspergillus oryzae* and Their Enzymatic Properties. *J. Biotechnol.* **1998**, *65*, 163–171.
198. Gorka-Niec, W.; Bankowska, R.; Palamarczyk, G.; Krotkiewski, H.; Kruszewska, J. S. Protein Glycosylation in pmt Mutants of *Saccharomyces cerevisiae*. Influence of Heterologously Expressed Cellobiohydrolase II of *Trichoderma reesei* and Elevated Levels of GDP-Mannose and *cis*-Prenyltransferase Activity. *Biochim. Biophys. Acta* **2007**, *1770*, 774–780.
199. Gorka-Niec, W.; Kania, A.; Perlinska-Lenart, U.; Smolenska-Sym, G.; Palamarczyk, G.; Kruszewska, J. S. Integration of Additional Copies of *Trichoderma reesei* Gene Encoding Protein *O*-Mannosyltransferase I Results in a Decrease of the Enzyme Activity and Alteration of Cell Wall Composition. *Fungal Biol.* **2011**, *115*, 124–132.
200. Gorka-Niec, W.; Perlinska-Lenart, U.; Zembek, P.; Palamarczyk, G.; Kruszewska, J. S. Influence of Sorbitol on Protein Production and Glycosylation and Cell Wall Formation in *Trichoderma reesei*. *Fungal Biol.* **2010**, *114*, 855–862.
201. Dubey, M. K.; Ubhayasekera, W.; Sandgren, M.; Jensen, D. F.; Karlsson, M. Disruption of the Eng18B ENGase Gene in the Fungal Biocontrol Agent *Trichoderma atroviride* Affects Growth, Conidiation and Antagonistic Ability. *PLoS One* **2012**, 7, e36152.
202. Jin, C. Protein Glycosylation in *Aspergillus fumigatus* Is Essential for Cell Wall Synthesis and Serves as a Promising Model of Multicellular Eukaryotic Development. *Int. J. Microbiol.* **2012**, *2012*, 654251.
203. Kotz, A.; Wagener, J.; Engel, J.; Routier, F. H.; Echtenacher, B.; Jacobsen, I.; Heesemann, J.; Ebel, F. Approaching the Secrets of N-Glycosylation in *Aspergillus fumigatus*: Characterization of the AfOch1 Protein. *PLoS One* **2010**, *5*, e15729.
204. Oka, T.; Sameshima, Y.; Koga, T.; Kim, H.; Goto, M.; Furukawa, K. Protein *O*-Mannosyltransferase A of *Aspergillus awamori* Is Involved in O-Mannosylation of Glucoamylase I. *Microbiology* **2005**, *151*, 3657–3667.
205. Argyros, R.; Nelson, S.; Kull, A.; Chen, M. T.; Stadheim, T. A.; Jiang, B. A Phenylalanine to Serine Substitution Within an O-Protein Mannosyltransferase Led to Strong Resistance to PMT-Inhibitors in *Pichia pastoris*. *PLoS One* **2013**, *8*, e62229.
206. Cantarel, B. L.; Coutinho, P. M.; Rancurel, C.; Bernard, T.; Lombard, V.; Henrissat, B. The Carbohydrate-Active EnZymes Database (CAZy): An Expert Resource for Glycogenomics. *Nucleic Acids Res.* **2009**, *37*, D233–D238.
207. Eneyskaya, E. V.; Kulmiskaya, A. A.; Savel'ev, A. N.; Shabalin, K. A.; Golubev, A. M.; Neustroev, K. N. α-Mannosidase from *Trichoderma reesei* Participates in the Postsecretory Deglycosylation of Glycoproteins. *Biochem. Biophys. Res. Commun.* **1998**, *245*, 43–49.
208. Maras, M.; Callewaert, N.; Piens, K.; Claeyssens, M.; Martinet, W.; Dewaele, S.; Contreras, H.; Dewerte, I.; Penttila, M.; Contreras, R. Molecular Cloning and ENZYMATIC Characterization of a *Trichoderma reesei* 1,2-α-D-Mannosidase. *J. Biotechnol.* **2000**, 77, 255–263.
209. Van Petegem, F.; Contreras, H.; Contreras, R.; Van Beeumen, J. *Trichoderma reesei* α-1,2-Mannosidase: Structural Basis for the Cleavage of Four Consecutive Mannose Residues. *J. Mol. Biol.* **2001**, *312*, 157–165.
210. Ichikawa, M.; Scott, D. A.; Losfeld, M. E.; Freeze, H. H. The Metabolic Origins of Mannose in Glycoproteins. *J. Biol. Chem.* **2014**, *289*, 6751–6761.
211. dos Santos Castro, L.; Pedersoli, W. R.; Antonieto, A. C. C.; Steindorff, A. S.; Silva-Rocha, R.; Martinez-Rossi, N. M.; Rossi, A.; Brown, N. A.; Goldman, G. H.; Faca, V. M.; Persinoti, G. F.; Silva, R. N. Comparative Metabolism of Cellulose, Sophorose and Glucose in *Trichoderma reesei* Using High-Throughput Genomic and Proteomic Analyses. *Biotechnol. Biofuels* **2014**, 7, 41.

212. Takegawa, K.; Miki, S.; Jikibara, T.; Iwahara, S. Purification and Characterization of *exo*-α-D-Mannosidase from a *Cellulomonas* sp. *Biochim. Biophys. Acta* **1989**, *991*, 431–437.
213. Athanasopoulos, V. I.; Niranjan, K.; Rastall, R. A. The Production, Purification and Characterisation of Two Novel α-D-Mannosidases from *Aspergillus phoenicis*. *Carbohydr. Res.* **2005**, *340*, 609–617.
214. Do Vale, L. H.; Gomez-Mendoza, D. P.; Kim, M. S.; Pandey, A.; Ricart, C. A.; Ximenes, F. F. E.; Sousa, M. V. Secretome Analysis of the Fungus *Trichoderma harzianum* Grown on Cellulose. *Proteomics* **2012**, *12*, 2716–2728.
215. Horta, M. A.; Vicentini, R.; Delabona Pda, S.; Laborda, P.; Crucello, A.; Freitas, S.; Kuroshu, R. M.; Polikarpov, I.; Pradella, J. G.; Souza, A. P. Transcriptome Profile of *Trichoderma harzianum* IOC-3844 Induced by Sugarcane Bagasse. *PLoS One* **2014**, *9*, e88689.
216. Benz, J. P.; Chau, B. H.; Zheng, D.; Bauer, S.; Glass, N. L.; Somerville, C. R. A Comparative Systems Analysis of Polysaccharide-Elicited Responses in *Neurospora crassa* Reveals Carbon Source-Specific Cellular Adaptations. *Mol. Microbiol.* **2014**, *91*, 275–299.
217. Liu, D.; Li, J.; Zhao, S.; Zhang, R.; Wang, M.; MIao, Y.; Shen, Y.; Shen, Q. Secretome Diversity and Quantitative Analysis of Celluloytic *Aspergillus fumigatus* Z5 in the Presence of Different Carbon Sources. *Biotechnol. Biofuels* **2013**, *6*, 149.
218. Sharma, M.; Soni, R.; Nazir, A.; Oberoi, H. S.; Chadha, B. S. Evaluation of Glycosyl Hydrolases in the Secretome of *Aspergillus fumigatus* and Saccharification of Alkali-Treated Rice Straw. *Appl. Biochem. Biotechnol.* **2011**, *163*, 577–591.
219. Liu, G.; Zhang, L.; Wei, X.; Zou, G.; Qin, Y.; Ma, L.; Li, J.; Zheng, H.; Wang, S.; Wang, C.; Xun, L.; Zhao, G. P.; Zhou, Z.; Qu, Y. Genomic and Secretomic Analyses Reveal Unique Features of the Lignocellulolytic Enzyme System of *Penicillium decumbens*. *PLoS One* **2013**, *8*. e55185.
220. Manavalan, A.; Adav, S. S.; Sze, S. K. iTRAQ-Based Quantitative Secretome Analysis of *Phanerochaete chrysosporium*. *J. Proteomics* **2011**, *75*, 642–654.
221. Hamaguchi, T.; Ito, T.; Inoue, Y.; Limpaseni, T.; Pongsawasdi, P.; Ito, K. Purification, Characterization and Molecular Cloning of a Novel *endo*-β-*N*-Acetylglucosaminidase from the Basidiomycete, Flammulina Velutipes. *Glycobiology* **2010**, *20*, 420–432.
222. Gerlach, J. Q.; Kilcoyne, M.; Farrell, M. P.; Kane, M.; Joshi, L. Differential Release of High Mannose Structural Isoforms by Fungal and Bacterial *endo*-β-*N*-Acetylglucosaminidases. *Mol. Biosyst.* **2012**, *8*, 1472–1481.
223. Tzelepis, G. D.; Melin, P.; Jensen, D. F.; Stenlid, J.; Karlsson, M. Functional Analysis of Glycoside Hydrolase Family 18 and 20 Genes in *Neurospora crassa*. *Fungal Genet. Biol.* **2012**, *49*, 717–730.
224. Tzelepis, G.; Hosomi, A.; Hossain, T. J.; Hirayama, H.; Dubey, M.; Jensen, D. F.; Suzuki, T.; Karlsson, M. *endo*-β-*N*-Acetylglucosamidases (ENGases) in the Fungus *Trichoderma atroviride*: Possible Involvement of the Filamentous Fungi-Specific Cytosolic ENGase in the ERAD Process. *Biochem. Biophys. Res. Commun.* **2014**, *449*, 256–261.
225. Fujita, K.; Kobayashi, K.; Iwamatsu, A.; Takeuchi, M.; Kumagai, H.; Yamamoto, K. Molecular Cloning of Mucor Hiemalis *endo* β-*N*-Acetylglucosaminidase and Some Properties of the Recombinant Enzyme. *Arch. Biochem. Biophys.* **2004**, *432*, 41–49.
226. Murakami, S.; Takaoka, Y.; Ashida, H.; Yamamoto, K.; Narimatsu, H.; Chiba, Y. Identification and Characterization of *endo*-β-*N*-Acetylglucosaminidase from Methylotrophic Yeast *Ogataea minuta*. *Glycobiology* **2013**, *23*, 736–744.
227. Fujita, K.; Yamamoto, K. A Remodeling System for the Oligosaccharide Chains on Glycoproteins with Microbial *endo*-β-*N*-Acetylglucosaminidases. *Biochim. Biophys. Acta* **2006**, *1760*, 1631–1635.

228. Tomabechi, Y.; Squire, M. A.; Fairbanks, A. J. endo-β-*N*-Acetylglucosaminidase Catalysed Glycosylation: Tolerance of Enzymes to Structural Variation of the Glycosyl Amino Acid Acceptor. *Org. Biomol. Chem.* **2014**, *12*, 942–955.
229. Wang, K.; Luo, H.; Bai, Y.; Shi, P.; Huang, H.; Xue, X.; Yao, B. A Thermophilic *endo*-1,4-Beta-Glucanase from *Talaromyces emersonii* CBS394.64 with Broad Substrate Specificity and Great Application Potentials. *Appl. Microbiol. Biotechnol.* **2014**, *98*, 7051–7060.
230. Wang, L. X.; Lomino, J. V. Emerging Technologies for Making Glycan-Defined Glycoproteins. *ACS Chem. Biol.* **2012**, 7, 110–122.
231. Hakkinen, M.; Arvas, M.; Oja, M.; Aro, N.; Penttila, M.; Saloheimo, M.; Pakula, T. M. Re-annotationof the CAZy Genes of *Trichoderma reesei* and Transcription in the Presence of Lignocellulosic Substrates. *Microb. Cell Fact.* **2012**, *11*.
232. Fekete, E.; Karaffa, L.; Karimi Aghcheh, R.; Nemeth, Z.; Fekete, E.; Orosz, A.; Paholcsek, M.; Stagel, A.; Kubicek, C. P. The Transcriptome of lae1 Mutants of *Trichoderma reesei* Cultivated at Constant Growth Rates Reveals New Targets of LAE1 Function. *BMC Genomics* **2014**, *15*, 447.
233. Ivanova, C.; Baath, J. A.; Seiboth, B.; Kubicek, C. P. Systems Analysis of Lactose Metabolism in *Trichoderma reesei* Identifies a Lactose Permease That Is Essential for Cellulase Induction. *PLoS One* **2013**, *8*, e62631.
234. Karimi-Aghcheh, R.; Bok, J. W.; Phatale, P. A.; Smith, K. M.; Baker, S. E.; Lichius, A.; Omann, M.; Zeilinger, S.; Seiboth, B.; Rhee, C.; Keller, N. P.; Freitag, M.; Kubicek, C. P. Functional Analyses of *Trichoderma reesei* LAE1 Reveal Conserved and Contrasting Roles of This Regulator. *G3 (Bethesda)* **2013**, *3*, 369–378.
235. Zhong, Y.; Liu, X.; Xiao, P.; Wei, S.; Wang, T. Expression and Secretion of the Human Erythropoietin Using an Optimized cbh1 Promoter and the Native CBH I Signal Sequence in the Industrial Fungus *Trichoderma reesei*. *Appl. Biochem. Biotechnol.* **2011**, *165*, 1169–1177.
236. Maras, M.; De Bruyn, A.; Vervecken, W.; Uusitalo, J.; Penttila, M.; Busson, R.; Herdewijn, P.; Contreras, R. In Vivo Synthesis of Complex *N*-Glycans by Expression of Human *N*-Acetylglucosaminyltransferase I in the Filamentous Fungus *Trichoderma reesei*. *FEBS Lett.* **1999**, *452*, 365–370.

CHAPTER FOUR

Human Milk Oligosaccharides (HMOS): Structure, Function, and Enzyme-Catalyzed Synthesis

Xi Chen
Department of Chemistry, University of California, Davis, California, USA

Contents

1. Introduction 115
2. Structures of HMOS 117
   2.1 HMOS Monosaccharide Building Blocks, Core Structures, and Glycosidic Linkages 117
   2.2 HMOS Structures 118
3. Biosynthesis of HMOS 151
4. Functions of HMOS 152
   4.1 Neutral Non-Fucosylated HMOS 153
   4.2 Fucosylated HMOS 155
   4.3 Sialylated HMOS 156
5. Production of HMOS by Enzyme-Catalyzed Processes 158
   5.1 2′FL 160
   5.2 3′SL and 3′SLN 161
   5.3 6′SL and 6′SLN 163
   5.4 LNT2, LNnT, LNnH, LNnO, LNnD, LSTd, and Disialyl Oligosaccharides 164
   5.5 Fucα1–2LNnT 165
   5.6 LNFP III, LNnFP V, and LNnDFH 166
   5.7 LNT 167
   5.8 3FL, LDFT, LNFP II, Le^{a} Tetrasaccharide, and Le^{X} Tetrasaccharide 168
   5.9 LNFP I and LNDFH I 168
   5.10 Other Oligosaccharides 168
6. Perspectives 169
Acknowledgments 170
References 170

Advances in Carbohydrate Chemistry and Biochemistry, Volume 72
ISSN 0065-2318
http://dx.doi.org/10.1016/bs.accb.2015.08.002

ABBREVIATIONS

2′FL 2′-fucosyllactose
3′S-3FL 3′-sialyl-3-fucosyllactose
3′SL 3′-sialyllactose
3′SLN 3′-sialyl-*N*-acetyllactosamine
3FL 3-fucosyllactose
6′SL 6′-sialyllactose
6′SLN 6′-sialyl-*N*-acetyllactosamine
ATP adenine 5′-triphosphate
B. bifidum *Bifidobacterium bifidum*
B. breve *Bifidobacterium breve*
B. infantis *Bifidobacterium longum* subsp. *infantis*
B. longum *Bifidobacterium longum* subsp. *longum*
BiNahK *B. infantis N*-acetylhexosamine-1-kinase
BLUSP *B. longum* UDP-sugar synthase
CDP cytidine 5′-diphosphate
CMP cytidine 5′-monophosphate
CMP-Neu5Ac cytidine 5′-monophosphate-*N*-acetylneuraminic acid
CstII *Campylobacter jejuni* α2–8-sialyltransferase II
CTP cytidine 5′-triphosphate
DSLNT disialyllacto-*N*-tetraose
E. coli *Escherichia coli*
EcGalK *E. coli* K-12 galactose kinase
EcNanA *E. coli* sialic acid aldolase
Fuc fucose
FucT fucosyltransferase
FUT2 human α1–2-fucosyltransferase encoded by the Secretor (*Se*) gene
FUT3 human α1–3/4-fucosyltransferase encoded by the Lewis (*Le*) gene
FutA *Helicobacter pylori* strain 26695 α1–3-fucosyltransferase
Gal galactose
GalNAc *N*-acetylgalactosamine
GalT galactosyltransferase
GDP guanidine 5′-diphosphate
Glc glucose
GlcNAc *N*-acetylglucosamine
GlcNAcT *N*-acetylglucosaminyltransferase
GMP guanidine 5′-monophosphate
GNB galacto-*N*-biose
GnT-I *N*-acetylglucosaminyltransferase I
GST glutathione S-transferase
GTP guanidine 5′-triphosphate
HMOS human milk oligosaccharides
HPAEC-PAD high-pH anion-exchange chromatography with pulsed amperometric detection
HPLC high-performance liquid chromatography
Lac lactose
LacNAc *N*-acetyllactosamine

LacY β-galactoside permease
LDFT lactodifucotetraose
Le Lewis
LNB lacto-*N*-biose
LNDFH lacto-*N*-difuco-hexoase
LNFP lacto-*N*-fucopentaose
LNnD lacto-*N*-neodecaose
LNnDFH lacto-*N*-neodifucohexaose
LNnFP lacto-*N*-neofucopentaose
LNnH lacto-*N*-neohexaose
LNnT lacto-*N*-neotetraose
LNT lacto-*N*-tetraose
LNT2 lacto-*N*-triose II
LPS lipopolysaccharides
LST sialyllacto-*N*-tetraose
MOS milk oligosaccharides
MS mass spectrometry
NanT Neu5Ac permease
NEC necrotizing enterocolitis
Neu4,5Ac$_2$ 4-*O*-acetyl-*N*-acetylneuraminic acid
Neu5Ac *N*-acetylneuraminic acid, the most abundant sialic acid form in nature
NK natural killer
NmCSS *Neisseria meningitidis* CMP-sialic acid synthetase
NmLgtA *Neisseria meningitidis* β1–3-*N*-acetylglucosaminyltransferase
NmLgtB *Neisseria meningitidis* β1–4-galactosyltransferase
NMR nuclear magnetic resonance spectroscopy
Nmα2–3ST *Neisseria meningitidis* α2–3-sialyltransferase
OPME "one-pot" multienzyme
Pd2,6ST *Photobacterium damselae* α2–6-sialyltransferase
PmGlmU *P. multocida* *N*-acetylglucosamine 1-phosphate uridylyltransferase
PmPpA *P. multocida* inorganic pyrophosphatase
PmST *Pasteurella multocida* α2–3-sialyltransferase (GenBank accession number AAK02272)
PmST1 multifunctional *Pasteurella multocida* α2–3-sialyltransferase 1
Psp2,6ST *Photobacterium* sp. JT-ISH-224 α2–6-sialyltransferase
Se Secretor
SiaT sialyltransferase
UTP uridine 5′-triphosphate

1. INTRODUCTION

Carbohydrates in human milk are presented in diverse forms including monosaccharides such as glucose and galactose, lactose (a disaccharide), oligosaccharides, glycoproteins, glycopeptides, and glycolipids.[1] Human milk oligosaccharides (HMOS) containing a diverse array of oligosaccharides with three or more monosaccharide units are the subject of this review.

HMOS are the third major component of human milk after lactose (55–70 g/L) and lipids (16–39 g/L).[2–4] Historically, purified HMOS were used to synthesize glycan antigens to obtain antibodies,[5,6] which were later used as important bioreagents to identify novel glycans and detect glycoconjugates.[7] The amounts of HMOS vary in lactation stages with 12–14 g/L in mature milk and 20–24 g/L in colostrum.[8–10] HMOS were found to be presented in higher concentrations in preterm human milk than those in term human milk.[11] The presence and the quantity of HMOS also vary among individuals and are related to the secretor status and the Lewis group type of the nursing mothers.[10–13] Four human milk groups have been classified based on the HMOS profiles controlled by the Secretor (*Se*) status and the Lewis (*Le*) blood type of the nursing mother.[11] Individuals (Se^+/Le^+) with both α1–2-fucosyltransferase (FUT2) encoded by the Secretor (*Se*) gene and α1–3/4-fucosyltransferase (FUT3) encoded by the Lewis (*Le*) gene represent about 70% of the European population and contain all types of fucosylated HMOS with α1–2/3/4-fucosyl linkages. Those (Se^-/Le^+) with no FUT2 but with FUT3 represent 20% of the population and do not have α1–2-fucosylated HMOS. Those (Se^+/Le^-) with FUT2 but no FUT3 represent 9% of the general population and do not have α1–4-fucosyl oligosaccharides. Finally, those (Se^-/Le^-) without FUT2 nor FUT3 that represent 1% of the general population contain α1–3-fucosylated HMOS but not other fucosylated HMOS due to the expression of a Lewis-independent α1–3-fucosyltransferase.[11,14]

Due to their structural complexity and the lack of efficient analytical methods, the presence of and the functions of HMOS were unknown in early times. For example, lactose was first isolated from milk in 1633.[15] In comparison, three centuries years later in the early 1930s, Polonowski and Lespagnol found and named nitrogen-containing "gynolactose,"[16,17] which was confirmed by two-dimensional paper chromatographic separation two decades later to be a mixture of more than 10 oligosaccharides.[18] In 1954, György *et al.* reported β-linked *N*-acetylglucosamine (GlcNAc)-containing oligosaccharides and polysaccharides in human milk as "bifidus factors"[19] that promote the growth of *Lactobacillus bifidus* var. *Penn* (now *Bifidobacterium bifidum*).[20–23] This ignited the efforts in elucidating the structures of HMOS. Several papers published in 1956 reported the structures of lacto-*N*-tetraose (LNT), 2′-fucosyllactose (2′FL), lacto-*N*-fucopentaose I (LNFP I), and 3-fucosyllactose (3FL).[24–27] By 1965, 14 HMOS structures had been reported, mainly by the groups of Kuhn and Montreuil.[28] Additional structures were soon elucidated by the efforts of Ginsburg, Kobata, and others. The introduction of mass spectrometry to the identification of HMOS[29] further speeded up the progress. Modern advances in separation and analysis methods allowed fast profiling of HMOS and the structural identification of additional HMOS. More than 200 HMOS species have now been reported,[30,31] and more than 100 HMOS structures have been elucidated.[28,32–34]

Unlike lactose, which is the primary component and the principal carbohydrate of human milk that is digestible by infants and provides them nutritional needs,[35] HMOS are not digestible by the infant.[1,36,37] Therefore, the direct physiological roles of HMOS are not clear. Accumulating evidence has shown that HMOS can survive the obstacles encountered upon suckling and reach the infant gut where they regulate the microbiota population, which in turn can affect the health of breastfed infants.[36–39] HMOS are believed to contribute significantly to the health of breast-fed infants in lowering their risk of diarrheal disease, respiratory infections, allergy, and other infectious diseases including otitis media.[15,40–42] The prebiotic (stimulating the growth and colonization of beneficial bacteria, mainly bifidobacteria, in the gut), antiadhesive antimicrobial (acting as decoys to inhibit specific pathogenic bacteria, viruses, or parasites binding to epithelial surface and translocation), immunomodulating, and brain development nutritional functions of HMOS have also been reported.[1,15,38,43–46] The enrichment of bifidobacteria in the gut also leads to the increased production of lactate and short-chain fatty acids, which lowers the pH and makes the environment less accommodating for the growth and colonization of some pathogens.[1] Additional mechanisms of pathogen inhibition may include the release of other antimicrobial substances by bifidobacteria.[47]

The functions of individual structures, however, are less clear. Only a handful of HMOS have known specific roles, and only a limited number of HMOS have been synthesized. The current knowledge about the structures, functions, and production of HMOS by enzyme-catalyzed processes is presented here.

2. STRUCTURES OF HMOS

2.1 HMOS Monosaccharide Building Blocks, Core Structures, and Glycosidic Linkages

Human milk is unique in containing a large number of oligosaccharides compared to the milk of other mammals.[48] Five monosaccharides have been found to be major building blocks for HMOS, which include D-glucose (Glc),

D-galactose (Gal), *N*-acetyl-D-glucosamine (GlcNAc), L-fucose (Fuc), and *N*-acetylneuraminic acid (Neu5Ac). These monosaccharide building blocks in HMOS are presented as six-membered ring pyranose (for Glc) or pyranoside (for Gal, GlcNAc, Fuc, and Neu5Ac) structures. HMOS are extended from lactose (Galβ1–4Glc with Glc at the reducing end) by *N*-acetylglucosaminylation and/or galactosylation with or without fucosylation and/or sialylation. Among the five major HMOS monosaccharide building blocks (Table 1), glucose (Glc) is at the reducing end with a mix of α and β anomers. While Gal and GlcNAc are always presented with β-D-glycosidic linkages, Fuc and Neu5Ac are always presented with α-L- and α-D-glycosidic linkages, respectively.

Other than lactose (which itself is not considered an HMOS), at least 15 neutral oligosaccharides (Table 2), including linear and branched structures, have been identified as the core structures of other HMOS.[32–34] It is interesting to observe that these structures rarely have GlcNAc as the terminal unit at the non-reducing end, indicating the high efficiency of galactosyltransferases (either β1–3- or β1–4-galactosyltransferase) in capping the GlcNAc residues in these HMOS. In addition, unlike *N*-acetyllactosamine (LacNAc, Galβ1–4GlcNAc), which can serve as both the internal and non-reducing-end terminal disaccharide unit(s), lacto-*N*-biose (LNB, Galβ1–3GlcNAc) can only serve as the non-reducing-end terminal disaccharide. Linear structures only contain the β1–3-linked GlcNAc residue, while any β1–6-linked GlcNAc generates branching.[38]

Other than the exceptions mentioned above, 12 glycosidic linkages constitute the structures of HMOS, and the list is shown in Table 3, which includes three types of galactosidic, two types of *N*-acetyl-glucosaminidic, four types of fucosidic, and three types of sialosidic linkages.

2.2 HMOS Structures

Over 200 individual HMOS molecular species have been found,[30,31] and the structures of more than 100 HMOS have been successfully elucidated (see Table 4).[28,32–34] These identifications and structural assignments were achieved using a variety of techniques, either alone or in combination. These include chromatographic separation, tritium labeling, derivatization with a chromophore or fluorophore, methylation analysis, glycosidase digestion, ^{1}H and ^{13}C nuclear magnetic resonance spectroscopy (NMR) spectral characterization, high-performance liquid chromatography (HPLC), high-pH anion-exchange chromatography with pulsed amperometric detection

Table 1 Major Monosaccharide Building Blocks for HMOS

HMOS Monosaccharide Building Blocks	Abbreviations	Symbols	Structures	Glycosidic Linkages in HMOS
N-Acetylneuraminic acid	Neu5Ac	◆	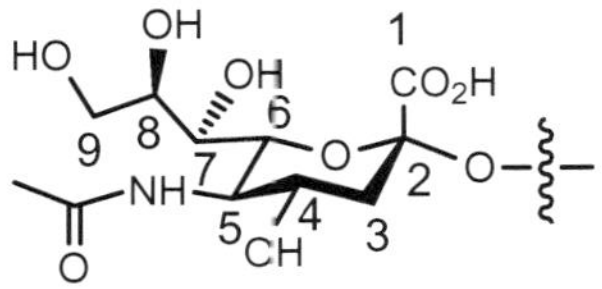	α-Linkages
Fucose	Fuc	▲	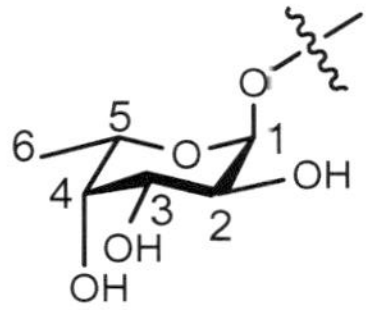	α-Linkages
Galactose	Gal	○	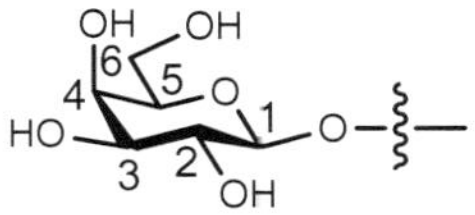	β-Linkages
N-Acetylglucosamine	GlcNAc	■	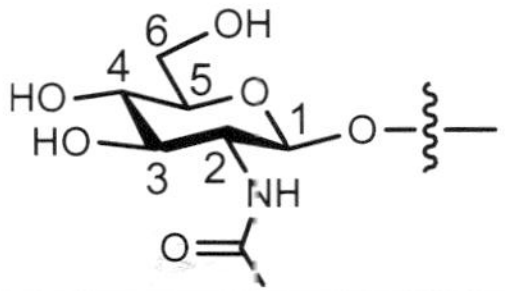	β-Linkages
Glucose	Glc	●	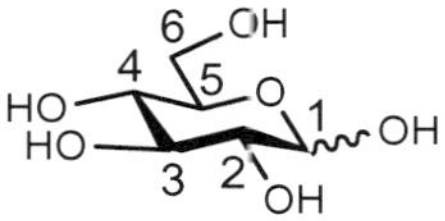	None (at the reducing end, a mix of α and β configurations at the anomeric carbon)

Table 2 Lactose and Neutral Non-Fucosylated HMOS That Can Serve as the Core Structures for Other HMOS[28,34]

Core #	Lactose and HMOS Core Structures	Abbreviations	Symbols	References
I	Lactose (not considered an HMOS itself)	Lac	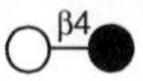 β4	49
II	Lacto-*N*-tetraose	LNT	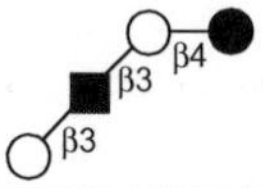 β4 β3 β3	24
III	Lacto-*N*-neotetraose	LNnT	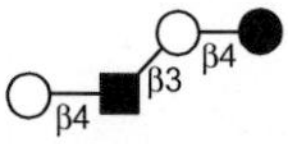 β4 β3 β4	50
IV	Lacto-*N*-hexaose	LNH	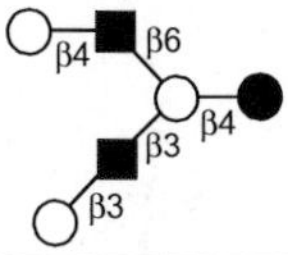 β4 β6 β4 β3 β3	51
V	Lacto-*N*-neohexaose	LNnH	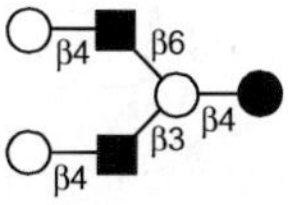 β4 β6 β4 β3 β4	52
VI	*para*-Lacto-*N*-hexaose	*p*LNH	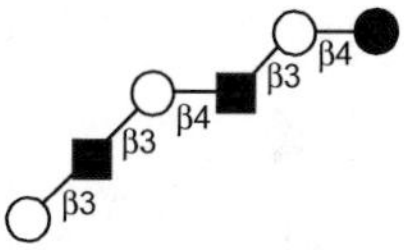 β4 β3 β4 β3 β3	53
VII	*para*-Lacto-*N*-neohexaose	*p*LNnH	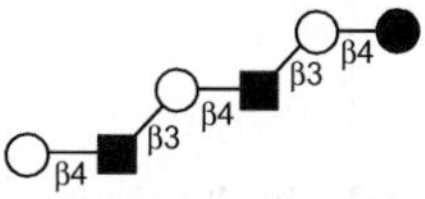 β4 β3 β4 β3 β4	53
VIII	Lacto-*N*-octaose	LNO	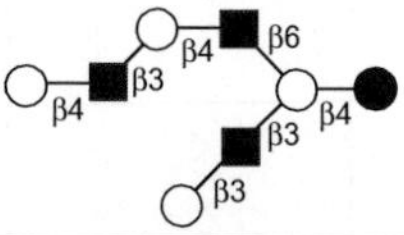 β4 β6 β4 β3 β4 β3 β3	54
IX	Lacto-*N*-neooctaose	LNnO	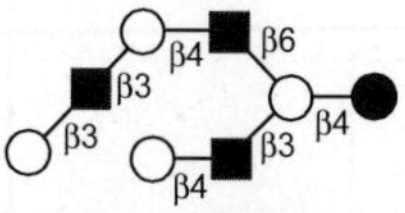 β4 β6 β3 β4 β3 β3 β4	54
X	*iso*-Lacto-*N*-octaose	*i*LNO	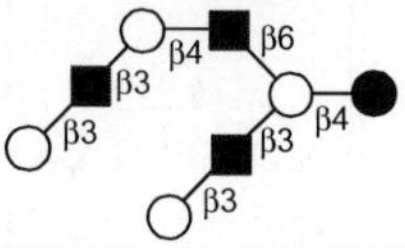 β4 β6 β3 β4 β3 β3 β3	55

Table 2 Lactose and Neutral Non-Fucosylated HMOS That Can Serve as the Core Structures for Other HMOS[28,34]—cont'd

Core #	Lactose and HMOS Core Structures	Abbreviations	Symbols	References
XI	*para*-Lacto-*N*-octaose	*p*LNO	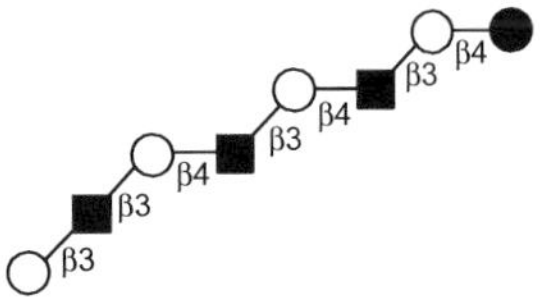	56
XII	*para*-Lacto-*N*-neooctaose	*p*LNnO	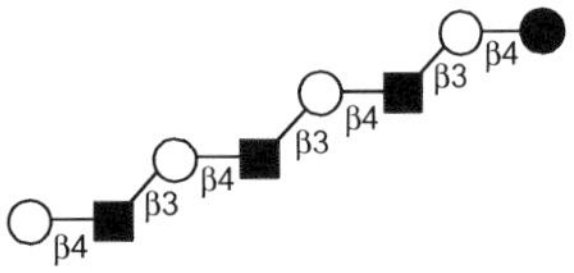	57
XIII	Lacto-*N*-decaose	LND	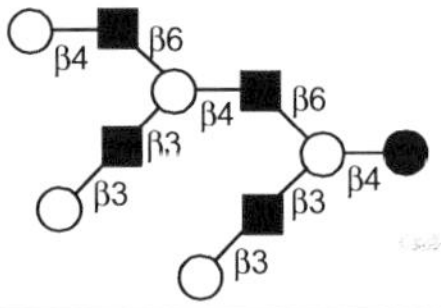	58
XIV	Lacto-*N*-neodecaose	LNnD	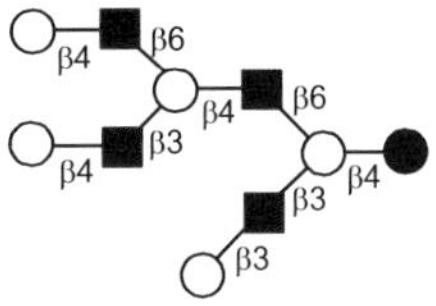	58
XV			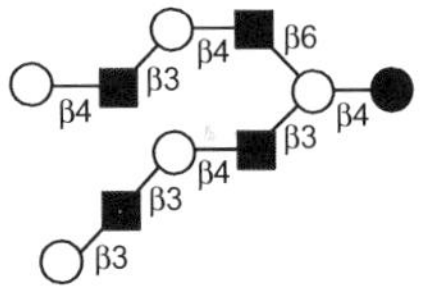	33
XVI			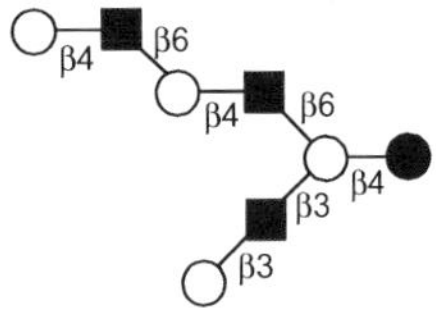	59

Symbols and abbreviations: ○ galactose (Gal), ■ *N*-acetylglucosamine (GlcNAc), ● glucose (Glc).

(HPAEC-PAD), capillary electrophoresis, and various mass spectrometric (MS) techniques.[8,11,30,58,60,61] The diversity of HMOS comes from 5 different monosaccharide building blocks (Table 1), the length, the size, the sequence (Table 2), and the 12 glycosidic linkages (Table 3) of the glycans.[15] Table 4 lists 144 structures (lactose is not counted) in the I–XVI categories

Table 3 Twelve Glycosidic Linkages That Constitute Diverse HMOS[33,34]

Glycosidic Linkages	Abbreviations	Symbol
Galactosidic bonds	Galβ1–4Glc	β4
	Galβ1–3GlcNAc	β3
	Galβ1–4GlcNAc	β4
N-Acetyl-glucosaminidic bond	GlcNAcβ1–3Gal	β3
	GlcNAcβ1–6Gal	β6
Fucosidic bond	Fucα1–2Gal	α2
	Fucα1–3Glc	α3
	Fucα1–3GlcNAc	α3
	Fucα1–4GlcNAc	α4
Sialidic bond	Neu5Acα2–3Gal	α3
	Neu5Acα2–6Gal	α6
	Neu5Acα2–6GlcNAc	α6

Symbols and abbreviations: *N*-acetylneuraminic acid (Neu5Ac), fucose (Fuc), galactose (Gal), *N*-acetylglucosamine (GlcNAc), glucose (Glc).

based on the differences of core structures and 12 structures in the XVII category of deviant structures.

Several exceptions have been found to the general structural features described above for HMOS. For example, a few HMOS containing a terminal *N*-acetylgalactosamine (GalNAc), such as the A-antigen tetrasaccharide, pentasaccharide, hexasaccharide,[62] and heptasaccharide,[63] were isolated from the urine or feces of blood group A breast-fed infants. In addition, several HMOS containing 6-*O*-sulfo monosaccharides have

Table 4 Structures of HMOS Grouped by Their Core Structures[28,32–34,61]

Core #	Lactose and HMOS	Abbreviations	Symbols	References
I	Lactose (not considered as HMOS itself)	Lac	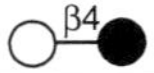	49
		LNTri II	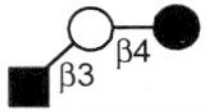	76
	[a]2′-Fucosyllactose	2′FL	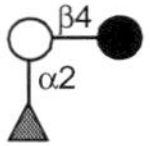	25
	3-Fucosyllactose	3FL	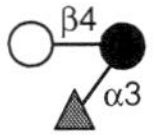	27
	[a]Lactodifucotetraose	LDFT	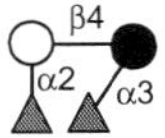	77
	3′-Sialyllactose	3′SL	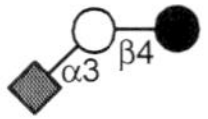	78
	6′-Sulfo-3′-sialyllactose	6′-Sulfo-3′SL	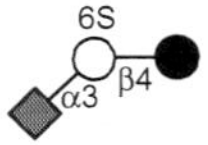	64
	6′-Sialyllactose	6′SL	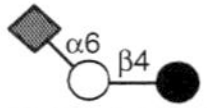	79
	3′-Sialyl-3-fucosyllactose	3′S3FL	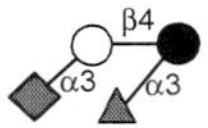	80

Continued

Table 4 Structures of HMOS Grouped by Their Core Structures[28,32–34,61]—cont'd

Core #	Lactose and HMOS	Abbreviations	Symbols	References
II	Lacto-*N*-tetraose	LNT		24
	[a]Lacto-*N*-fucopentaose I	LNFP I		26
	[b]Lacto-*N*-fucopentaose II	LNFP II		81
	Lacto-*N*-fucopentaose V	LNFP V		82
	[a,b]Lacto-*N*-difuco-hexaose I	LNDFH I		83
	[b]Lacto-*N*-difuco-hexaose II	LNDFH II		84
	Sialyllacto-*N*-tetraose a	LSTa		66

Sialyllacto-*N*-tetraose b	LSTb	α6 β3 β4 β3	66
Disialyllacto-*N*-tetraose	DSLNT	α6 β3 β4 β3 α3	85
Sialylfucosyllacto-*N*-tetraose	S-LNF II or F-LSTa	α4 β3 β4 β3 α3	86
Fucosylsialyllacto-*N*-tetraose	S-LNF I or F-LSTb	α6 β3 β4 β3 α2	86
Fucosyldisialyllacto-*N*-tetraose	DS-LNF II or FDS-LNT I	α6 α4 β3 β4 β3 α3	87
Disialylfucosyllacto-*N*-tetraose	DS-LNF V or FDS-LNT II	α6 β3 β4 α3 β3 α3	87

Continued

Table 4 Structures of HMOS Grouped by Their Core Structures[28,32–34,61]—cont'd

Core #	Lactose and HMOS	Abbreviations	Symbols	References
III	Lacto-*N*-neotetraose	LNnT	β4 β3 β4	50
	Lacto-*N*-fucopentaose III	LNFP III	β4 β3 β4 α3	88
	Lacto-*N*-neofucopentaose V	LNnFP V	β4 β3 β4 α3	89
	Lacto-*N*-neodifucohexaose II	LNnDFH II (LNDFH III)	β4 β3 β4 α3 α3	67
	Sialyllacto-*N*-neotetraose c	LSTc	α6 β4 β3 β4	90
	Fucosylsialyllacto-*N*-neotetroase	F-LSTc	α6 β4 β3 β4 α3	91
IV	Lacto-*N*-hexaose	LNH	β4 β6 β4 β3 β3	51
	Fucosyllacto-*N*-hexaose I	FLNH I	β4 β6 β4 β3 β3 α2	92

	Fucosyllacto-*N*-hexaose II	FLNH II	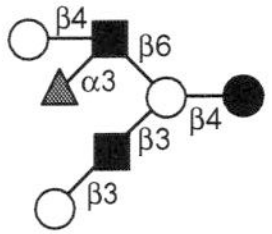 β4 β6 α3 β3 β4 β3	93
	[c]4120a		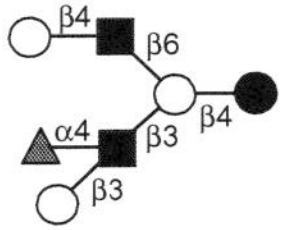 β4 β6 α4 β3 β4 β3	33
	Difucosyllacto-*N*-hexaose a	DF-LNH a	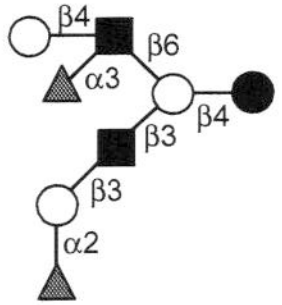 β4 β6 α3 β3 β4 β3 α2	92
	Difucosyllacto-*N*-hexaose b	DF-LNH b	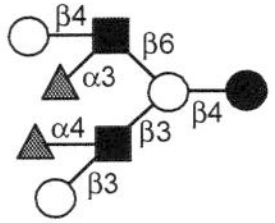 β4 β6 α3 α4 β3 β4 β3	51
	Difucosyllacto-*N*-hexaose c	DF-LNH c	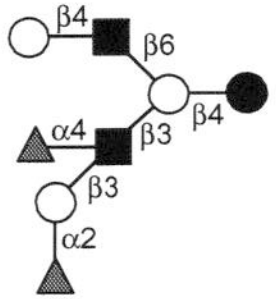 β4 β6 α4 β3 β4 β3 α2	33
	Trifucosyllacto-*N*-hexaose	TF-LNH	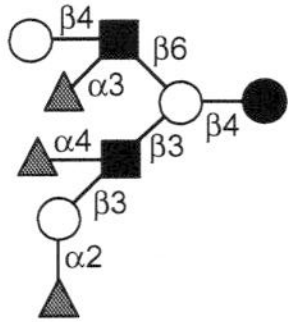 β4 β6 α3 α4 β3 β4 β3 α2	55

Continued

Table 4 Structures of HMOS Grouped by Their Core Structures[28,32–34,61]—cont'd

Core #	Lactose and HMOS	Abbreviations	Symbols	References
	Sialyllacto-*N*-hexaose	S-LNH	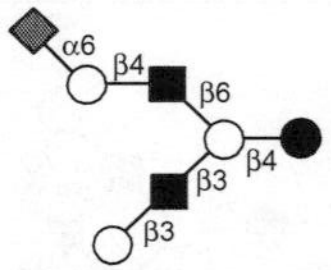	51
	[c]4021a		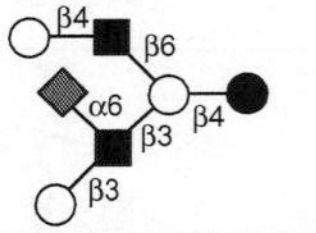	34
	Disialyllacto-*N*-hexaose I	DS-LNH I	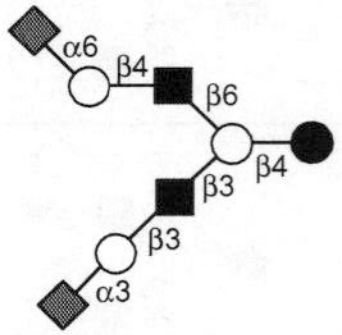	94
	Disialyllacto-*N*-hexaose II	DS-LNH II	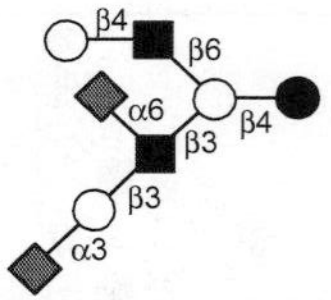	94
	Trisialyllacto-*N*-hexaose	TS-LNH	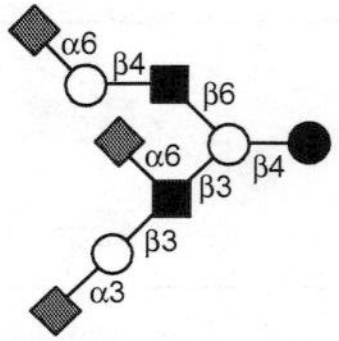	95

	Fucosylsialyllacto-*N*-hexaose	FS-LNH	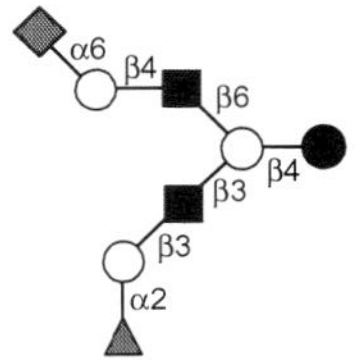 α6 β4 β6 β4 β3 β3 α2	92
	Fucosylsialyllacto-*N*-hexaose I	FS-LNH I	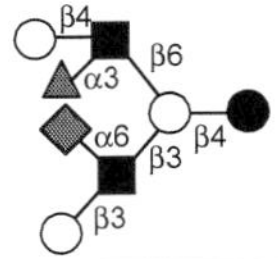 β4 β6 α3 α6 β4 β3 β3	96
	Fucosylsialyllacto-*N*-hexaose II	FS-LNH II	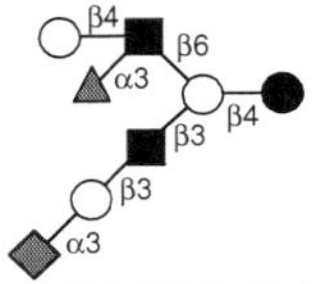 β4 β6 α3 β4 β3 β3 α3	96
	Fucosylsialyllacto-*N*-hexaose III	FS-LNH III	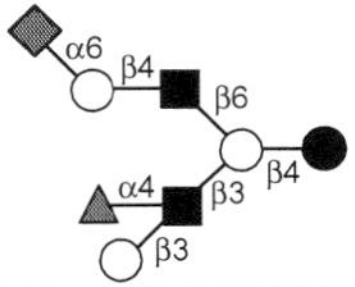 α6 β4 β6 β4 α4 β3 β3	96
	Fucosylsialyllacto-*N*-hexaose IV	FS-LNH IV	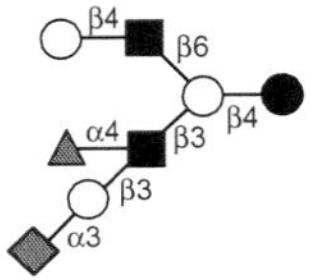 β4 β6 β4 α4 β3 β3 α3	97

Continued

Table 4 Structures of HMOS Grouped by Their Core Structures[28,32–34,61]—cont'd

Core #	Lactose and HMOS	Abbreviations	Symbols	References
	Difucosylsialyllacto-*N*-hexaose I	DFS-LNH I	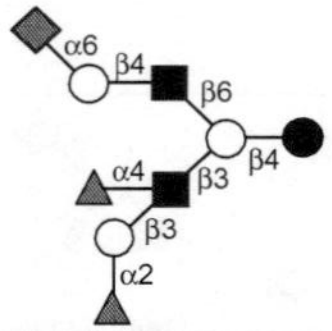	96
	Difucosylsialyllacto-*N*-hexaose II	DFS-LNH II	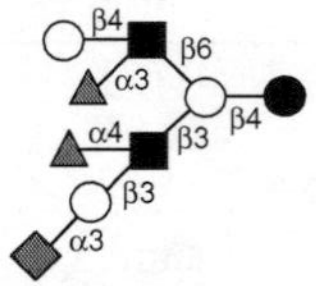	97
	Fucosyldisialyllacto-*N*-hexaose I	FDS-LNH I	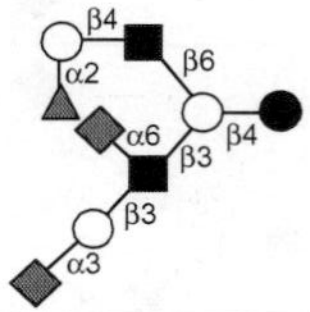	98
	Fucosyldisialyllacto-*N*-hexaose II	FDS-LNH II	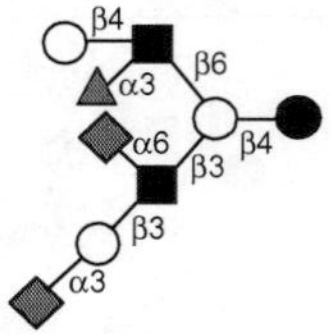	98
	Fucosyldisialyllacto-*N*-hexaose III	FDS-LNH III	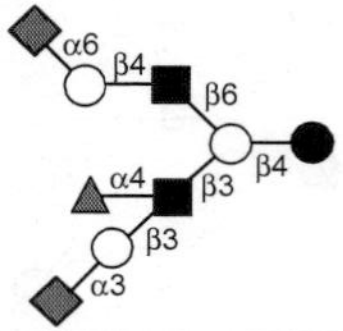	97

V	Lacto-*N*-neohexaose	LNnH		52
	Fucosyllacto-*N*-neohexaose	F-LNnH		52
	Difucosyllacto-*N*-neohexaose	DF-LNnH		56
	Sialyllacto-*N*-neohexaose I	S-LNnH I		51
	Sialyllacto-*N*-neohexaose II	S-LNnH II		80
	Disialyllacto-*N*-neohexaose	DS-LNnH		87
	Fucosylsialyllacto-*N*-neohexaose I	FS-LNnH I		80

Continued

Table 4 Structures of HMOS Grouped by Their Core Structures[28,32–34,61]—cont'd

Core #	Lactose and HMOS	Abbreviations	Symbols	References
	Fucosylsialyllacto-*N*-neohexaose II	FS-LNnH II	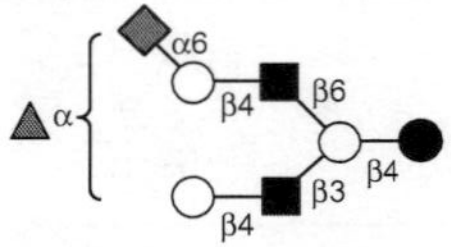 α6 β4 β6 α β4 β3 β4	52
	Difucosylsialyllacto-*N*-neohexaose	DFS-LNnH	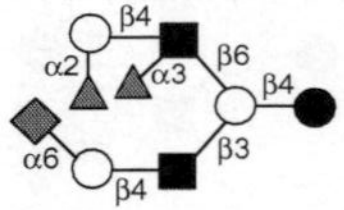 β4 α2 α3 β6 β4 α6 β3 β4	80
	Fucosyldisialyllacto-*N*-neohexaose	FDS-LNnH	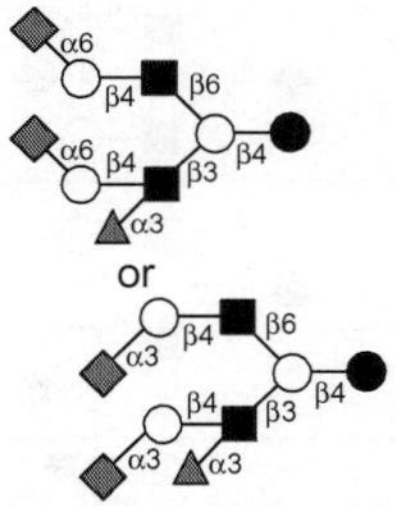 α6 β4 β6 α6 β4 β3 β4 α3 or β4 β6 α3 β4 β4 β3 α3 α3	98
VI	*para*-Lacto-*N*-hexaose	*p*LNH	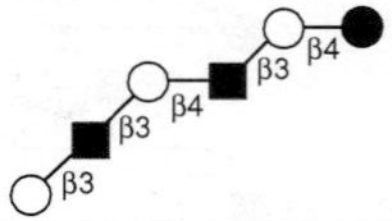 β3 β4 β4 β3 β3 β3	53
	Fucosyl-*para*-lacto-*N*-hexaose I	F-*p*LNH I	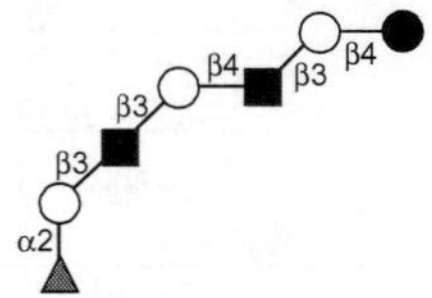 β4 β3 β4 β3 β3 α2	99

	Fucosyl-*para*-lacto-*N*-hexaose II	F-*p*LNH II	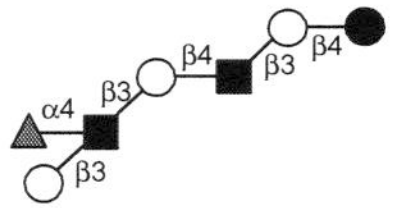	100
	Fucosyl-*para*-lacto-*N*-hexaose IV	F-*p*LNH IV	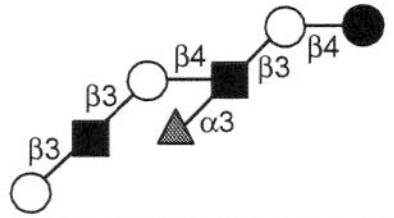	100
	Difucosyl-*para*-lacto-*N*-hexaose	DF-*p*LNH	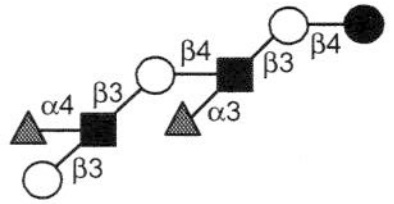	53
	Trifucosyl-*para*-lacto-*N*-hexaose I	TF-*p*LNH I	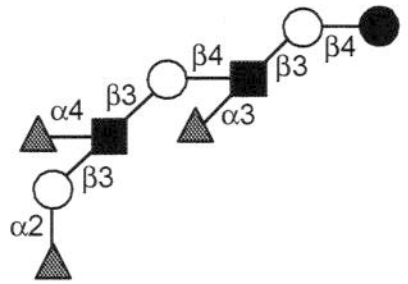	101
	Trifucosyl-*para*-lacto-*N*-hexaose II	TF-*p*LNH II	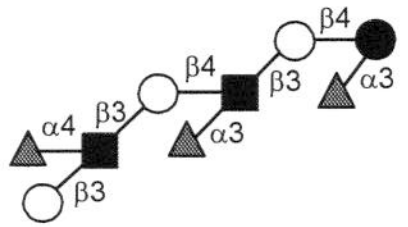	100
	Difucosyl-*para*-lacto-*N*-hexaose sulfate I	DF-*p*LNH sulfate I	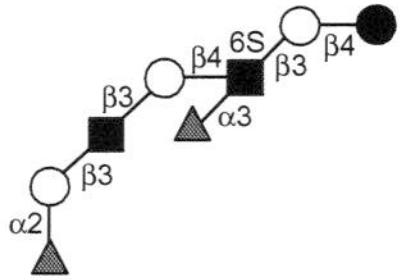	65

Continued

Table 4 Structures of HMOS Grouped by Their Core Structures[28,32–34,61]—cont'd

Core #	Lactose and HMOS	Abbreviations	Symbols	References
	Difucosyl-*para*-lacto-*N*-hexaose sulfate II	DF-*p*LNH sulfate II	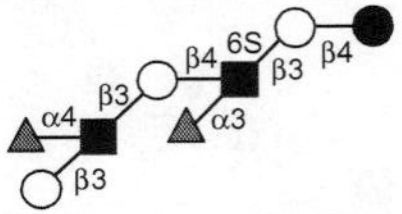	65
	Difucosyl-*para*-lacto-*N*-hexaose sulfate III	DF-*p*LNH sulfate III	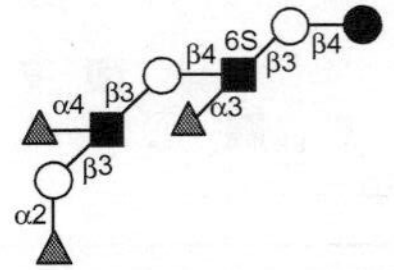	65
VII	*para*-Lacto-*N*-neohexaose	*p*LNnH	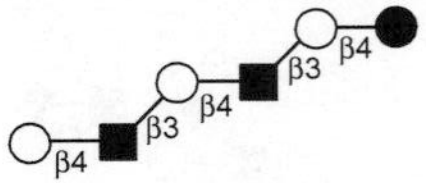	53
	Fucosyl-*para*-lacto-*N*-neohexaose	F-*p*LNnH or IFLNH III	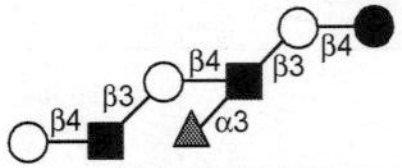	99
	Difucosyl-*para*-lacto-*N*-neohexaose	DF-*p*LNnH	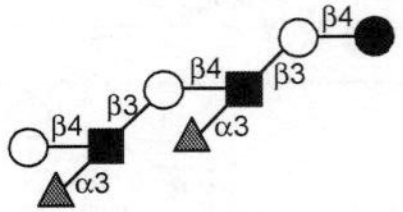	53
	Trifucosyl-*para*-lacto-*N*-neohexaose	TF-*p*LNnH	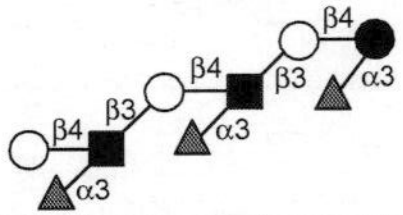	100

	[c]4021b		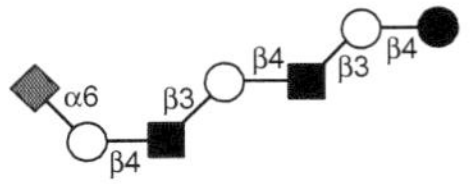	34
	[c]4121a		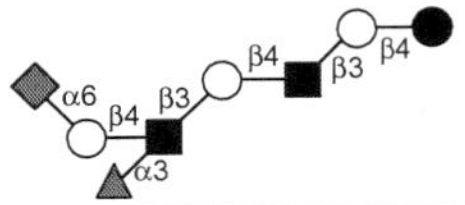	34
	[c]4121b		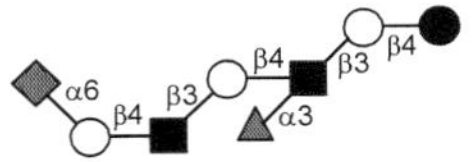	34
VIII	Lacto-*N*-octaose	LNO	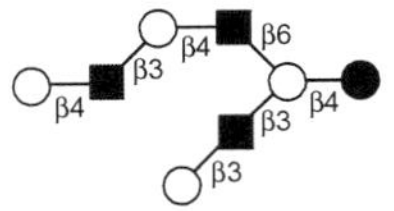	54
	[c]5130c		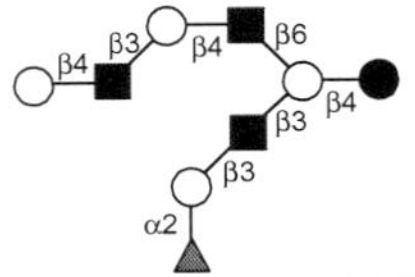	33
	[c]5130b		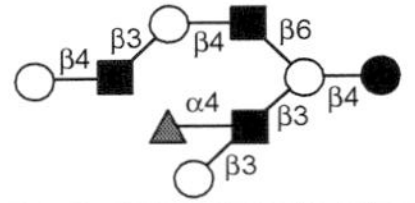	33
	Fucosyllacto-*N*-octaose	F-LNO	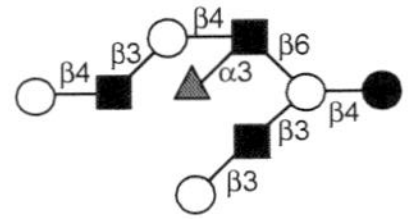	102

Continued

Table 4 Structures of HMOS Grouped by Their Core Structures[28,32–34,61]—cont'd

Core #	Lactose and HMOS	Abbreviations	Symbols	References
	Difucosyllacto-*N*-octaose I	DF-LNO I	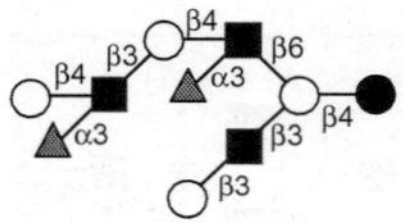 β4 β3 β4 β6 α3 α3 β4 β3 β3	54
	Difucosyllacto-*N*-octaose II	DF-LNO II	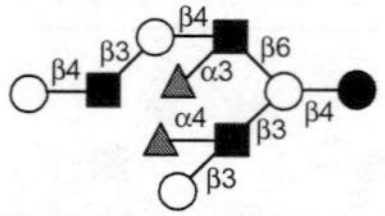 β4 β3 β4 β6 α3 α4 β4 β3 β3	54
	Trifucosyllacto-*N*-octaose	TF-LNO	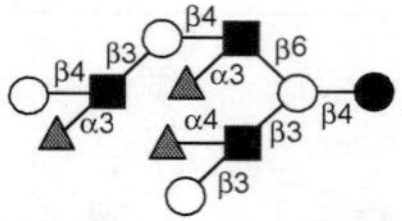 β4 β3 β4 β6 α3 α3 α4 β4 β3 β3	54
	[c]5031a		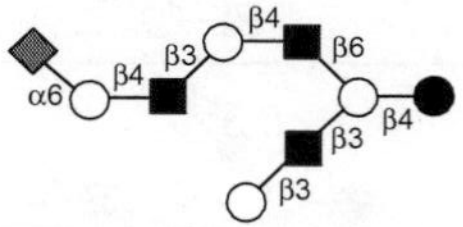 α6 β4 β3 β4 β6 β4 β3 β3	34
	Fucosylsialyllacto-*N*-octaose	FS-LNO	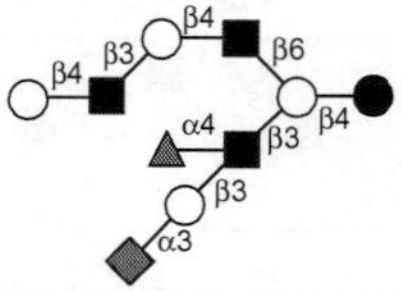 β4 β3 β4 β6 α4 β4 β3 β3 α3	103
	[c]5131a		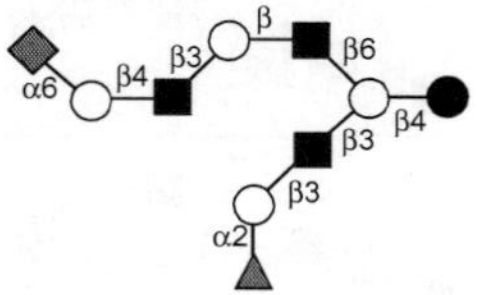 α6 β4 β3 β β6 β4 β3 β3 α2	34

	[c]5231a		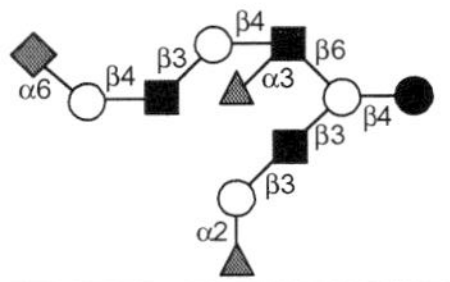	34
	[c]5231b		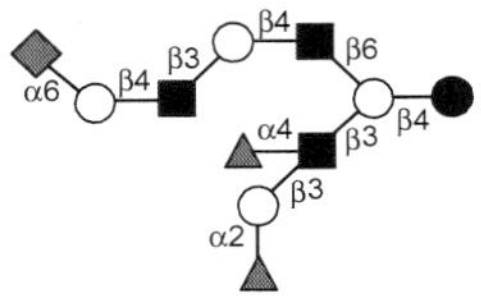	34
	Difucosylsialyllacto-*N*-octaose	DFS-LNO	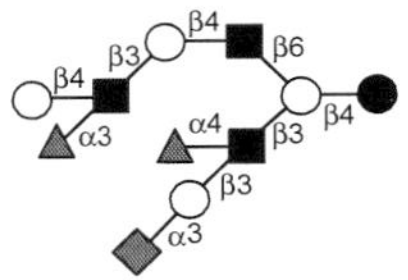	104
	[c]5331a		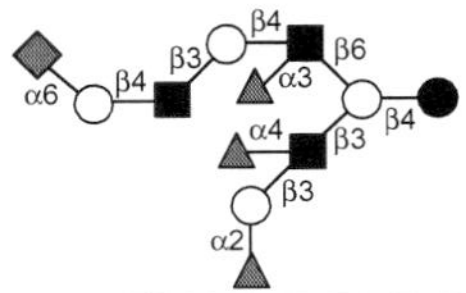	34
IX	Lacto-*N*-neooctaose	LNnO	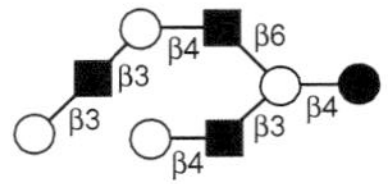	54
	Fucosyllacto-*N*-neooctaose	F-LNnO	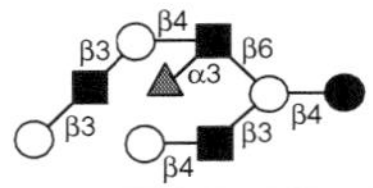	54

Continued

Table 4 Structures of HMOS Grouped by Their Core Structures[28,32–34,61]—cont'd

Core #	Lactose and HMOS	Abbreviations	Symbols	References
	Difucosyllacto-*N*-neooctaose I	DF-LNnO I		54
	Difucosyllacto-*N*-neooctaose II	DF-LNnO II		54
	Trifucosyllacto-*N*-neooctaose I	TF-LNnO I		54,55
	Trifucosyllacto-*N*-neooctaose II	TF-LNnO II		55
X	*iso*-Lacto-*N*-octaose	*i*LNO		55
	[c]5130a Fucosyl-*iso*-Lacto-*N*-octaose	F-*i*LNO		33,105

	Difucosyl-*iso*-lacto-*N*-octaose I	DF-*i*LNO I	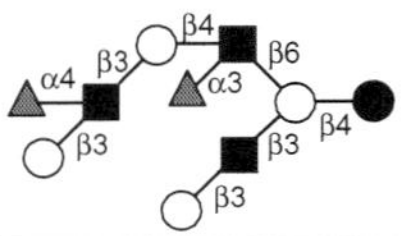 β4 β6 α4 β3 α3 β3 β3 β4 β3	106
	Difucosyl-*iso*-lacto-*N*-octaose II	DF-*i*LNO II	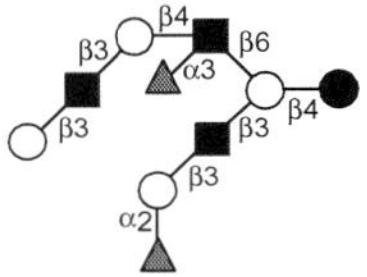 β4 β6 β3 α3 β3 β3 β4 β3 α2	106
	[c]5230a		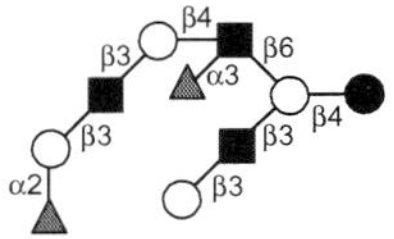 β4 β6 β3 α3 β3 β3 β4 α2 β3	33
	Trifucosyl-*iso*-lacto-*N*-octaose I	TF-*i*LNO I	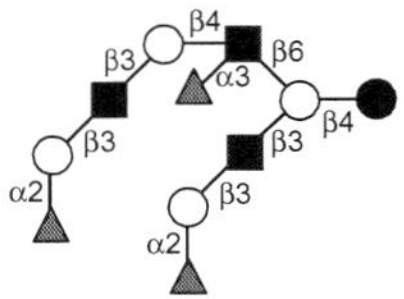 β4 β6 β3 α3 β3 β3 β4 α2 β3 α2	55
	Trifucosyl-*iso*-lacto-*N*-octaose II	TF-*i*LNO II	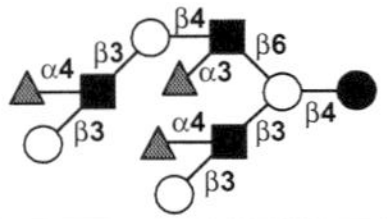 β4 β6 α4 β3 α3 β3 α4 β3 β4 β3	56,105
	[c]5330a		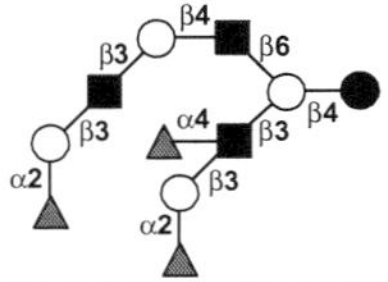 β4 β6 β3 β3 α4 β3 β4 α2 β3 α2	33

Continued

Table 4 Structures of HMOS Grouped by Their Core Structures[28,32–34,61]—cont'd

Core #	Lactose and HMOS	Abbreviations	Symbols	References
	Tetrafucosyl-*iso*-lacto-*N*-octaose	TetraF-*i*LNO	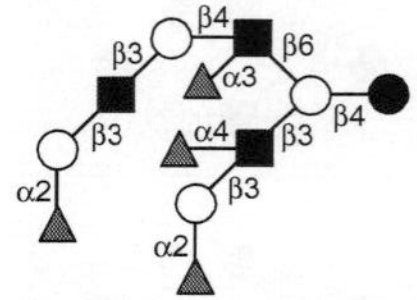	56
	Pentafucosyl-*iso*-lacto-*N*-octaose	PentaF-*i*LNO	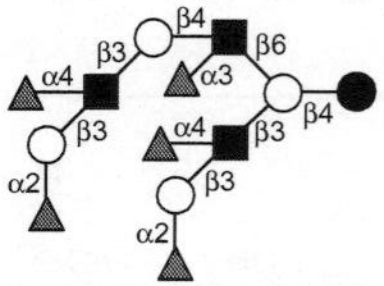	56
	Fucosylsialyl-*iso*-lacto-*N*-octaose	FS-*i*LNO	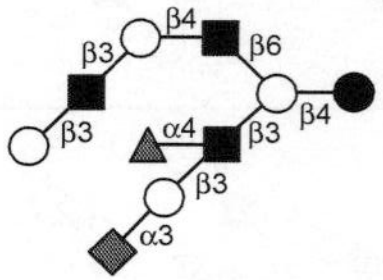	104
	Difucosylsialyl-*iso*-lacto-*N*-octaose I	DFS-*i*LNO I	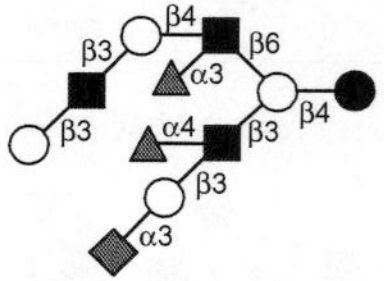	104
	Difucosylsialyl-*iso*-lacto-*N*-octaose II	DFS-*i*LNO II	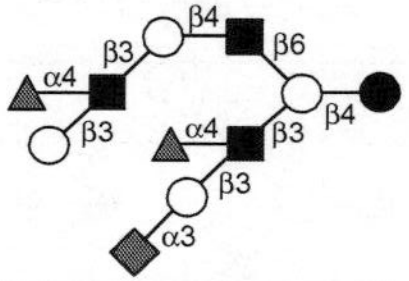	104

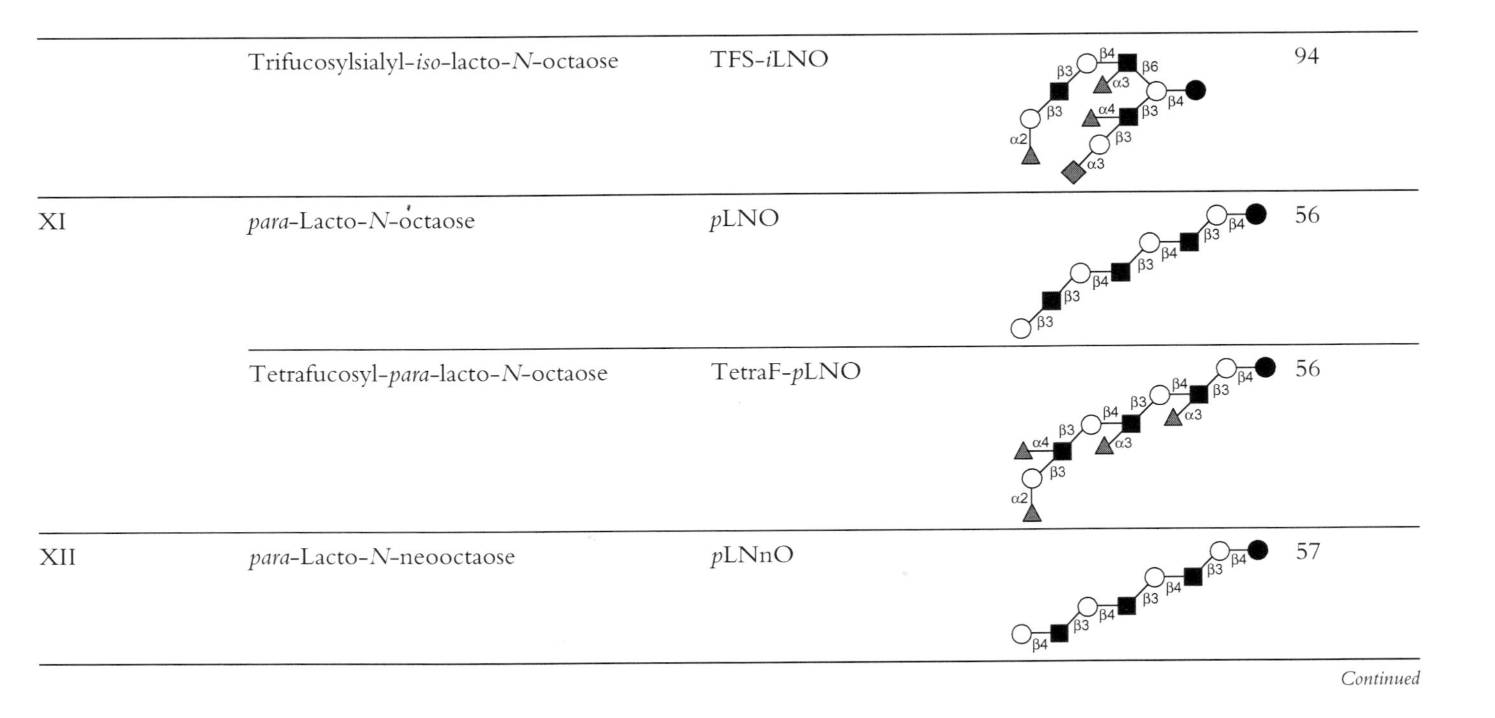

	Trifucosylsialyl-*iso*-lacto-*N*-octaose	TFS-*i*LNO		94
XI	*para*-Lacto-*N*-octaose	*p*LNO		56
	Tetrafucosyl-*para*-lacto-*N*-octaose	TetraF-*p*LNO		56
XII	*para*-Lacto-*N*-neooctaose	*p*LNnO		57

Continued

Table 4 Structures of HMOS Grouped by Their Core Structures[28,32–34,61]—cont'd

Core #	Lactose and HMOS	Abbreviations	Symbols	References
XIII	Lacto-*N*-decaose	LND	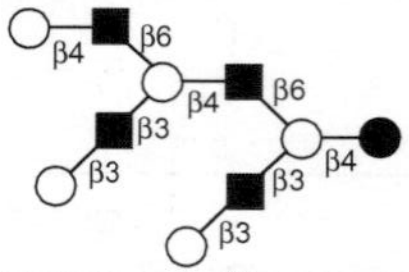	58,107
	Fucosyl-lacto-*N*-decaose I	F-LND I	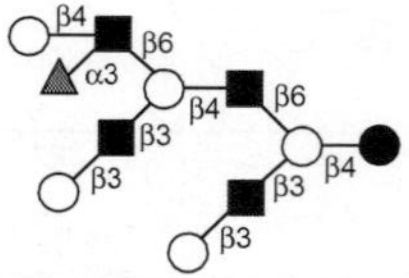	107
	Difucosyl-lacto-*N*-decaose I	DF-LND I	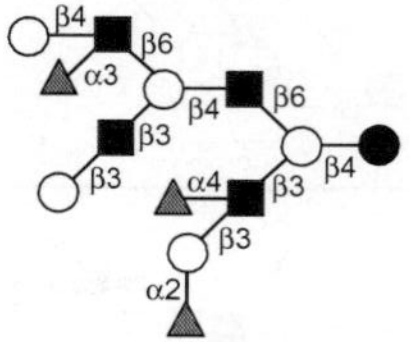	58
	Difucosyl-lacto-*N*-decaose II	DF-LND II	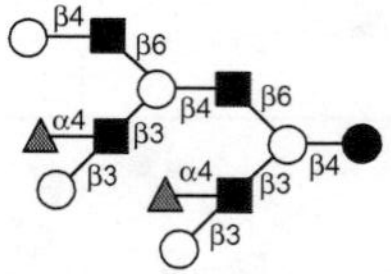	58
	Difucosyl-lacto-*N*-decaose III	DF-LND III	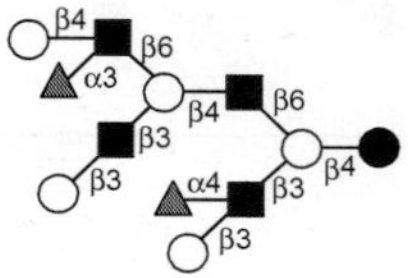	58

	Difucosyl-lacto-*N*-decaose IV	DF-LND IV	β4 β6 α3 α4 β3 β3 β4 β6 β4 β3 β3	58
	Difucosyl-lacto-*N*-decaose V	DF-LND V	β4 β6 α3 β4 β6 β3 β3 β4 α2 β3 β3	58
	Difucosyl-lacto-*N*-decaose VI	DF-LND VI	β4 β6 β4 β6 α4 β3 β3 β4 β3 β3 α2	58
			β4 β6 β4 β6 α4 β3 β3 β4 α2 β3 β3	58
			β4 β6 β4 β6 β3 β3 α4 β3 β4 β3 α2	58

Continued

Table 4 Structures of HMOS Grouped by Their Core Structures[28,32–34,61]—cont'd

Core #	Lactose and HMOS	Abbreviations	Symbols	References
	Trifucosyl-lacto-*N*-decaose I	TriF-LND I	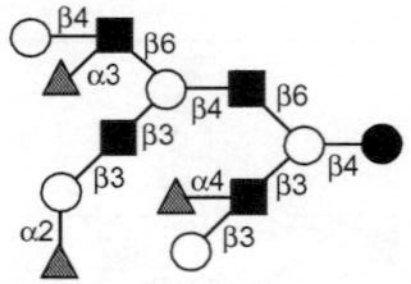 β4 α3 β6 β4 β6 β3 β3 α2 α4 β3 β3 β4	58
	Trifucosyl-lacto-*N*-decaose II	TriF-LND II	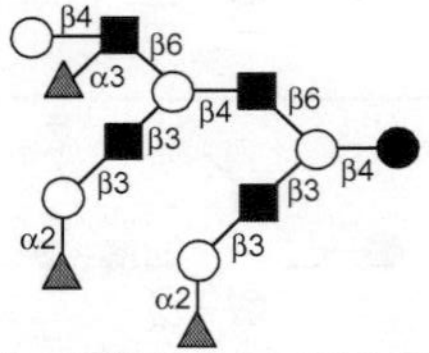 β4 α3 β6 β4 β6 β3 β3 α2 β3 β4 β3 α2	58
	Trifucosyl-lacto-*N*-decaose III	TriF-LND III	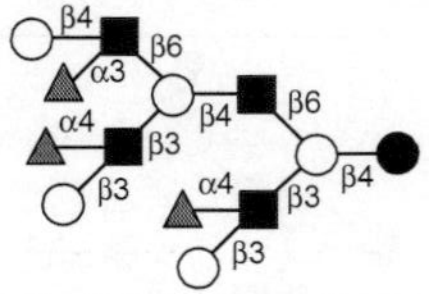 β4 α3 β6 α4 β4 β6 β3 β3 α4 β3 β4 β3	58
	Trifucosyl-lacto-*N*-decaose IV	TriF-LND IV	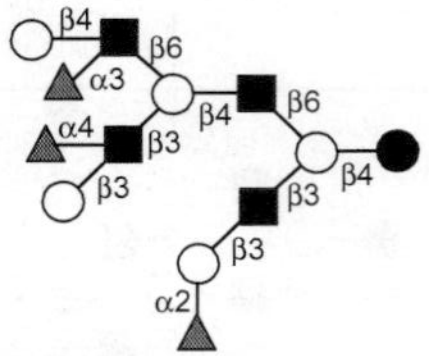 β4 α3 β6 α4 β4 β6 β3 β3 β3 β4 β3 α2	58
			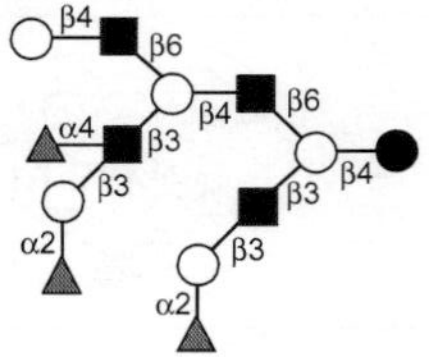 β4 β6 β4 β6 α4 β3 β3 α2 β3 β4 β3 α2	58

			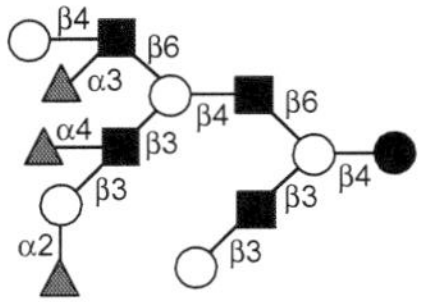	58
	[c]6340a		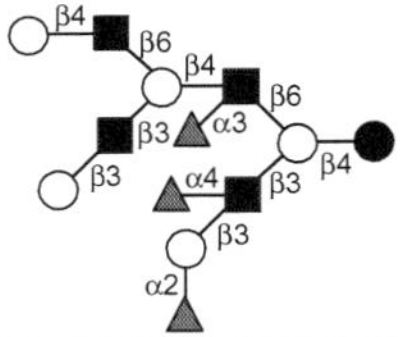	33
	Tetrafucosyl-lacto-*N*-decaose I	TetraF-LND I	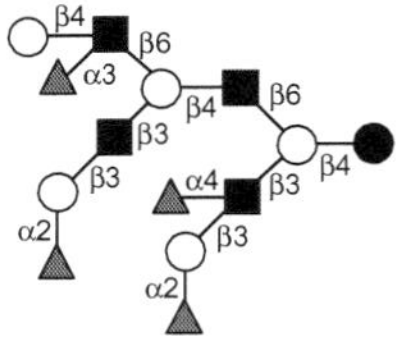	58
	Tetrafucosyl-lacto-*N*-decaose II	TetraF-LND II	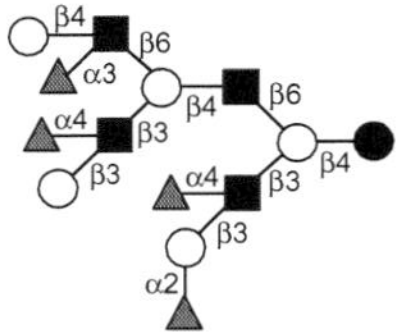	58

Continued

Table 4 Structures of HMOS Grouped by Their Core Structures[28,32–34,61]—cont'd

Core #	Lactose and HMOS	Abbreviations	Symbols	References
	Tetrafucosyl-lacto-*N*-decaose III	TetraF-LND III	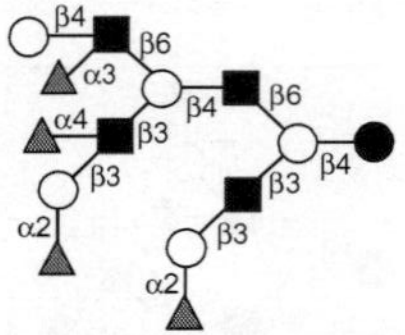	58
			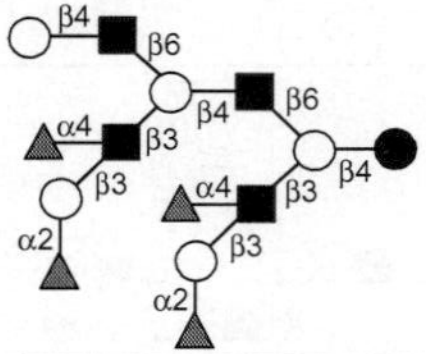	58
XIV	Lacto-*N*-neodecaose	LNnD	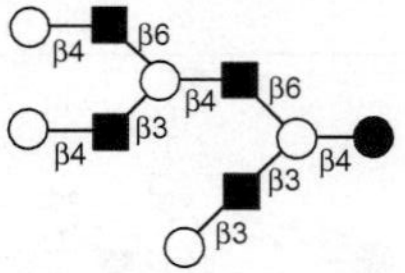	58
	Fucosyllacto-*N*-neodecaose I	F-LNnD I	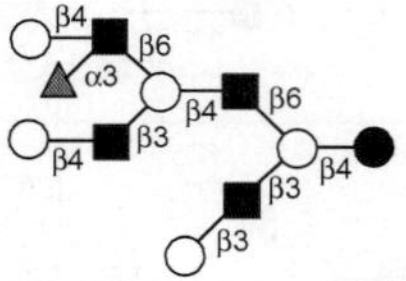	58

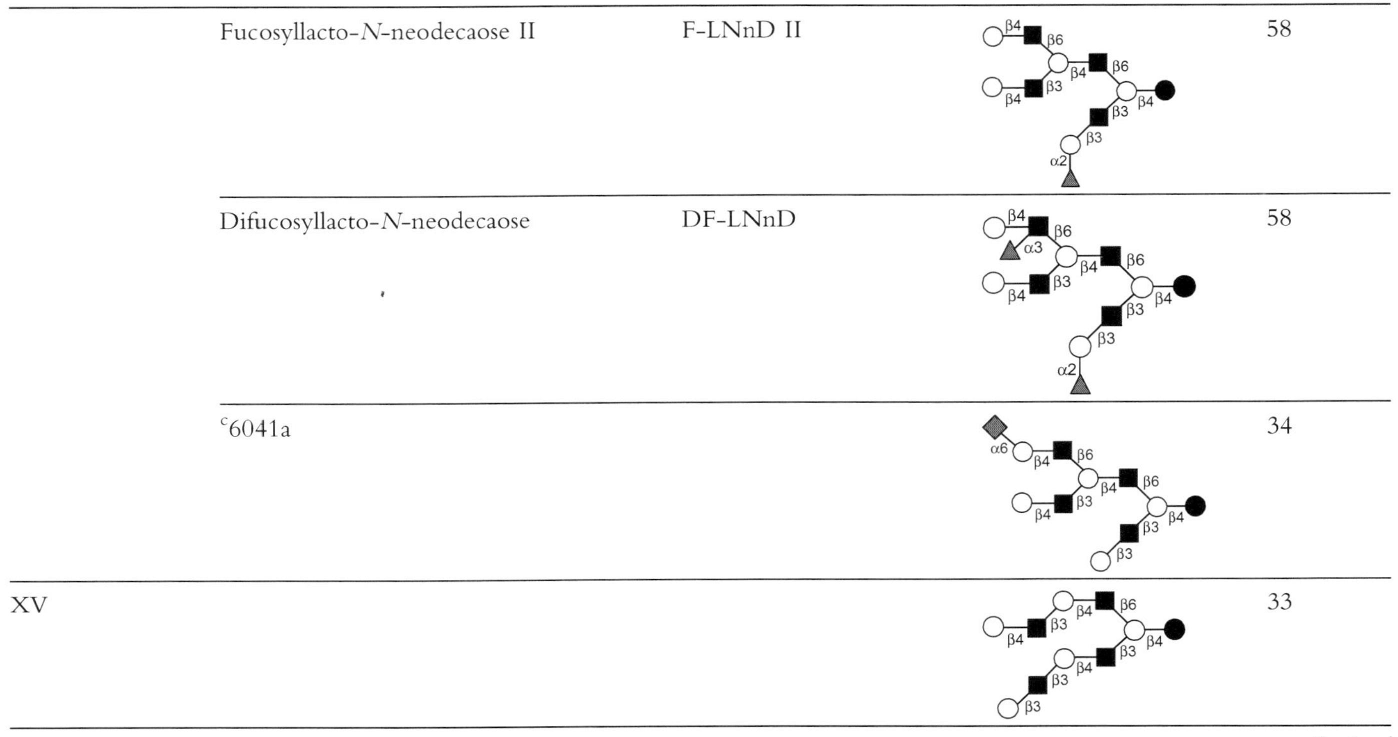

	Fucosyllacto-*N*-neodecaose II	F-LNnD II		58
	Difucosyllacto-*N*-neodecaose	DF-LNnD		58
	[c]6041a			34
XV				33

Continued

Table 4 Structures of HMOS Grouped by Their Core Structures[28,32–34,61]—cont'd

Core #	Lactose and HMOS	Abbreviations	Symbols	References
XVI			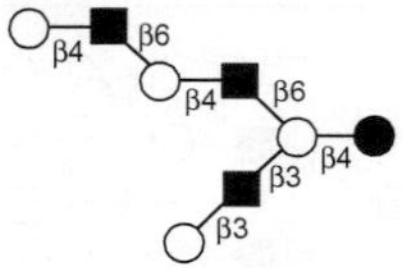	59
			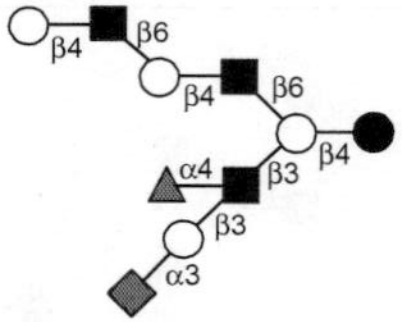	59
			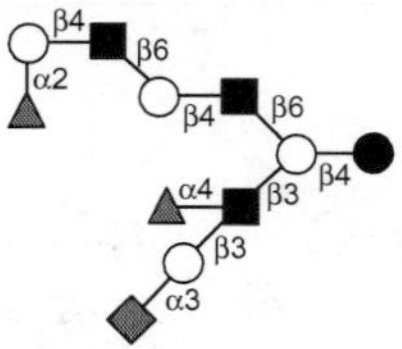	59
XVII Deviant structures	A antigen-tetrasaccharide	A-Tri	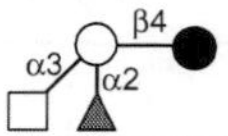	62,108
	A antigen-pentasaccharide	A-Penta	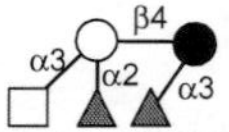	62,108–110
	A antigen-hexasaccharide	A-Hexa	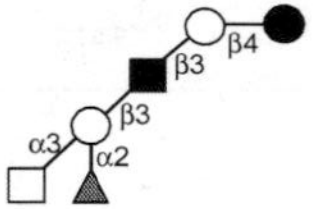	62

	Name	Abbreviation	Structure	References
	A antigen-heptasaccharide	A-Hepta	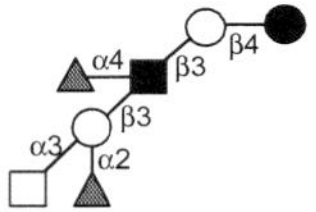	62,108,109
			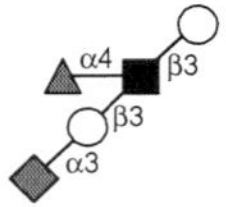	111
	3′-Sialyl Lewis a	3′SLe[a]	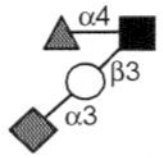	111
	3′-Sialyl-*N*-acetyllactosamine	3′SLN	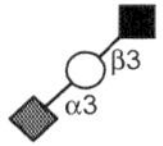	34
	6′-Sialyl-*N*-acetyllactosamine	6′SLN	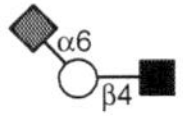	66
	Fucosylsialyl-*novo*-lacto-*N*-pentaose I	FS-*novo*-LNP I	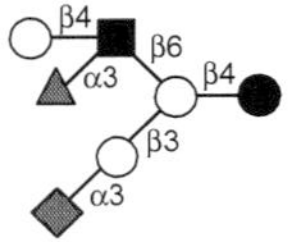	96
	3′-Galactosyllactose	β3′GL	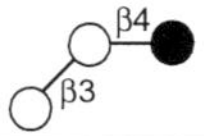	67
	4′-Galactosyllactose	β4′GL	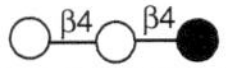	68
	6′-Galactosyllactose	β6′GL	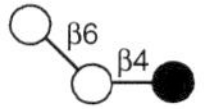	69

[a]Missing in the milk of Le^{a+b-} nonsecretors.[73]
[b]Missing in the milk of Lewis negative (Le^{a-b-}) individuals.[74]
[c]Indicate the number of Hexose, Fucose, HexNAc, and Neu5Ac in the oligosaccharide.[33,34]
Symbols and abbreviations: ◆ *N*-acetylneuraminic acid (Neu5Ac), ▲ fucose (Fuc), ○ galactose (Gal), ■ *N*-acetylglucosamine (GlcNAc), ● glucose (Glc), □ *N*-acetylgalactosamine (GalNAc).

been identified,[64,65] as well as several HMOS missing the glucose[59] or lactose[34,59,66] at the reducing end. Also, an unusual Galβ1–3Gal,[59,67] Galβ1–4Gal,[68] or Galβ1–6Gal[69] component has been found in several HMOS (see Core No. XVII, Deviant Structures, in Table 4).

The presence of some HMOS is related to the secretor status and the Lewis blood type of the mother.[70] The milk produced by Le^{a+b+} secretors, presenting in 70% of the general population, has the highest diversity of HMOS.[14] Fucα1–2Gal-containing HMOS such as 2′-fucosyllactose (2′FL),[71,72] lactodifucotetraose (LDFT), lacto-*N*-fucopentaose I (LNFP I), and lacto-*N*-difuco-hexoase I (LNDFH I) are missing in the milk of Le^{a+b-} nonsecretors.[73] Fucα1–4GlcNAc-containing HMOS including LNFP II, LNDFH I, and LNDFH II are missing in the milk of Lewis negative (Le^{a-b-}) individuals[74] (Table 4). It has been shown that in the absence of blood samples, the ratios of 2′FL versus 3′FL; LNFP I, LDFT, and LNDFH I versus LNT; and 6′SL versus 3′SL in human milk can be used as specific and sensitive markers for determining the secretor status of individuals.[75]

High-molecular-weight HMOS (M_r 2242–8000),[112] complex neutral HMOS with up to 10 fucose residues on a core structure containing 7 LacNAc units,[60] HMOS with up to 32 monosaccharide units,[113] neutral HMOS with up to 35 monosaccharides, and HMOS with more than 50 monosaccharide units[114] have all been observed. Nevertheless, the HMOS identified containing more than 14 monosaccharide units have not been structurally elucidated and are not presented in Table 4.

Overall, about 70% of HMOS in pooled milk are fucosylated and about 20% are sialylated.[30] The major components of HMOS are lacto-*N*-tetraose (LNT), lacto-*N*-neotetraose (LNnT), as well as monofucosylated, monosialylated, difucosylated, and disialylated lactose, LNT, and LNnT (Table 4).[57] Oligosaccharides with both sialic acid and fucose are also presented in HMOS. The top 10 most abundant HMOS species are responsible for about 46% of the HMOS mass. The top 50 most abundant HMOS species in a pooled human milk sample account for 83% of the total intensity, and the least abundant half of the total constitute only 8% of the entire intensities.[30] Among HMOS, 20–25 of them are considered to be the major components.[11] Most of them contain 3–9 monosaccharide units.[8] 2′FL, LNFP I, LNDFH I, and LNT were shown to be the most abundant HMOS in the colostra of Japanese women (85% are secretors) and in mature human milk.[10,61,115–117] The most abundant acidic HMOS were LSTc, DSLNT, 6′SL, 3′SL and LSTa.[61,115]

Compared to the milk oligosaccharides (MOS) characterized for some domestic animals[118–120] and other primates,[48,121] HMOS are higher in quantities and complexity with more diversity and longer structures. In general, HMOS are high in fucosylation, which is rare in the MOS of cows and pigs. In comparison, sialylation is more abundant in the MOS of cows and pigs.[118–120] Furthermore, *N*-glycolylneuraminic acid (Neu5Gc)-containing MOS found in the milk of cows, pigs, and primates[48,118–121] and 4-*O*-acetyl-*N*-acetylneuraminic acid (Neu4,5Ac$_2$)-containing MOS found in the milk of monotremes, including echidna[122,123] and platypus,[124–126] have not been observed in human milk. Except for 3′-galactosyllactose (3′GL), other oligo-β1–3-galactoside structures abundant in the milk of metatheria (marsupials),[127–129] such as the common brushtail possum, tammar wallaby, red kangaroo, and koala, have not been found in HMOS.

3. BIOSYNTHESIS OF HMOS

Our understanding on the biosynthesis of HMOS is limited. Most, if not all, HMOS are extended from lactose at the non-reducing ends and are believed to be catalyzed by glycosyltransferases in the mammary gland.[130–133] Glucose and galactose can be synthesized *de novo* in the mammary gland by a process named hexoneogenesis, although plasma glucose is the major carbon source of milk lactose.[134,135] Lactose and other MOS are most likely accumulated in the secretory vesicle and secreted by exocytotic fusion with the apical plasma membrane.[136] Lactose itself is produced in the mammary gland by β1–4-galactosyltransferase 1 (β1–4GalT1) bound to α-lactalbumin in a lactose synthase complex.[137–139] However, most of the specific glycosyltransferases that are responsible for the formation of HMOS structures with specific glycosidic linkages have not been identified. The best understood examples are human α1–2-fucosyltransferase (FUT2) encoded by the secretor (*Se*) gene[71,140] and α1–3/4-fucosyltransferase (FUT3) encoded by the Lewis (*Le*) gene[74] that are responsible for the formation of α1–2- and α1–3/4-linked fucosides, respectively, in human mammary glands.[130,141] Lewis (*Le*) gene-independent α1–3-fucosyltransferase presented in all women has also been described.[11,14] Transgenic introduction of a human α1–2-fucosyltransferase gene to mice was shown to allow the mice to express large quantities of 2′-fucosyllactose,[142] which is a good indication of the ability of the mammary gland in producing corresponding oligosaccharides in the presence of suitable glycosyltransferases. Similar success was achieved for the transgenic manipulation of mice, but not rabbit, using other glycosyltransferases, including a homologous galactosyltransferase and different fucosyltransferases.[143]

The presence of several glycosyltransferases in human milk has been confirmed. For example, β1–4GalT1,[144,145] α1–3- and α1–4-fucosyltransferases,[146] as well as an α1–3/4-fucosyltransferase[147–149] have been purified from human milk. The activity of β1–3-*N*-acetylglucosaminyltransferase was identified in human colostrums but not in bovine (Holstein and Jersey cow) colostrums studied.[150] In addition to the presence of fucosyltransferase activity in human milk, α-fucosidase activity has also been identified.[151]

Enzymes purified from human milk have been used for synthesis. For example, partially purified α1–3/4-fucosyltransferase from human milk was used for synthesizing sialyl Lewis^{a},[152] Lewis^{a} and Lewis^{X},[153] including their deoxy analogs,[154] sulfated Lewis^{X},[155] and multivalent tyrosinamide-tagged Lewis^{X} structures.[156,157] Purified human milk fucosyltransferase preparation was also used for the synthesis of tumor-associated trimeric Lewis^{X}[158] and its sialylated structures,[159] sialyl Lewis^{a} and sialyl Lewis^{X} tetrasaccharide structures that are modified at the C-2 position of the glucose residue at the reducing end.[160] Fucosylation of lacto-*N*-neohexaose (LNnH) by a partially purified human milk α1–3-fucosyltransferase was found to add fucose at the LacNAc units of LNnH at the non-reducing end.[161,162] Human milk α1–3FucTs were also shown to fucosylate chitin oligosaccharides containing 2–4 GlcNAc units.[163] Purified human milk β1–4GalT was used together with a partially purified rat liver β1–3GalT, a recombinant core 2 β1–6GlcNAcT, and a recombinant human α1–3FucT in synthesizing a sialyl Lewis^{X} hexasaccharide.[164] ^{13}C-labeled linear *N*-acetylpolylactosamines $(LacNAc)_n$ were enzymatically synthesized on the 10–100 μmol scale using the partially purified and immobilized bovine milk β1–4GalT and human serum β1–3GlcNAcT.[165] *N*-Acetylglucosaminyltransferase I (GnT-I) purified from human milk was shown to be able to catalyze the transfer of deoxy derivatives of GlcNAc.[166,167] These synthetic applications of human milk enzymes provide important information about their properties. Nevertheless, the syntheses were limited to small-scale preparations and were mostly used for HMOS derivatives instead of the natural HMOS structures with a free reducing end.

4. FUNCTIONS OF HMOS

The functional studies of HMOS were usually carried out using mixtures of HMOS that were isolated from human milk pools. The benefits of breast-feeding were observed as early as the end of the 19th century.[38] Increasing evidence has now shown that HMOS contribute significantly to the health of breast-fed infants via several mechanisms by serving as listed in the following:[1,2,15,35,38,44,45,57,133,168–177]

1. Prebiotics: HMOS are carbon and energy sources preferably used by beneficial bacteria such as probiotic bifidobacteria, thus promoting their growth, which in turn produce lactic acid and short-chain fatty acid to decrease the pH of the gut, making it less desirable for the growth of pathogens. The predominant growth and colonization of bifidobacteria allow them to compete well with pathogens for the limited nutrients available in the gut. Bifidobacteria also occupy the epithelial binding sites and make them less available for the binding of pathogens. Some antimicrobial substances released by bifidobacteria also generate an unfavorable environment for pathogens.[47]
2. Antiadhesive antimicrobials: HMOS mimic the glycan structures presented on the surface of the gut epithelium and serve as soluble decoy receptors to pathogenic bacteria to decrease their binding to the infant gut surface for colonization, thus lowering the risk for viral, bacterial, and protozoan parasite infections. HMOS can also serve as inhibitors for toxins released by pathogenic bacteria.
3. Immunomodulators: Evidence has shown that HMOS can modulate epithelial and immune cell responses. Some HMOS can directly influence the functions of the gut epithelium[178] and reduce excessive mucosal leukocyte infiltration and activation, which can lower the risk for necrotizing enterocolitis (NEC), one of the most common and fatal intestinal disorders in preterm infants. *Bifidobacterium infantis* grown on HMOS can also change the functions of intestinal cells.[43]
4. Nutrient providers for the brain: Some HMOS, mainly sialylated ones, may also be providers of sialic acid for the synthesis of sialic acid-containing glycolipids (gangliosides) and glycoproteins important for the development of the brain and cognition of infants.

These functions have been discussed quite thoroughly in several excellent reviews recently published.[1,35,38,44,45,61,133,168–172] The functional roles of individual HMOS species, however, are less clear. This is mainly due to the unavailability of sufficient amounts of pure HMOS for detailed functional studies. Only a handful of examples have been shown. These are discussed briefly in the sections that follow as three categories.

4.1 Neutral Non-Fucosylated HMOS

Neutral non-fucosylated HMOS constitute the core structures or the backbones of all HMOS. Despite earlier studies on identifying β-GlcNAc-containing oligosaccharides and polysaccharides in human milk as "bifidus factors,"[19,179] their identities have not been conclusively elucidated. The discovery about 10 years ago of a novel galactose operon, which is responsible for the assembly of the GNB/LNB pathway in *Bifidobacterium longum* JCM1217 for galacto-*N*-biose (GNB) and lacto-*N*-biose (LNB) consumption, pointed to lacto-*N*-biose (LNB, Galβ1–3GlcNAc), presented at the non-reducing end of many neutral non-fucosylated HMOS, as a potential "bifidus factor."[9] This was further supported by the property of LNB in selectively stimulating the growth of bifidobacteria, but not *Clostridia*, *Enterococci*, and *Lactobacillus*.[180–182] The extracellular lacto-*N*-biosidase, α1–2-fucosidase, α1–3/4-fucosidase, and sialidase of *B. bifidum*[183–185] can decap fucosylated and/or sialylated HMOS to release their core structures, which can then be used by its extracellular lacto-*N*-biosidase to produce LNB.[186] LNB can be transported into the bacterium by the GNB/LNB transporter in the GNB/LNB pathway and be metabolized by other enzymes involved in the GNB/LNB pathway.[187] On the other hand, LacNAc-terminated core HMOS can be broken down by extracellular β-galactosidase and β-*N*-acetylhexosaminidase of *B. bifidum*.[188]

Among bifidobacteria species commonly found in breast-fed infants such as *B. longum* subsp. *longum*, *B. longum* subsp. *infantis*, *B. bifidum*, and *B. breve*,[175] *B. longum* subsp. *infantis* (e.g., JCM1222) and *B. bifidum* (e.g., JCM1254) were both found to consume both type I and type II HMOS core structures equally well. The other two species tested had preference toward LNT, but not LNnT. The *B. longum* subsp. *infantis* strain tested also consumed quite well the mono- and difucosylated LNT/LNnT, disaccharides and monosaccharides monitored in the experiment.[182] LNnT was further confirmed to provide advantages for *B. infantis* versus *B. thetaiotaomicron* in both in growth studies *in vitro* and in germ-free mice studies.[189]

Different from *B. bifidum* that expresses extracellular glycosidases, *B. longum* subsp. *infantis* expresses internal glycosidases[190–192] and relies on its glycan ABC transporters[193] for internalization of the corresponding HMOS.[175] The extracellular glycosidases on some bacteria such as *B. bifidum* could be used as a mechanism to release components from HMOS, which can be readily transported into *B. longum* subsp. *infantis* or other bacteria for consumption. The symbiotic sharing of HMOS and components could be one of the mechanisms used to shape the gut microbiota. The enrichment of bifidobacteria in the infant gut could be the result of coevolution of the bacteria and milk ingredients including HMOS.[175,182]

LNT is a carbon source that can be used by most bifidobacteria.[182] LNT in HMOS, and maybe in other HMOS with Gal at the non-reducing end, was shown to reduce *Entamoeba histolytica* (a protozoan parasite infecting ~50 million people and causing ~100,000 deaths per year[194]) attachment and cytotoxicity toward human intestinal epithelial HT-29 cells in a dose-dependent manner.[194] Further studies *in vivo* are needed to show the prebiotic and antimicrobial potentials of LNT.

In comparison, LNnT was shown to be a selective carbon source for certain bifidobacteria such as *B. longum* subsp. *infantis* and *B. bifidum*. LNnT was also shown to have immunosuppressive functions[195] and can inhibit the binding of *Streptococcus pneumoniae* to ciliated chinchilla tracheal epithelium.[196] Higher concentrations of LNnT in the milk of HIV-infected women were found to be associated with reduced postnatal transmission via breastfeeding.[197] Therefore, LNnT is a potential candidate for developing prebiotics and therapeutics against infectious disease.[198]

4.2 Fucosylated HMOS

Fucosylated HMOS are the most abundant HMOS species.[33] Their prebiotic, antiadhesive antimicrobial, and immunomodulation activities have been demonstrated.

2′FL, 3FL, and LDFT were shown to selectively promote the growth of bifidobacteria.[199] Fucosylated HMOS, including 2′FL, 3FL, LDFT, LNFP I/II/III, LNDFH I, and LNDFHII, showed preferred consumption by *B. longum* subsp. *infantis* and *B. bifidum* compared to *B. longum* subsp. *longum* or *B. breve*.[182] In fact, five fucosidases have been identified from *B. longum* subsp. *infantis* strain ATCC 15697 and characterized. Their ability in using fucosylated HMOS was confirmed.[190]

Several examples have been shown for the antiadhesive antimicrobial functions of fucosylated HMOS including antibacterial, antiyeast, and antiviral activities. Fucosylated HMOS were also found to bind norovirus[200] and inhibit the adhesion of an enteropathogenic *Escherichia coli* (EPEC) to HEp-2 cells.[201] A minor neutral fucosylated HMOS component was shown to protect suckling mice from the diarrheagenic effects caused by the heat-stable enterotoxin of *E. coli*.[202] α1–2-Fucosylated HMOS were shown to inhibit the adherence of Std fimbriated *Salmonella enterica* serotype *Typhimurium* to Caco-2 cells.[203] They also inhibit the binding of *Campylobacter jejuni* to intestinal H(O) antigen and lower the chance of infection[204] and potentially protect infants against diarrhea caused by campylobacter or calicivirus.[205]

More specifically, high levels of 2′FL in mother's milk corresponded to lower occurrences of campylobacter diarrhea of the infants. LDFH I was also shown to correlate to lower incidences of calicivirus diarrhea.[205] In addition, α1–2-fucosylated HMOS, but not those of the Lewis blood group type, were found to inhibit the binding of *Candida albicans* yeasts to human buccal epithelial cells.[206] On the other hand, Lewis blood group antigen-containing HMOS bind well to dendritic cell-specific, ICAM3-grabbing, non-integrin (DC-SIGN), competing against human immunodeficiency virus (HIV) surface glycoprotein gp120 binding to DC-SIGN *in vitro*.[207] Indeed, breastfeeding with human milk with high concentrations of α1–2-fucosylated and α1–3-fucosylated HMOS was found to be protective against mortality for HIV-exposed uninfected (HEU) children during breastfeeding.[208] Lewis^b (Le^b) antigens including Le^b-terminated LNDFH I that was synthesized enzymatically were shown to bind to *Helicobacter pylori*.[209,210]

The immunomodulating function of fucosylated HMOS was represented by Lewis^X-type LNFP III, which was shown to have immunosuppressive functions.[195] It was able to activate macrophages *in vitro*, which can further activate natural killer (NK) cells.[211] HMOS containing Lewis blood group antigens were also shown to reduce selectin-mediated cell–cell interactions.[176,212] 2′FL and 3FL were shown to decrease colon motor contractions in a dose-dependent fashion with a better activity observed for 3FL than for 2′FL.[178]

The understanding of the important roles of α1–2- and α1–3/4-fucosylated HMOS for infant health is greatly facilitated by the presence of nursing mothers with differences on the secretor status [determined by α1–2-fucosyltransferase (FUT2)] and Lewis blood type [determined by α1–3/4-fucosyltransferase (FUT3)]. Bifidobacteria were shown to be established earlier and more often in infants fed by secretor mothers.[213] Mother's milk with a higher ratio of α1–2-fucosylated versus non-α 1–2-fucosylated HMOS was shown to provide protection for breast-fed infants against diarrhea.[214]

The secretor status of premature infants was also shown to be a predictor for the outcome of infants on their survival or susceptibility to diseases. Low or nonsecretor status was associated with a higher death rate, a higher incident of necrotizing enterocolitis (NEC), and Gram-negative sepsis.[215]

4.3 Sialylated HMOS

Sialylated HMOS are charged species and represent about 20% of HMOS.[30] Their prebiotic, antiadhesive antimicrobial, and immunomodulating activities, as well as their nutritional value for infant brain development, have been demonstrated.

A sialylated HMOS fraction was shown to inhibit the adhesion of *Escherichia coli* serotype O119, *Vibrio cholerae*, and *Salmonella fyris* to differentiated Caco-2 cells.[216] As hemagglutinins on the surface of influenza viruses bind to sialylated glycans on the host-cell surface, it is not a surprise that sialylated HMOS bind to the influenza virus or inhibit the viral hemagglutinin binding to its ligand.[217]

Sialylated HMOS have been shown to influence lymphocyte maturation[218] and have shown anti-infective and immunomodulating effects.[38] Sialylated HMOS, but not non-sialylated HMOS, reduce leukocyte rolling and adhesion in a dose-dependent manner.[176] In fact, a sialylated HMOS fraction in a physiological range (12.5–125 μg/mL) was shown to be even better than soluble sialyl Lewis^X in inhibiting leukocyte rolling and adhesion. 3′SL and 3′S-3FL were further identified to be the key ingredients and were suggested to contribute to the lower incidence of inflammatory diseases in breast-fed infants.[176] Similarly, sialylated HMOS reduce platelet-neutrophil complex (PNC) formation and subsequent neutrophil activation in an ex vivo model with whole human blood.[212]

Sialylated HMOS may also be used as source of sialic acid for the synthesis of sialic acid-containing glycolipids (gangliosides) and glycoproteins important for the development of brain and cognition of infants.[45]

The simplest and the most well-studied sialylated HMOS are the sialyllactoses, including 6′SL and 3′SL. Sialyllactose inhibited cholera toxin-induced fluid accumulation in a rabbit intestinal loop model. These effects are believed to be responsible for the activity of human milk and its low-molecular-weight fraction in inhibiting the cholera toxin B subunit binding to monosialoganglioside (GM1).[219] Sialyllactose was also shown to inhibit the binding of *Aspergillus fumigatus* conidia to laminin extracted from mouse sarcoma tumor[220] and the binding of *Pseudomonas aeruginosa* 8830 to immobilized asialo GM1 in a microtiter plate assay,[221] although the mechanism for the latter is unknown. Sialyllactoses were also shown to induce differentiation in transformed human intestinal cells (HT-29) and human intestinal epithelial cells (HIEC).[222] 6′SL alone or with 3′SL, but not 3′SL alone or oligofructose alone, was shown to enhance the adhesion of *B. longum* subsp. *infantis* strain ATCC15697 to HT-29 human intestinal cells.[223] 3′SL was shown to bind to polyomarvirus.[224] It inhibited the binding of S fimbriated *E. coli* to endothelial and epithelial cells.[225,226] It also inhibited the adhesion of *Helicobacter pylori* binding to human epithelial cells in vitro and was shown to decrease *Helicobacter pylori* colonization in a rhesus monkey antiadhesive therapy model.[227] 3′SL was shown to inhibit the binding of some sialyl oligosaccharides to *Helicobacter pylori*,[228] *E. coli S-fimbriate*,[229] and influenza viruses.[173]

The 3′SL level in human milk, however, can also be an indicator of HIV infection. Higher relative abundances of 3′SL were shown in the milk of HIV-infected mothers[230] and in the milk of mothers who transmit HIV to their babies via breastfeeding.[197]

Another exciting example about the potential use of sialylated HMOS is their potential application in treating necrotizing enterocolitis, one of the most common and fatal intestinal disorders in preterm infants[231,232] that does not currently have an ideal therapeutic outcome.[233–235] A single sialylated HMOS, disialyllacto-*N*-tetraose (DSLNT), but not its non-sialylated or monosialylated analog, was identified as a specific HMOS component that is effective for preventing necrotizing enterocolitis (NEC) in a neonatal rat model.[236] Low concentrations of DSLNT in mother's milk are corresponding to an increased risk of NEC in the preterm very-low-birth-weight infants.[230]

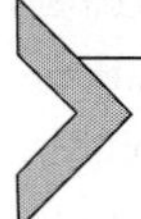

5. PRODUCTION OF HMOS BY ENZYME-CATALYZED PROCESSES

The chemical syntheses of more than 15 different HMOS and derivatives with 3–11 monosaccharide units have been reported. These include 2′FL,[237] 3FL,[237] LDFT,[237] LNT,[238–240] LNnT,[238,241] LNFP I,[239,242] LNFP III[242–245] and its protected form,[246] LSTa and LSTd (Neu5Acα2–3LNnT, not found in human milk) with an aglycon,[247,248] LNDFH I with a β-linked aglycon,[249] *p*LNnH,[241,250] LNnH[251] and its protected forms,[252,253] *i*LNO,[254] *p*LNnO[241] and its protected form,[255] trifucosylated *p*LNnO in its protected form,[255] DF-LNH II,[256,257] and DF-LNnH.[256,257] Recent successes in the synthesis of LNFP I and its α1–2-fucosylated LNnT analog using "one-pot" glycosylation techniques have been reported.[258] These chemical synthetic efforts are out of the scope of this review. The focus of this section will be a survey on enzyme-catalyzed processes for the production of HMOS.

The production of only a handful of HMOS has been reported using enzyme-catalyzed processes,[259] and the synthesized HMOS are limited to those with relatively simple structures. Despite the success on the characterization of mammalian enzymes and purification of several glycosyltransferases from human milk, their application in synthesis has been limited due to the difficulties in obtaining them in large amounts and in an economically efficient manner. On the other hand, bacteria express a wide array of glycosyltransferases that are responsible for the construction of diverse lipopolysaccharides (LPS) and capsular polysaccharide structures. Some of these glycan structures mimic those found on human cell surfaces and those in HMOS.[260,261] Therefore, bacteria are a rich source of glycosyltransferases that can be used for the synthesis of HMOS as well as the glycans and glycoconjugates presented on the surface of human cells.[262–265] Recombinant bacterial glycosyltransferases have been increasingly used for the synthesis of several HMOS structures in enzymatic, chemoenzymatic, whole-cell, and living-cell approaches.

Early enzymatic methods used expensive sugar nucleotides as donor substrates for glycosyltransferases for the production of HMOS. Glycosyltransferase-catalyzed reactions with in situ donor regeneration cycles that were applied to the preparative scale synthesis of oligosaccharides[266–268] can also be used for the synthesis of HMOS. Recently, highly efficient "one-pot" multienzyme (OPME) systems have been established for the synthesis of HMOS.[264,269–271] These systems use inexpensive, free-hydroxy monosaccharides as starting materials, which are enzymatically converted to sugar nucleotides with or without the formation of sugar-1-phosphate intermediates. The activated sugars in the forms of sugar nucleotides are supplied to the corresponding glycosyltransferases in a single reaction vessel for the formation of the corresponding oligosaccharides. Multiple OPME systems can be used in sequence to build up more complex oligosaccharides.[269,272] The high efficiency of the systems is facilitated by the elucidation of novel salvage pathways of sugar nucleotide biosynthesis, as well as the identification and characterization of new bacterial glycosyltransferases and mutants with high expression levels in *E. coli*, good solubility and stability, and high activity.

Much progress has been made recently in identifying glycosyltransferase mutants with improved functions, and many of these successes are based on protein crystal structure-based rational design, and some are from directed evolution coupled with high-throughput screening methods.[262,273] These are effective approaches for obtaining additional or better catalysts that are not readily available from nature.

If not all glycosyltransferases that are responsible for formation of desired HMOS are available, enzymatically synthesized oligosaccharide derivatives can be used as building blocks (or synthons) for chemical synthesis of more complex HMOS and derivatives. Such chemoenzymatic methods have been explored for the synthesis of sialyl galactosides,[274,275] sialyl Lewis^{X}

tetrasaccharides,[276,277] protected sialyllacto-*N*-tetraose a (LSTa, Neu5-Acα2–3LNT) and LSTd (Neu5Acα2–3LNnT,[126,263] has not been found in human milk),[278] and an LNT derivative.[279] The LNT derivative so obtained was further used as a glycosyltransferase acceptor for the production of LSTa derivatives by OPME enzymatic sialylation.[279]

A limited number of HMOS have also been synthesized by whole-cell synthesis and engineered *E. coli* living-cell fermentation approaches. Both approaches make good use of the microorganism's own metabolic machinery for the production of some components (such as nucleotides, monosaccharides, and/or sugar nucleotides) from less expensive materials (simple carbon and energy sources such as glycerol or glucose). One of the limitations of the living-cell system is the restriction of the oligosaccharide transporter systems for transfer acceptors from external sources into the cells for product formation.[262]

Alternative enzymatic synthetic strategies using glycosidases, transglycosidases, and glycosidase mutants designed for synthesizing carbohydrates (e.g., glycosynthase[280]) have also been developed for obtaining HMOS. These methods require the use of glycosylated donors that may not be readily available. The synthetic donors used have to be chemically synthesized and may not be stable.[281,282] Low yields and poor regioselectivity are also common problems for glycosidase-catalyzed reactions. Strategies to improve the trans-glycosylation reactions of glycosidases including controlling acceptor/donor ratio and reaction time, removing product continuously, enzyme immobilization and recycling, using cosolvents, and enzyme engineering have been recently reviewed.[283]

Examples of HMOS that have been synthesized as their natural free-hydroxy, oligosaccharide forms with a free reducing end using enzymatic, whole-cell, and living-cell approaches are shown in the following sections.

5.1 2′FL

Enzymatic production of 2′FL (18 mg, 65% yield) from lactose was achieved using a reaction catalyzed by *Helicobacter pylori* NCTC 364 α1–2-fucosyltransferase (glutathione S-transferase (GST) fusion was shown to improve the expression of soluble protein)[284] using GDP-L-fucose (78 mg, 78% yield) produced from GDP-D-mannose by enzymatic reactions catalyzed by *E. coli* K-12 GDP-D-mannose 4,6-dehydratase and GDP-4-keto-6-deoxy-D-mannose 3,5-epimerase-4-reductase.[285]

Living-cell biosynthesis of 2′FL (1.23 g/L, 20% yield) from lactose (14.5 g/L) in batch fermentation was achieved using *E. coli* JM109(DE3) cells engineered to overexpress *Helicobacter pylori* 26695 strain (ATCC 700392) α1–2-fucosyltransferase[286] and overproduce GDP-fucose.[287] The production of 2′FL (6.4 g/L) was further improved using an engineered *Helicobacter pylori* α1–2-fucosyltransferase by adding three aspartate residues at its N-terminus in an alternative expression host obtained by engineering the *E. coli* BL21star(DE) strain to delete its endogenous lactose operon and to introduce a *lacZΔM15*-containing modified lactose operon from *E. coli* K-12.[288]

An improved large-scale production of 2′FL (20 g/L) from lactose and glycerol was achieved using an antibiotic-free, fed-batch fermentation (13 L) of engineered *E. coli* JM109 (*lacY*$^+$, *lacZ*$^-$) cells. The cells were engineered by chromosome incorporation of genes involved in the *de novo* GDP-L-fucose biosynthetic pathway, two copies of *Helicobacter pylori* alpha1-2-fucosyltransferase *futC* gene, and *Bacteroides fragilis* bifunctional fucokinase/GDP-fucose pyrophosphorylase *fkp* gene involved in the salvage pathway of GDP fucose formation.[289]

2′FL production was also achieved in the milk of transgenic mice by introducing to mice a fusion gene containing a human α1–2-fucosyltransferase gene downstream of a murine whey acidic protein promoter and upstream of a polyadenylation signal.[142] The same transgenic manipulation on rabbits seemed to interfere with their lactation process.[290] The presence of glycoconjugates containing the Fucα1–2Gal epitope reduces the rate and duration of pathogen colonization in pups inoculated with pathogenic strains of *Campylobacter jejuni*.[204]

Several α1–2-fucosynthases were obtained from *Bifidobacterium bifidum* α1–2-fuocisdase (AfcA), an inverting glycosidase, by mutating the amino acid residues involved in catalysis (N421G, N423G, or D766G).[183,291] The D766G mutant was found to be the most effective enzyme in catalyzing the synthesis of 2′FL from β-L-fucosyl fluoride (10 mM) and lactose (30 mM). A 6% yield was obtained based on the β-L-fucosyl fluoride donor substrate used.[292]

5.2 3′SL and 3′SLN

Neu5Acα2–3Lac (3′SL) and Neu5Acα2–3LacNAc (3′SLN) were synthesized using a "one-pot" three-enzyme (OP3E) system containing an *E. coli* sialic acid aldolase (EcNanA),[262,293] *Neisseria meningitidis* CMP-sialic acid synthetase (NmCSS),[293] and a multifunctional *Pasteurella multocida* α2–3-sialyltransferase 1 (PmST1).[294] The amount of the enzyme used and the reaction time both needed to be controlled to allow for the optimal production of the product due to the multifunctionality of PmST1.

The synthesis can be improved by replacing the wild-type PmST1 with a PmST1 E271F/R313Y double mutant, which has retained its α2–3-sialyltransferase activity with >6000-fold decreased α2–3-sialidase activity.[295] A PmST1 M144D mutant with decreased donor hydrolysis and lowered α2–3-sialidase activities[296] can also be used for the highly efficient synthesis of 3′SL and 3′SLN. The sialosides can also be synthesized from Neu5Ac and a suitable acceptor using a "one-pot" two-enzyme system containing NmCSS and a sialyltransferase.

Production of 3′SL and 3′SLN has also been reported from CMP-Neu5-Ac and lactose catalyzed by a *Pasteurella multocida* α2–3-sialyltransferase[297] or *Pasteurella dagmatis* α2–3-sialyltransferase.[298,299]

The α2–3-trans-sialidase activity of *Pasteurella multocida* α2–3-sialyltransferase (GenBank accession number AAK02272) (PmST), which differs from the PmST1 protein sequence by three amino acid residues (N105D, Q135R, and E295G) and has α2–3- and α2–6-dual trans-sialidase activities, was used for the synthesis of 3′SL from lactose and casein glycomacropeptide (whey protein). The product 3′SL was accumulated up to 2.75 mM from lactose (100 mM) and 5% (w/v) casein glycomacropeptide (containing 9 mM bound sialic acid) under optimal conditions at pH 6.4 and 40 °C for 6 h.[300]

The trans-sialidase activities of *Bacteroides fragilis* sialidase,[301] *Arthrobacter ureafaciens* or *Bifidobacterium infantis* sialidase,[302] and *Trypanosoma cruzi* α2–3-trans-sialidase have also been explored for the synthesis of 3′SL.[303,304] Low or moderate yields were achieved.

A fusion protein of NmCSS and *Neisseria meningitidis* α2–3-sialyltransferase (Nmα2–3ST) was used in a sugar nucleotide regeneration reaction for the synthesis of 3′SL (68 g in a partially purified solid form, 68% yield) on a 100-gram scale from lactose, Neu5Ac, phosphoenolpyruvate, and catalytic amounts of ATP and CMP.[305]

Large-scale production of 3′SL was also achieved using a whole-cell approach.[306] In this process, *Corynebacterium ammoniagenes* DN510 cells (for the production of UTP from inexpensive orotic acid and converting CMP to CDP) and three recombinant *E. coli* strains (containing *E. coli* K12 CTP synthetase, *E. coli* K1 CMP-Neu5Ac synthetase, and *Neisseria gonorrhoeae* α2–3-sialyltransferase, respectively) were permeabilized by treating cell pellets with polyoxyethylene octadecylamine (Nymeen S-215) and dimethylbenzenes (xylene). Multiple grams of 3′SL (0.99 g, 36% yield and 72 g, 44% yield) were synthesized from lactose, Neu5Ac, and orotic acid at 32 °C for 11 h.[306]

3′SL (2.6 g/L, 49% yield) has also been produced from Neu5Ac and lactose fed to living *E. coli* (*lacY*$^+$, *lacZ*$^-$, *nanT*$^+$, *nanA*$^-$) cells engineered to express *N. meningitidis* CMP-Neu5Ac synthetase (NmCSS) and an *N. meningitidis* L3 strain MC58 α2–3-sialyltransferase (Nm2–3ST). The knockout of *lacZ*$^-$ and *nanA*$^-$ genes was to ensure that the lactose and Neu5Ac fed to the cells were not broken down by the β-galactosidase and sialic acid aldolase, respectively. Neu5Ac permease NanT and β-galactoside permease LacY, endogenous to the *E. coli* host cells, were responsible for transporting exogenous Neu5Ac and lactose, respectively, into *E. coli* cells for the production of 3′SL.[307]

To decrease the cost for 3′SL production, the engineered 3′SL biosynthetic *E. coli* K12 cells were modified further by deleting ManNAc kinase *nanK* gene and incorporating plasmids for the expression of *Campylobacter jejuni* strain ATCC43438 *neuABC* genes encoding GlcNAc-6-phosphate 2 epimerse, sialic acid synthase, and CMP-Neu5Ac synthetase to produce CMP-Neu5Ac from endogenous UDP-GlcNAc and avoid the need of exogenous Neu5Ac. Using this improved, engineered bacterial strain, a higher concentration (25 g/L) of 3′SL was obtained.[308]

5.3 6′SL and 6′SLN

Neu5Acα2–6LacNAc (6′SLN) was synthesized using an OP3E sialylation system similar to that described above for the synthesis of 3′SL and 3′SLN, except for replacing the PmST1 with *Photobacterium damselae* α2–6-sialyltransferase (Pd2,6ST).[309] Neu5Acα2–6Lac (6′SL) can also be synthesized similarly using the same OP3E system as shown for the synthesis of 6′SL derivatives.

Both 6′SL and 6′SLN have been synthesized from CMP-Neu5Ac and lactose using a *Pasteurella dagmatis* α2–3-sialyltransferase P7H/M117A double mutant that was completely switched to an α2–6-sialyltransferase.[298]

More recently, 6′SL (3.33 mM) was synthesized from lactose (100 mM) and casein glycomacropeptide (containing 9 mM bound sialic acid) at pH 5.4 and 40 °C for 8 h using the α2–6-trans-sialidase activity of PmST (GenBank accession number AAK02272), which has dual α2–3- and α2–6-trans-sialidase activities.[300] A PmST1 P34H mutant with enhanced α2–6-trans-sialidase activity was used to further improve the regioselective production of 6′SL versus 3′SL.

6′SL was also produced, together with its disialylated derivative, 6,6′-disialyllactose, using a living-cell system engineered to overexpress

Photobacterium sp. JT-ISH-224 α2–6-sialyltransferase (Psp2,6ST)[304,310] with *Campylobacter jejuni* strain ATCC43438 *neuABC* genes encoding GlcNAc-6-phosphate 2-epimerse, sialic acid synthase, and CMP-Neu5Ac synthetase.[311] A 6′SL derivative Kdoα2–6Lac was also produced using a similar system with the Psp2,6ST gene under the control of a strong Ptrc promoter and *neuABC* genes under the control of a weaker Plac promoter.[311]

5.4 LNT2, LNnT, LNnH, LNnO, LNnD, LSTd, and Disialyl Oligosaccharides

Recently, two β-*N*-acetylhexosaminidases, HEX1 and HEX2, identified from soil-derived, metagenomic library screening, were found to catalyze trans-glycosylation reactions using chitin oligosaccharides as donor substrates and lactose as the acceptor for the formation of lacto-*N*-triose II (LNT2, GlcNAcβ1–3Lac),[312] the precursor for the synthesis of LNT and LNnT. Although the yields are low (2% and 8%, respectively), they have the potential for improvement by mutagenesis.

LNT2 (106.3 mg) was also synthesized from lactose and UDP-GlcNAc catalyzed by bovine serum β1–3-*N*-acetylglucosaminyltransferase. LNnT (12 mg) was subsequently produced from LNT2 and *ortho*-nitrophenyl β-galactoside by a commercially available *Bacillus circulans* β-D-galactosidase.[313]

Large-scale production of LNT2 trisaccharide and LNnT in several hundred gram quantities in a 100-L reactor has been reported. LNT2 trisaccharide (250 g) was synthesized from lactose and UDP-GlcNAc using *E. coli* cells expressing β1–3-*N*-acetylglucosaminyltransferase (LgtA). LNnT (300 g, >85% yield) was subsequently synthesized from LNT2 and UDP-galactose using *E. coli* cells expressing β1–4GalT (LgtB). Sialyllacto-*N*-tetraose d (LSTd, Neu5Acα2–3LNnT, which has not been identified in human milk) was further produced in 50-g quantity with a 90% yield from LNnT and Neu5Acα2–3Lac using a recombinant *Trypanosoma cruzi* α2–3-trans-sialidase expressed in *E. coli*.[263]

LNT2 and LNnT were reported to be produced in kilogram quantities in a fermentation-based system to allow the conduct of clinical trials.[290,314] At the tested concentration, LNnT was proven to be stable and safe to use as a component of infant formula, although it did not reduce oropharyngeal colonization of *Streptococcus pneumoniae* in children of age 6 months or older.[315]

LNnT was also synthesized from 1-thio-β-LNT2 conjugated to a polyethylene glycol (PEG)-based dendrimeric support and UDP-Glc using reactions catalyzed by UDP-Gal 4-epimerase and bovine milk β1–4GalT. In this system, the UDP-Gal 4-epimerase was responsible for the formation of UDP-Gal from less expensive UDP-Glc, thus providing a donor substrate for the bovine milk β1–4GalT for the formation of LNnT. The thio-linked PEG support was readily cleaved off using mercury(II) trifluoroacetate ($(CF_3CO_2)_2Hg$, 2 equiv) in acetic acid (0.05 M) at room temperature to release free LNnT (18 mg).[316]

Large-scale production of LNT2 (6 g/L, 73% yield) and LNnT (>5 g/L), and lower level formation of lacto-*N*-neohexaose (LNnH), lacto-*N*-neooctaose (LNnO), and even lacto-*N*-neodecaose (LNnD) were reported using living *E. coli* JM109 cells (*lacY*+ *lacZ*$^-$) engineered to overexpress *Neisseria meningitidis* β1–3-*N*-acetylglucosaminyltransferase (NmLgtA) and *Neisseria meningitidis* β1–4GalT (NmLgtB).[307]

Enzymatic syntheses of LNT2 (1.36 g, 95% yield), LNnT (1.19 g, 92% yield), and disialyl glycans were successfully achieved using sequential "one-pot" multienzyme (OPME) systems as shown in Figs. 1 and 2.[269] In these systems, free monosaccharides were added one by one to each of the "one-pot" systems containing multiple enzymes responsible for catalyzing monosaccharide activation, followed by transfer processes. Multiple OMPE systems were used sequentially for building up complex HMOS structures. The combination of several OPME systems was used for the synthesis of disialyl oligosaccharides milk including DSLNnT (236 mg, 99% yield), GD3 tetraose (239 mg, 82% yield), DSLac (112 mg, 93% yield), and DS′LNT (268 mg, 98% yield), all of which are analogs of disialyl lacto-*N*-tetraose (DSLNT), a hexaose commonly found in human milk. A monosialylpentaose LSTd (or 3‴-sLNnT) (138 mg, 98% yield) was synthesized similarly using sequential OPME systems.[269] Similar to DSLNT and the HMOS pool,[236] both synthetic DSLNnT and DS′LNT were shown to protect neonatal rats from necrotizing enterocolitis.[269]

5.5 Fucα1–2LNnT

Fucα1–2LNnT, a monofucosylated pentaose that has not been identified from human milk, was produced together with 2′FL using *E. coli* living cells engineered to overproduce GDP-fucose[325] and LNnT[307] with an additional introduction of a modified *H. pylori* strain 26695 α1–2-fucosyltransferase.[326]

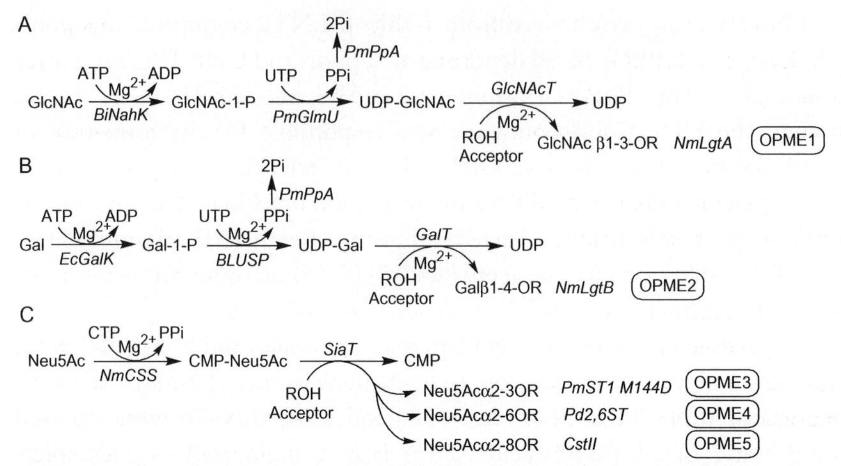

Figure 1 "One-pot" multienzyme (OPME) GlcNAc (A), Gal (B), and Neu5Ac (C) activation and transfer systems.[269,270] Enzyme abbreviations: BiNahK, *B. infantis* *N*-acetylhexosamine-1-kinase[317]; PmGlmU, *P. multocida* *N*-acetylglucosamine 1-phosphate uridylyltransferase[318]; PmPpA, *P. multocida* inorganic pyrophosphatase[318]; GlcNAcT, *N*-acetylglucosaminyltransferase; NmLgtA, *Neisseria meningitidis* β1–3-*N*-acetylglucosaminyltransferase[319]; EcGalK, *E. coli* K-12 galactose kinase[320]; BLUSP: *B. longum* UDP-sugar synthase[320]; GalT, galactosyltransferase; NmLgtB, *Neisseria meningitidis* β1–4-galactosyltransferase[321]; NmCSS, *Neisseria meningitidis* CMP-sialic acid synthetase[293]; SiaT, sialyltransferases; PmST1 M144D, *Pasteurella multocida* α2–3-sialyltransferase 1 M144D mutant[296]; Pd2,6ST, *Photobacterium damselae* α2–6-sialyltransferase[309,322]; CstII, *Campylobacter jejuni* α2–8-sialyltransferase II.[323,324]

5.6 LNFP III, LNnFP V, and LNnDFH

Lacto-*N*-neofucopentaose (LNnFP V), lacto-*N*-neodifucohexaose (LNnDFH), and a lacto-*N*-neodifucooctaose [Galβ1–4GlcNAcβ1–3Galβ1–4(Fucα1–3)GlcNAcβ1–3Galβ1–4(Fucα1–3)Glc] have been synthesized from lactose using living *E. coli* cells engineered to inactivate the *wcaJ* gene involved in colanic acid synthesis and to express NmLgtA, NmLgtB, *Helicobacter pylori* strain 26695 α1–3-fucosyltransferase FutA (encoded by the *HP0379* gene), and RcsA (a positive regulator of the colanic acid operon). Glucose was used as the carbon source.[325] The construct was further modified to improve the yield for the synthesis of LNnDFH (1.7 g/L). In addition, the living-cell system containing another *Helicobacter pylori* strain 26695 α1–3-fucosyltransferase *futB* gene (*HP0651*) was shown to produce both lacto-*N*-neofucopentaose III (LNFP III) (260 mg/L) and LNnFP V (280 mg/L).[327]

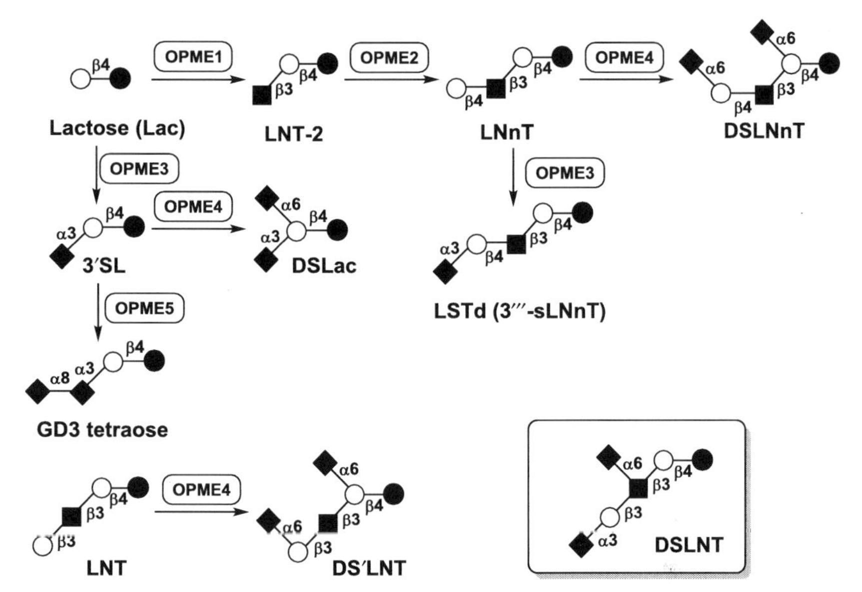

Figure 2 OPME and sequential OPME systems for the synthesis of disialyl oligosaccharides including DSLNnT, GD3 tetraose, DSLac, DS′LNT, and a monosialylpentaose LSTd (3‴-sLNnT).[269] The structure of DSLNT found in human milk is shown for comparison purposes.

5.7 LNT

LNT was enzymatically synthesized from LNT2 and *ortho*-nitrophenyl β-D-galactoside using a *Bacillus circulans* ATCC31382 β-galactosidase-catalyzed trans-glycosylation reaction. Alternatively, LNT (7.1 mg) was synthesized from lactose and Galβ1–3GlcNAcβ*p*NP using an *Aureobacterium* sp. L-101 lacto-*N*-biosidase-catalyzed trans-glycosylation reaction.[313] Inherent low yields (19–26%) were observed for typical glycosidase-catalyzed trans-glycosylation reactions.

An LNT benzyl glycoside was efficiently produced from LNT2 benzyl glycoside (synthesized by an NmLgtA-catalyzed glycosylation reaction from lactose benzyl glycoside and UDP-GlcNAc) and UDP-Gal using a GST-tagged *Escherichia coli* O55:H7 β1–3-*N*-acetylglucosaminyltransferase WbgO fusion protein.[313]

Large-scale production of LNT was not achieved until recently using *E. coli* strain LJ110 (with intact LacY but with *lacZ* knockout) chromosomally integrated with *Neisseria meningitidis* β1–3-*N*-acetylglucosaminyltransferase *lgtA* and *Escherichia coli* O55:H7 β1–3-*N*-acetylglucosaminyltransferase *wbgO* genes.[328] Nevertheless, when glucose was used as the carbon source, LNT2 was the major product and only about 5% of the lactose was converted to LNT (219 mg/L).[328] By substituting the glucose with galactose, the yield of LNT production (811 mg/L) was improved by 3.6-fold. Fed-batch cultivation with galactose further improved the efficiency and produced 173 g (12.72 g/L) of LNT.[329]

5.8 3FL, LDFT, LNFP II, Le^a Tetrasaccharide, and Le^X Tetrasaccharide

Several α1–3/4-fucosynthases were obtained from *Bifidobacterium bifidum* α1–3/4-fuocisdase (BbAfcB), a retaining glycosidase, by mutating the amino acid residue that was predicted to serve as a nucleophile (D703). Among the four mutants, D703A, D703C, D703G, and D703S, the D703S mutant was found to be the best α1–3/4-fucosynthase and was used for the production of several fucosylated HMOS and derivatives using β-L-fucosyl fluoride (40 mM) and a suitable acceptor (100 mM) such as Lac, 2′FL, LNT, LNB, and LacNAc. HMOS and derivatives 3FL, LDFT, LNFP II, Le^a tetrasaccharide, and Le^X tetrasaccharide were obtained in 13%, 5.5%, 41%, 47%, and 55% yields, respectively, based on the β-L-fucosyl fluoride donor substrate used. Increasing the LNB concentration to 200 mM was able to improve the yield to 56%.[330]

5.9 LNFP I and LNDFH I

LNFP I (7% yield) and LNDFH I (6% yield) were synthesized from lactose using several glycosyltransferase-catalyzed reactions with the corresponding sugar nucleotides and a galactosidase-catalyzed reaction with a corresponding synthetic donor. LNT2 (44% yield) was initially synthesized from lactose and UDP-GlcNAc catalyzed by a β1–3GlcNAcT that was partially purified from bovine blood. LNT (22% yield) was then produced from LNT2 and *o*-nitrophenyl-β-galactoside (Galβ*o*NP) using a recombinant *Bacillus circulans* β1–3-galactosidase. The production of LNFP I (71% yield) was achieved from LNT and GDP-Fuc using a recombinant human α1–2-fucosyltransferase 1 (FUT1) expressed in a baculovirus system. Finally, LNDFH I (85% yield) was produced from LNFP I and GDP-Fuc by a FUT3-catalyzed reaction using a commercial enzyme.[210]

5.10 Other Oligosaccharides

Gram-scale production of globotriose (Gb_3) and globotetraose (Gb_4) oligosaccharides was achieved using bacterial glycosyltransferases and sugar nucleotides. Gb_3 trisaccharide (5 g, 75% yield) was synthesized from lactose and UDP-galactose using *Neisseria gonorrhoeae* α1–4-galactosyltransferase (NgLgtC). Gb_4 tetrasaccharide (1.5 g, 60% yield) was synthesized from Gb_3 trisaccharide and UDP-GalNAc using *Neisseria gonorrhoeae* β1–3-*N*-acetylgalactosaminyltransferase (NgLgtD).[263]

UDP-galactose (44 g/L) and globotriose Galα1–4Lac (188 g/L) were also produced using permeabilized *Corynebacterium ammoniagenes* cells (for the production of UTP from orotic acid) and *E. coli* cells engineered to overexpress UDP-Gal biosynthetic genes with or without *Neisseria gonorrhoeae* α1–4-galactosyltransferase.[331] A whole-cell approach using permeabilized *Corynebacterium ammoniagenes* cells (for the production of GTP from GMP) and *E. coli* cells engineered to overexpress de novo GDP-Fuc biosynthetic enzymes, phosphoglucomutase, phosphofructokinase, and *Helicobacter pylori* α1–3-fucosyltransferase was also developed for the production of $Lewis^X$ trisaccharide (21 g/L on a 30-mL scale, purification yield 32%) from GMP, mannose, and *N*-acetyllactosamine (LacNAc).[332]

The engineered *E. coli* living-strategy can be used for the synthesis of other HMOS by introducing plasmids for the expression of necessary glycosyltransferases and sugar nucleotide biosynthetic enzymes. For example, the gram-scale synthesis of $Lewis^X$ tetrasaccharides Galβ1–4(Fucα1–3)GlcNAcβ1–4GlcNAc (0.62 g) and Galβ1–4(Fucα1–3)GlcNAcβ1–3Gal (1.84 g) was achieved by placing four de novo GDP-Fuc biosynthetic genes under the *Plac* promoter without knocking out *wcaJ* involved in colanic acid synthesis or introducing RcsA (a positive regulator of the colanic acid operon).[333] The engineered living-cell strategy to produce the oligosaccharide components of gangliosides GM2 and GM1, GalNAcβ1–4(Neu5-Acα2–3)Lac and Galβ1–3GalNAcβ1–4(Neu5Acα2–3)Lac, has been achieved.[334]

6. PERSPECTIVES

Future efforts in HMOS research should be focused on elucidating the structures of more complex and longer chain HMOS, including those with more than 14 monosaccharide residues that are not currently presented in Table 4. These more complex structures are unlikely the by-products of the biosynthesis of HMOS. Although less abundant, they could have significant biological functions as yet unknown. Profiling HMOS in a high-throughput format will also help to find correlation of disease states and the roles of certain populations of HMOS. Various enzyme-catalyzed synthetic methods that have been successfully used in production of relatively simple HMOS include glycosyltransferase-catalyzed reactions with or without cofactor recycling, sialidase- and trans-glycosidase-catalyzed reactions, "one-pot" multienzyme (OPME) and sequential OPME systems, whole-cell approaches, and living-cell strategies. Such efforts, however, have not been applied for the production of more complex structures, especially the ones with branches. With the limited access to all enzymes that are needed in research, especially essential glycosyltransferases, a combination of chemical and enzymatic methods can be used. Such methods have been developed but have not been broadly used for the synthesis of HMOS. The strategy of chemoenzymatic synthesis of asymmetrically branched *N*-glycan structures[335,336] can be readily applicable for the synthesis of branched HMOS structures. Efficient purification methods for large-scale production of HMOS, either by chemical, enzymatic, or cell-based systems, are also in a great demand. The availability of structurally defined, more complex, individual HMOS species will greatly facilitate their functional studies and help to explore their prebiotic and therapeutic potentials. Other than functional studies using pooled HMOS and individual pure synthetic HMOS, the synergistic effect of a mixture of two or more structurally defined HMOS should also be investigated as, most likely, a single compound will not provide the desired protection against multiple pathogens.[15]

ACKNOWLEDGMENTS

X.C. expresses her sincerest appreciation for the invitation to write on the subject from the late Prof. Derek Horton, who provided much-needed instructions, but unfortunately could not see the submission of this work. X.C. is also grateful for the editing provided by Prof. David C. Baker. Research activities on milk oligosaccharide projects in the Chen group were supported, in part, by the National Institutes of Health (NIH) Grant R01HD065122.

REFERENCES

1. Smilowitz, J. T.; Lebrilla, C. B.; Mills, D. A.; German, J. B.; Freeman, S. L. Breast Milk Oligosaccharides: Structure–Function Relationships in the Neonate. *Annu. Rev. Nutr.* **2014**, *34*, 143–169.
2. Newburg, D. S. Oligosaccharides and Glycoconjugates in Human Milk: Their Role in Host Defense. *J. Mammary Gland Biol. Neoplasia* **1996**, *1*, 271–283.
3. Jenness, R. The Composition of Human Milk. *Semin. Perinatol.* **1979**, *3*, 225–239.
4. Jensen, R. G.; Blanc, B.; Patton, S. The Structure of Milk: Implications for Sampling and Storage. B. Particulate Constituents in Human and Bovine Milks. In: *Handbook of Milk Composition*; Jensen, R. G., Ed.; Academic Press, Inc: San Diego, CA, 1995; pp 50–62.

5. Zopf, D. A.; Smith, D. F.; Drzeniek, Z.; Tsai, C. M.; Ginburg, V. Affinity Purification of Antibodies Using Oligosaccharide-Phenethylamine Derivatives Coupled to Sepharose. *Methods Enzymol.* **1978**, *50*, 171–175.
6. Smith, D. F.; Prieto, P. A.; Torres, B. V. Rabbit Antibodies Against the Human Milk Sialyloligosaccharide Alditol of LS-Tetrasaccharide a (NeuAcα2–3Galβ1–3GlcNAcβ1–3Galβ1–4GlcOH). *Arch. Biochem. Biophys.* **1985**, *241*, 298–303.
7. Prieto, P. A.; Smith, D. F. A New Ganglioside in Human Meconium Detected with Antiserum Against Human Milk Sialyltetrasaccharide a. *Arch. Biochem. Biophys.* **1986**, *249*, 243–253.
8. Coppa, G. V.; Pierani, P.; Zampini, L.; Carloni, I.; Carlucci, A.; Gabrielli, O. Oligosaccharides in Human Milk During Different Phases of Lactation. *Acta Paediatr. Suppl.* **1999**, *88*, 89–94.
9. Montreuil, J.; Mullet, S. Etude des variations des constituants glucidiques du lait de femme au cours de la lactation. *Bull. Soc. Chim. Biol.* **1960**, *42*, 365–377.
10. Chaturvedi, P.; Warren, C. D.; Altaye, M.; Morrow, A. L.; Ruiz-Palacios, G.; Pickering, L. K.; Newburg, D. S. Fucosylated Human Milk Oligosaccharides Vary Between Individuals and over the Course of Lactation. *Glycobiology* **2001**, *11*, 365–372.
11. Gabrielli, O.; Zampini, L.; Galeazzi, T.; Padella, L.; Santoro, L.; Peila, C.; Giuliani, F.; Bertino, E.; Fabris, C.; Coppa, G. V. Preterm Milk Oligosaccharides During the First Month of Lactation. *Pediatrics* **2011**, *128*, e1520–e1531.
12. Viverge, D.; Grimmonprez, L.; Cassanas, G.; Bardet, L.; Bonnet, H.; Solere, M. Variations of Lactose and Oligosaccharides in Milk from Women of Blood Types Secretor A or H, Secretor Lewis, and Secretor H/Nonsecretor Lewis During the Course of Lactation. *Ann. Nutr. Metab.* **1985**, *29*, 1–11.
13. Kobata, A.; Yamashita, K.; Tachibana, Y. Oligosaccharides from Human Milk. *Methods Enzymol.* **1978**, *50*, 216–220.
14. Thurl, S.; Henker, J.; Siegel, M.; Tovar, K.; Sawatzki, G. Detection of Four Human Milk Groups with Respect to Lewis Blood Group Dependent Oligosaccharides. *Glycoconjugate J.* **1997**, *14*, 795–799.
15. Newburg, D. S.; Ruiz-Palacios, G. M.; Morrow, A. L. Human Milk Glycans Protect Infants Against Enteric Pathogens. *Annu. Rev. Nutr.* **2005**, *25*, 37–58.
16. Polonowski, M.; Lespagnol, A. Nouvelles acquisitions sur les composes glucidiques du lai de femme. *Bull. Soc. Chim. Biol.* **1933**, *15*, 320–349.
17. Polonowski, M.; Lespagnol, A. Sur la Nature Glucidique de la Substance Levogyre du Lait de Femme. *Bull. Soc. Biol.* **1929**, *101*, 61–63.
18. Polonowski, M.; Montreuil, J. Etude Chromatographique des Polyosides du Lai de Femme. *C. R. Acad. Sci.* **1954**, *238*, 2263–2264.
19. Schönfeld, H. Über die Beziehungen der Einzelnen Bestandteile der Frauenmilch zur Bifidusflora. *Jahrb. Kinderheilkd.* **1929**, *113*, 19–69.
20. Rose, C. S.; Kuhn, R.; Zilliken, F.; György, P. Bifidus Factor. V. The Activity of α- and β-Methyl-*N*-Acetyl-D-Glucosaminides. *Arch. Biochem. Biophys.* **1954**, *49*, 123–129.
21. György, P.; Norris, R. F.; Rose, C. S. Bifidus Factor. I. A Variant of *Lactobacillus bifidus* Requiring a Special Growth Factor. *Arch. Biochem. Biophys.* **1954**, *48*, 193–201.
22. György, P.; Kuhn, R.; Rose, C. S.; Zilliken, F. Bifidus Factor. II. Its Occurrence in Milk from Different Species and in Other Natural Products. *Arch. Biochem. Biophys.* **1954**, *48*, 202–208.
23. Gauhe, A.; György, P.; Hoover, J. R.; Kuhn, R.; Rose, C. S.; Ruelius, H. W.; Zilliken, F. Bifidus Factor. IV. Preparations Obtained from Human Milk. *Arch. Biochem. Biophys.* **1954**, *48*, 214–224.
24. Kuhn, R.; Baer, H. H. Die Konstitution der Lacto-*N*-tetraose. *Chem. Ber.* **1956**, *89*, 504–511.

25. Kuhn, R.; Baer, H. H.; Gauhe, A. Kristallisierte Fucosido-Lactose. *Chem. Ber.* **1956**, *89*, 2513.
26. Kuhn, R.; Baer, H. H.; Gauhe, A. Kristallisation und Konstituionsermittlung der Lacto-*N*-fucopentaose I. *Chem. Ber.* **1956**, *89*, 2514–2523.
27. Montreuil, J. Structure de Deux Triholosides Isoles du Lai de Femme. *C. R. Hebd. Seances Acad. Sci.* **1956**, *242*, 192–193.
28. Kobata, A. Structures and Application of Oligosaccharides in Human Milk. *Proc. Jpn. Acad. Ser. B Phys. Biol. Sci.* **2010**, *86*, 731–747.
29. Egge, H.; Dell, A.; Von Nicolai, H. Fucose Containing Oligosaccharides from Human Milk. I. Separation and Identification of new Constituents. *Arch. Biochem. Biophys.* **1983**, *224*, 235–253.
30. Ninonuevo, M. R.; Park, Y.; Yin, H.; Zhang, J.; Ward, R. E.; Clowers, B. H.; German, J. B.; Freeman, S. L.; Killeen, K.; Grimm, R.; Lebrilla, C. B. A Strategy for Annotating the Human Milk Glycome. *J. Agric. Food Chem.* **2006**, *54*, 7471–7480.
31. Dai, D.; Nanthkumar, N. N.; Newburg, D. S.; Walker, W. A. Role of Oligosaccharides and Glycoconjugates in Intestinal Host Defense. *J. Pediatr. Gastroenterol. Nutr.* **2000**, *30*(Suppl 2), S23–S33.
32. Newburg, D. S.; Neubauer, S. H. Carbohydrates in Milks: Analysis, Quantities, and Significance. In: *Handbook of Milk Composition*; Jensen, R. G., Ed.; Academic Press: San Diego, 1995; pp 273–349.
33. Wu, S.; Tao, N.; German, J. B.; Grimm, R.; Lebrilla, C. B. Development of an Annotated Library of Neutral Human Milk Oligosaccharides. *J. Proteome Res.* **2010**, *9*, 4138–4151.
34. Wu, S.; Grimm, R.; German, J. B.; Lebrilla, C. B. Annotation and Structural Analysis of Sialylated Human Milk Oligosaccharides. *J. Proteome Res.* **2011**, *10*, 856–868.
35. Newburg, D. S. Glycobiology of Human Milk. *Biochem. Mosc.* **2013**, *78*, 771–785.
36. Engfer, M. B.; Stahl, B.; Finke, B.; Sawatzki, G.; Daniel, H. Human Milk Oligosaccharides Are Resistant to Enzymatic Hydrolysis in the Upper Gastrointestinal Tract. *Am. J. Clin. Nutr.* **2000**, *71*, 1589–1596.
37. Gnoth, M. J.; Kunz, C.; Kinne-Saffran, E.; Rudloff, S. Human Milk Oligosaccharides Are Minimally Digested In Vitro. *J. Nutr.* **2000**, *130*, 3014–3020.
38. Bode, L. Human Milk Oligosaccharides: Every Baby Needs a Sugar Mama. *Glycobiology* **2012**, *22*, 1147–1162.
39. Hinde, K.; Lewis, Z. T. MICROBIOTA. Mother's Littlest Helpers. *Science* **2015**, *348*, 1427–1428.
40. Grulee, C.; Sanford, H.; Schwartz, H. Breast and Artificially fed Infants; Study of the Age Incidence in Morbidity and Mortality in 20,000 Cases. *JAMA* **1935**, *104*, 1986–1988.
41. Morrow, A. L.; Ruiz-Palacios, G. M.; Jiang, X.; Newburg, D. S. Human-Milk Glycans That Inhibit Pathogen Binding Protect Breast-Feeding Infants Against Infectious Diarrhea. *J. Nutr.* **2005**, *135*, 1304–1307.
42. Stepans, M. B.; Wilhelm, S. L.; Hertzog, M.; Rodehorst, T. K.; Blaney, S.; Clemens, B.; Polak, J. J.; Newburg, D. S. Early Consumption of Human Milk Oligosaccharides Is Inversely Related to Subsequent Risk of Respiratory and Enteric Disease in Infants. *Breastfeed. Med.* **2006**, *1*, 207–215.
43. Chichlowski, M.; De Lartigue, G.; German, J. B.; Raybould, H. E.; Mills, D. A. Bifidobacteria Isolated from Infants and Cultured on Human Milk Oligosaccharides Affect Intestinal Epithelial Function. *J. Pediatr. Gastroenterol. Nutr.* **2012**, *55*, 321–327.
44. Etzold, S.; Bode, L. Glycan-Dependent Viral Infection in Infants and the Role of Human Milk Oligosaccharides. *Curr. Opin. Virol.* **2014**, 7, 101–107.
45. Wang, B. Molecular Mechanism Underlying Sialic Acid as an Essential Nutrient for Brain Development and Cognition. *Adv. Nutr.* **2012**, *3*, 465S–472S.

46. Yatsunenko, T.; Rey, F. E.; Manary, M. J.; Trehan, I.; Dominguez-Bello, M. G.; Contreras, M.; Magris, M.; Hidalgo, G.; Baldassano, R. N.; Anokhin, A. P.; Heath, A. C.; Warner, B.; Reeder, J.; Kuczynski, J.; Caporaso, J. G.; Lozupone, C. A.; Lauber, C.; Clemente, J. C.; Knights, D.; Knight, R.; Gordon, J. I. Human Gut Microbiome Viewed Across Age and Geography. *Nature* **2012**, *486*, 222–227.
47. Gibson, G. R.; Wang, X. Regulatory Effects of Bifidobacteria on the Growth of Other Colonic Bacteria. *J. Appl. Bacteriol.* **1994**, 77, 412–420.
48. Tao, N.; Wu, S.; Kim, J.; An, H. J.; Hinde, K.; Power, M. L.; Gagneux, P.; German, J. B.; Lebrilla, C. B. Evolutionary Glycomics: Characterization of Milk Oligosaccharides in Primates. *J. Proteome Res.* **2011**, *10*, 1548–1557.
49. Levene, P. A.; Sobotka, H. Lactone Formation of Lacto- and Maltobionic Acids and Its Bearing on the Structure of Lactose and Maltose. *J. Biol. Chem.* **1927**, *71*, 471–475.
50. Kuhn, R.; Gauhe, A. Die Konstitution der Lacto-*N*-neotetraose. *Chem. Ber.* **1962**, *95*, 518–522.
51. Kobata, A.; Ginsburg, V. Oligosaccharides of Human Milk. 3. Isolation and Characterization of a New Hexasaccharide, Lacto-*N*-hexaose. *J. Biol. Chem.* **1972**, *247*, 1525–1529.
52. Kobata, A.; Ginsburg, V. Oligosaccharides of Human Milk. IV. Isolation and Characterization of a New Hexasaccharide, Lacto-*N*-neohexaose. *Arch. Biochem. Biophys.* **1972**, *150*, 273–281.
53. Yamashita, K.; Tachibana, Y.; Kobata, A. Oligosaccharides of Human Milk. Structural Studies of Two New Octasaccharides, Difucosyl Derivatives of *para*-Lacto-*N*-hexaose and *para*-Lacto-*N*-neohexaose. *J. Biol. Chem.* **1977**, *252*, 5408–5411.
54. Tachibana, Y.; Yamashita, K.; Kobata, A. Oligosaccharides of Human Milk: Structural Studies of Di-and Trifucosyl Derivatives of Lacto-*N*-octaose and Lacto-*N*-neoctaose. *Arch. Biochem. Biophys.* **1978**, *188*, 83–89.
55. Strecker, G.; Fievre, S.; Wieruszeski, J. M.; Michalski, J. C.; Montreuil, J. Primary Structure of Four Human Milk Octa-, Nona-, and Undeca-Saccharides Established by ^{1}H- and ^{13}C-Nuclear Magnetic Resonance Spectroscopy. *Carbohydr. Res.* **1992**, *226*, 1–14.
56. Haeuw-Fievre, S.; Wieruszeski, J. M.; Plancke, Y.; Michalski, J. C.; Montreuil, J.; Strecker, G. Primary Structure of Human Milk Octa-, Dodeca- and Tridecasaccharides Determined by a Combination of ^{1}H-NMR Spectroscopy and Fast-Atom-Bombardment Mass Spectrometry. Evidence for a New Core Structure, the ^{para}-Lacto-*N*-octaose. *Eur. J. Biochem.* **1993**, *215*, 361–371.
57. Kunz, C.; Rudloff, S.; Baier, W.; Klein, N.; Strobel, S. Oligosaccharides in Human Milk: Structural, Functional, and Metabolic Aspects. *Annu. Rev. Nutr.* **2000**, *20*, 699–722.
58. Amano, J.; Osanai, M.; Orita, T.; Sugahara, D.; Osumi, K. Structural Determination by Negative-Ion MALDI-QIT-TOFMSn After Pyrene Derivatization of Variously Fucosylated Oligosaccharides with Branched Decaose Cores from Human Milk. *Glycobiology* **2009**, *19*, 601–614.
59. Kobata, A. Possible Application of Milk Oligosaccharides for Drug Development. *Chang Gung Med. J.* **2003**, *26*, 620–636.
60. Pfenninger, A.; Chan, S. Y.; Karas, M.; Finke, B.; Stahl, B.; Costello, C. E. Mass Spectrometric Detection of Multiple Extended Series of Neutral Highly Fucosylated Acetyllactosamine Oligosaccharides in Human Milk. *Int. J. Mass Spectrom.* **2008**, *278*, 129–136.
61. Urashima, T.; Kitaoka, M.; Terabayashi, T.; Fukuda, K.; Ohnishi, M.; Kobata, A. Milk Oligosaccharides. In: *Oligosaccharides: Sources, Properties, and Applications;* Gordon, N. G. Ed.; Nova Science Publishers: New York, 2011; pp 1–58.

62. Sabharwal, H.; Nilsson, B.; Chester, M. A.; Sjoblad, S.; Lundblad, A. Blood Group Specific Oligosaccharides from Faeces of a Blood Group A Breast-Fed Infant. *Mol. Immunol.* **1984**, *21*, 1105–1112.
63. Strecker, G.; Trentesaux-Chauvet, C.; Riazi-Farzad, T.; Fournet, B.; Bouquelet, S.; Montreuil, J. Demonstration of Oligosaccharidosuria Associated with Various Meliturias, and Determination of the Structure of the Oligosaccharides Excreted. Hypothesis Concerning the Origins of Oligosaccharides in Biological Fluids. *C.R. Hebd. Seances Acad. Sci. Ser. D Sci. Nat.* **1973**, *277*, 1569–1572.
64. Sturman, J. A.; Lin, Y. Y.; Higuchi, T.; Fellman, J. H. *N*-Acetylneuramin Lactose Sulfate: A Newly Identified Nutrient in Milk. *Pediatr. Res.* **1985**, *19*, 216–219.
65. Guerardel, Y.; Morelle, W.; Plancke, Y.; Lemoine, J.; Strecker, G. Structural Analysis of Three Sulfated Oligosaccharides Isolated from Human Milk. *Carbohydr. Res.* **1999**, *320*, 230–238.
66. Kuhn, R.; Gauhe, A. Bestimmung der Bindungsstelle von Sialinsaureresten in Oligosaccharides mit Hilfe von Periodat. *Chem. Ber.* **1965**, *98*, 395–413.
67. Donald, A. S.; Feeney, J. Separation of Human Milk Oligosaccharides by Recycling Chromatography. First Isolation of Lacto-*N*-*neo*-difucohexaose II and 3′-Galactosyllactose from This Source. *Carbohydr. Res.* **1988**, *178*, 79–91.
68. Sugawara, M.; Idota, T. A New Oligosaccharide 4′-Galactosyllactose in Human Milk. In: *Proc. Ann. Meet. Jpn. Soc. Biosci. Biotech. Agrochem., Sappro, Japan* 1995; p 132.
69. Yamashita, K.; Kobata, A. Oligosaccharides of Human Milk V. Isolation and Characterization of a New Trisaccharide, 6′-Galactosyllactose. *Arch. Biochem. Biophys.* **1974**, *161*, 164–170.
70. Erney, R.; Hilty, M.; Pickering, L.; Ruiz-Palacios, G.; Prieto, P. Human Milk Oligosaccharides: A Novel Method Provides Insight into Human Genetics. *Adv. Exp. Med. Biol.* **2001**, *501*, 285–297.
71. Shen, L.; Grollman, E. F.; Ginsburg, V. An Enzymatic Basis for Secretor Status and Blood Group Substance Specificity in Humans. *Proc. Natl. Acad. Sci. U.S.A.* **1968**, *59*, 224–230.
72. Grollman, E. F.; Ginsburg, V. Correlation Between Secretor Status and the Occurrence of 2′-Fucosyllactose in Human Milk. *Biochem. Biophys. Res. Commun.* **1967**, *28*, 50–53.
73. Kobata, A.; Ginsburg, V.; Tsuda, M. Oligosaccharides of Human Milk. I. Isolation and Characterization. *Arch. Biochem. Biophys.* **1969**, *130*, 509–513.
74. Grollman, E. F.; Kobata, A.; Ginsburg, V. An Enzymatic Basis for Lewis Blood Types in Man. *J. Clin. Invest.* **1969**, *48*, 1489–1494.
75. Totten, S. M.; Zivkovic, A. M.; Wu, S.; Ngyuen, U.; Freeman, S. L.; Ruhaak, L. R.; Darboe, M. K.; German, J. B.; Prentice, A. M.; Lebrilla, C. B. Comprehensive Profiles of Human Milk Oligosaccharides Yield Highly Sensitive and Specific Markers for Determining Secretor Status in Lactating Mothers. *J. Proteome Res.* **2012**, *11*, 6124–6133.
76. Dabrowski, U.; Egge, H.; Dabrowski, J. Proton–Nuclear Magnetic Resonance Study of Peracetylated Derivatives of Ten Oligosaccharides Isolated from Human Milk. *Arch. Biochem. Biophys.* **1983**, *224*, 254–260.
77. Kuhn, R.; Gauhe, A. Uber die Lacto-Difuco-Tetraose der Frauenmilch. *Justus Liebigs Ann. Chem.* **1958**, *611*, 249–252.
78. Kuhn, R.; Brossmer, R. Uber das Durch Viren der Influenza-Gruppe Spaltbare Trisaccharid der Milch. *Chem. Ber.* **1959**, *92*, 1667–1671.
79. Kuhn, R. Biochemie der Rezeptoren und Resistenzfaktoren. Von der Widerstandsfähigkeit der Lebewesen Gegen Einwirkungen der Umwelt. *Naturwissenschaften* **1959**, *46*, 43–50.

80. Gronberg, G.; Lipniunas, P.; Lundgren, T.; Erlansson, K.; Lindh, F.; Nilsson, B. Isolation of Monosialyated Oligosaccharides from Human Milk and Structural Analysis of Three New Compounds. *Carbohydr. Res.* **1989**, *191*, 261–278.
81. Kuhn, R.; Baer, H. H.; Gauhe, A. Die Konstitution der Lacto-*N*-fucopentaose II. *Chem. Ber.* **1958**, *91*, 364.
82. Ginsburg, V.; Zopf, D. A.; Yamashita, K.; Kobata, A. Oligosaccharides of Human Milk. Isolation of a New Pentasaccharide, Lacto-*N*-fucopentaose V. *Arch. Biochem. Biophys.* **1976**, *175*, 565–568.
83. Kuhn, R.; Baer, H. H.; Gauhe, A. 2-α-L-Fucopyranosyl-D-galaktose und 2-α-L-Fucopyranosyl-D-talose. *Justus Liebigs Ann. Chem.* **1958**, *611*, 242–249.
84. Kuhn, R.; Gauhe, A. Uber ein Kristallisiertes, Lea-Aktives Hexasaccharid aus Frauenmilch. *Chem. Ber.* **1960**, *93*, 647–651.
85. Grimmonprez, L.; Montreuil, J. Physico-Chemical Study of 6 New Oligosides Isolated from Human Milk. *Bull. Soc. Chim. Biol.* **1968**, *50*, 843–855.
86. Wieruszeski, J. M.; Chekkor, A.; Bouquelet, S.; Montreuil, J.; Strecker, G.; Peter-Katalinic, J.; Egge, H. Structure of Two New Oligosaccharides Isolated from Human Milk: Sialylated Lacto-*N*-fucopentaoses I and II. *Carbohydr. Res.* **1985**, *137*, 127–138.
87. Gronberg, G.; Lipniunas, P.; Lundgren, T.; Lindh, F.; Nilsson, B. Isolation and Structural Analysis of Three New Disialylated Oligosaccharides from Human Milk. *Arch. Biochem. Biophys.* **1990**, *278*, 297–311.
88. Kobata, A.; Ginsburg, V. Oligosaccharides of Human Milk. II. Isolation and Characterization of a New Pentasaccharide, Lacto-*N*-fucopentaose 3. *J. Biol. Chem.* **1969**, *244*, 5496–5502.
89. Perret, S.; Sabin, C.; Dumon, C.; Pokorna, M.; Gautier, C.; Galanina, O.; Ilia, S.; Bovin, N.; Nicaise, M.; Desmadril, M.; Gilboa-Garber, N.; Wimmerova, M.; Mitchell, E. P.; Imberty, A. Structural Basis for the Interaction Between Human Milk Oligosaccharides and the Bacterial Lectin PA-IIL of *Pseudomonas aeruginosa*. *Biochem. J.* **2005**, *389*, 325–332.
90. Kuhn, R.; Gauhe, A. Uber Drei Saure Pentasaccharide aus Frauenmilch. *Chem. Ber.* **1962**, *95*, 513–517.
91. Smith, D. F.; Prieto, P. A.; McCrumb, D. K.; Wang, W. C. A Novel Sialylfucopentaose in Human Milk. Presence of This Oligosaccharide Is not Dependent on Expression of the Secretor or Lewis Fucosyltransferases. *J. Biol. Chem.* **1987**, *262*, 12040–12047.
92. Yamashita, K.; Tachibana, Y.; Kobata, A. Oligosaccharides of Human Milk: Structures of Three Lacto-*N*-hexaose Derivatives with H-Haptenic Structure. *Arch. Biochem. Biophys.* **1977**, *182*, 546–555.
93. Dua, V. K.; Goso, K.; Dube, V. E.; Bush, C. A. Characterization of Lacto-*N*-hexaose and two Fucosylated Derivatives from Human Milk by High-Performance Liquid Chromatography and Proton NMR Spectroscopy. *J. Chromatogr.* **1985**, *328*, 259–269.
94. Kitagawa, H.; Takaoka, M.; Nakada, H.; Fukui, S.; Funakoshi, I.; Kawasaki, T.; Tate, S.; Inagaki, F.; Yamashina, I. Isolation and Structural Studies of Human Milk Oligosaccharides That Are Reactive with a Monoclonal Antibody MSW 113. *J. Biochem.* **1991**, *110*, 598–604.
95. Fievre, S.; Wieruszeski, J. M.; Michalski, J. C.; Lemoine, J.; Montreuil, J.; Strecker, G. Primary Structure of a Trisialylated Oligosaccharide from Human Milk. *Biochem. Biophys. Res. Commun.* **1991**, *177*, 720–725.
96. Gronberg, G.; Lipniunas, P.; Lundgren, T.; Lindh, F.; Nilsson, B. Structural Analysis of Five New Monosialylated Oligosaccharides from Human Milk. *Arch. Biochem. Biophys.* **1992**, *296*, 597–610.
97. Kitagawa, H.; Nakada, H.; Kurosaka, A.; Hiraiwa, N.; Numata, Y.; Fukui, S.; Funakoshi, I.; Kawasaki, T.; Yamashina, I.; Shimada, I.; Inagaki, F. Three Novel

Oligosaccharides with the Sialyl-Lea Structure in Human Milk: Isolation by Immunoaffinity Chromatography. *Biochemistry* **1989**, *28*, 8891–8897.

98. Yamashita, K.; Tachibana, Y.; Kobata, A. Oligosaccharides of Human Milk. Isolation and Characterization of Three New Disialyfucosyl Hexasaccharides. *Arch. Biochem. Biophys.* **1976**, *174*, 582–591.
99. Pfenninger, A.; Karas, M.; Finke, B.; Stahl, B. Structural Analysis of Underivatized Neutral Human Milk Oligosaccharides in the Negative Ion Mode by Nano-Electrospray MS(n) (Part 2: Application to Isomeric Mixtures). *J. Am. Soc. Mass Spectrom.* **2002**, *13*, 1341–1348.
100. Bruntz, R.; Dabrowski, U.; Dabrowski, J.; Ebersold, A.; Peter-Katalinic, J.; Egge, H. Fucose-Containing Oligosaccharides from Human Milk from a Donor of Blood Group 0 Le^a Nonsecretor. *Biol. Chem. Hoppe Seyler* **1988**, *369*, 257–273.
101. Strecker, G.; Wieruszeski, J. M.; Michalski, J. C.; Montreuil, J. Structure of a New Nonasaccharide Isolated Form Human Milk: VI^2-Fuc, V^4Fuc, III^3Fuc-*p*-Lacto-*N*-hexaose. *Glycoconjugate J.* **1988**, *5*, 385–396.
102. Yamashita, K.; Tachibana, Y.; Kobata, A. Oligosaccharides of Human Milk: Isolation and Characterization of Two New Nonasaccharides, Monofucosyllacto-*N*-octaose and Monofucosyllacto-*N*-neooctaose. *Biochemistry* **1976**, *15*, 3950–3955.
103. Kitagawa, H.; Nakada, H.; Fukui, S.; Funakoshi, I.; Kawasaki, T.; Yamashina, I.; Tate, S.; Inagaki, F. Novel Oligosaccharides with the Sialyl-Le^a Structure in Human Milk. *J. Biochem.* **1993**, *114*, 504–508.
104. Kitagawa, H.; Nakada, H.; Fukui, S.; Funakoshi, I.; Kawasaki, T.; Yamashina, I.; Tate, S.; Inagaki, F. Novel Oligosaccharides with the Sialyl-Le^a Structure in Human Milk. *Biochemistry* **1991**, *30*, 2869–2876.
105. Kogelberg, H.; Piskarev, V. E.; Zhang, Y.; Lawson, A. M.; Chai, W. Determination by Electrospray Mass Spectrometry and ^{1}H-NMR Spectroscopy of Primary Structures of Variously Fucosylated Neutral Oligosaccharides Based on the iso-Lacto-*N*-octaose Core. *Eur. J. Biochem.* **2004**, *271*, 1172–1186.
106. Strecker, G.; Wieruszeski, J. M.; Michalski, J. C.; Montreuil, J. Primary Structure of Human Milk Nona- and Decasaccharides Determined by a Combination of Fast-Atom Bombardment Mass Spectrometry and ^{1}H-/^{13}C-Nuclear Magnetic Resonance Spectroscopy. Evidence for a New Core Structure, iso-Lacto-*N*-octaose. *Glycoconjugate J.* **1989**, *6*, 169–182.
107. Chai, W.; Piskarev, V. E.; Zhang, Y.; Lawson, A. M.; Kogelberg, H. Structural Determination of Novel Lacto-N-decaose and Its Monofucosylated Analogue from Human Milk by Electrospray Tandem Mass Spectrometry and ^{1}H NMR Spectroscopy. *Arch. Biochem. Biophys.* **2005**, *434*, 116–127.
108. Lundblad, A.; Hallgren, P.; Rudmark, A.; Svensson, S. Structures and Serological Activities of Three Oligosaccharides Isolated from Urines of Nonstarved Secretors and from Secretors on Lactose Diet. *Biochemistry* **1973**, *12*, 3341–3345.
109. Strecker, G.; Montruil, J. Isolation and Structural Study of 16 Oligosaccharides Isolated from Human Urine. *C.R. Hebd. Seances Acad. Sci. Ser. D Sci. Nat.* **1973**, *277*, 1393–1396.
110. Lundblad, A.; Svensson, S. Letters: The Structure of a Urinary Difucosyl Pentasaccharide, Characteristic of Secretors with the Blood-Group A Gene. *Carbohydr. Res.* **1973**, *30*, 187–189.
111. Kitagawa, H.; Nakada, H.; Numata, Y.; Kurosaka, A.; Fukui, S.; Funakoshi, I.; Kawasaki, T.; Shimada, I.; Inagaki, F.; Yamashina, I. Occurrence of Tetra- and Pentasaccharides with the Sialyl-Le^a Structure in Human Milk. *J. Biol. Chem.* **1990**, *265*, 4859–4862.
112. Stahl, B.; Thurl, S.; Zeng, J.; Karas, M.; Hillenkamp, F.; Steup, M.; Sawatzki, G. Oligosaccharides from Human Milk as Revealed by Matrix-Assisted Laser Desorption/Ionization Mass Spectrometry. *Anal. Biochem.* **1994**, *223*, 218–226.

113. Ballard, O.; Morrow, A. L. Human Milk Composition: Nutrients and Bioactive Factors. *Pediatr. Clin. North Am.* **2013**, *60*, 49–74.
114. Boehm, G.; Stahl, B. Oligosaccharides. In: *Functional Dairy Products*; Mattila-Sandholm, T. Ed.; Woodhead Publishers: Cambridge, UK, 2003; pp 203–243.
115. Thurl, S.; Muller-Werner, B.; Sawatzki, G. Quantification of Individual Oligosaccharide Compounds from Human Milk Using High-pH Anion-Exchange Chromatography. *Anal. Biochem.* **1996**, *235*, 202–206.
116. Asakuma, S.; Urashima, T.; Akahori, M.; Obayashi, H.; Nakamura, T.; Kimura, K.; Watanabe, Y.; Arai, I.; Sanai, Y. Variation of Major Neutral Oligosaccharides Levels in Human Colostrum. *Eur. J. Clin. Nutr.* **2008**, *62*, 488–494.
117. Asakuma, S.; Akahori, M.; Kimura, K.; Watanabe, Y.; Nakamura, T.; Tsunemi, M.; Arai, I.; Sanai, Y.; Urashima, T. Sialyl Oligosaccharides of Human Colostrum: Changes in Concentration During the First Three Days of Lactation. *Biosci. Biotechnol. Biochem.* **2007**, *71*, 1447–1451.
118. Tao, N.; DePeters, E. J.; Freeman, S.; German, J. B.; Grimm, R.; Lebrilla, C. B. Bovine Milk Glycome. *J. Dairy Sci.* **2008**, *91*, 3768–3778.
119. Tao, N.; DePeters, E. J.; German, J. B.; Grimm, R.; Lebrilla, C. B. Variations in Bovine Milk Oligosaccharides During Early and Middle Lactation Stages Analyzed by High-Performance Liquid Chromatography-Chip/Mass Spectrometry. *J. Dairy Sci.* **2009**, *92*, 2991–3001.
120. Tao, N.; Ochonicky, K. L.; German, J. B.; Donovan, S. M.; Lebrilla, C. B. Structural Determination and Daily Variations of Porcine Milk Oligosaccharides. *J. Agric. Food Chem.* **2010**, *58*, 4653–4659.
121. Kawai, Y.; Nakamura, T.; Arai, I.; Saito, T.; Namiki, M.; Yamaoka, K.; Kawahawa, K.; Messer, M. Chemical Characterisation of Six Oligosaccharides in a Sample of Colostrum of the Brown Capuchin, *Cebus apella* (Cebidae: Primates). *Comp. Biochem. Physiol. C Pharmacol. Toxicol. Endocrinol.* **1999**, *124*, 295–300.
122. Oftedal, O. T.; Nicol, S. C.; Davies, N. W.; Sekii, N.; Taufik, E.; Fukuda, K.; Saito, T.; Urashima, T. Can an Ancestral Condition for Milk Oligosaccharides Be Determined? Evidence from the Tasmanian echidna (*Tachyglossus aculeatus setosus*). *Glycobiology* **2014**, *24*, 826–839.
123. Messer, M.; Kerry, K. R. Milk Carbohydrates of the Echidna and the Platypus. *Science* **1973**, *180*, 201–203.
124. Messer, M.; Gadiel, P. A.; Ralston, G. B.; Griffiths, M. Carbohydrates of the Milk of the Platypus. *Aust. J. Biol. Sci.* **1983**, *36*, 129–137.
125. Amano, J.; Messer, M.; Kobata, A. Structures of the Oligosaccharides Isolated from Milk of the Platypus. *Glycoconjugate J.* **1985**, *2*, 121–135.
126. Urashima, T.; Inamori, H.; Fukuda, K.; Saito, T.; Messer, M.; Oftedal, O. T. 4-O-Acetyl-Sialic Acid (Neu4,5Ac2) in Acidic Milk Oligosaccharides of the Platypus (*Ornithorhynchus anatinus*) and Its Evolutionary Significance. *Glycobiology* **2015**, *25*, 683–697.
127. Urashima, T.; Fujita, S.; Fukuda, K.; Nakamura, T.; Saito, T.; Cowan, P.; Messer, M. Chemical Characterization of Milk Oligosaccharides of the Common Brushtail Possum (*Trichosurus vulpecula*). *Glycoconjugate J.* **2014**, *31*, 387–399.
128. Anraku, T.; Fukuda, K.; Saito, T.; Messer, M.; Urashima, T. Chemical Characterization of Acidic Oligosaccharides in Milk of the Red Kangaroo (*Macropus rufus*). *Glycoconjugate J.* **2012**, *29*, 147–156.
129. Urashima, T.; Taufik, E.; Fukuda, R.; Nakamura, T.; Fukuda, K.; Saito, T.; Messer, M. Chemical Characterization of Milk Oligosaccharides of the Koala (*Phascolarctos cinereus*). *Glycoconjugate J.* **2013**, *30*, 801–811.
130. Grollman, E. F.; Kobata, A.; Ginsburg, V. Enzymatic Basis of Blood Types in Man. *Ann. N. Y. Acad. Sci.* **1970**, *169*, 153–160.

131. Grimmonprez, L.; Montreuil, J. Isolation and Physico-Chemical Properties of Oligosaccharides of Human Milk. *Biochimie* **1975**, *57*, 695–771.
132. Newburg, D. S. Human Milk Glycoconjugates That Inhibit Pathogens. *Curr. Med. Chem.* **1999**, *6*, 117–127.
133. Bertino, E.; Peila, C.; Giuliani, F.; Martano, C.; Cresi, F.; Di Nicola, P.; Occhi, L.; Sabatino, G.; Fabris, C. Metabolism and Biological Functions of Human Milk Oligosaccharides. *J. Biol. Regul. Homeost. Agents* **2012**, *26*, 35–38.
134. Sunehag, A. L.; Louie, K.; Bier, J. L.; Tigas, S.; Haymond, M. W. Hexoneogenesis in the Human Breast During Lactation. *J. Clin. Endocrinol. Metab.* **2002**, *87*, 297–301.
135. Mohammad, M. A.; Maningat, P.; Sunehag, A. L.; Haymond, M. W. Precursors of Hexoneogenesis Within the Human Mammary Gland. *Am. J. Physiol. Endocrinol. Metab.* **2015**, *308*, E680–E687.
136. Sasaki, M.; Eigel, W. N.; Keenan, T. W. Lactose and Major Milk Proteins Are Present in Secretory Vesicle-Rich Fractions from Lactating Mammary Gland. *Proc. Natl. Acad. Sci. U. S. A.* **1978**, *75*, 5020–5024.
137. Ramakrishnan, B.; Qasba, P. K. Crystal Structure of Lactose Synthase Reveals a Large Conformational Change in Its Catalytic Component, the β1,4-Galactosyltransferase-I. *J. Mol. Biol.* **2001**, *310*, 205–218.
138. Brodbeck, U.; Denton, W. L.; Tanahashi, N.; Ebner, K. E. The Isolation and Identification of the B Protein of Lactose Synthetase as α-Lactalbumin. *J. Biol. Chem.* **1967**, *242*, 1391–1397.
139. Brew, K.; Hill, R. L. Lactose Biosynthesis. *Rev. Physiol. Biochem. Pharmacol.* **1975**, *72*, 105–158.
140. Kumazaki, T.; Yoshida, A. Biochemical Evidence That Secretor Gene, Se, Is a Structural Gene Encoding a Specific Fucosyltransferase. *Proc. Natl. Acad. Sci. U. S. A.* **1984**, *81*, 4193–4197.
141. Ceppellini, R. On the Genetics of Secretor and Lewis Characters: A Family Study. *Proc. Fifth Intern. Congr. Blood Transfusion, Paris* **1955**, *1954*, 207–211.
142. Prieto, P. A.; Mukerji, P.; Kelder, B.; Erney, R.; Gonzalez, D.; Yun, J. S.; Smith, D. F.; Moremen, K. W.; Nardelli, C.; Pierce, M.; Li, Y.; Chen, X.; Wagner, T. E.; Cummings, R. D.; Kopchick, J. J. Remodeling of Mouse Milk Glycoconjugates by Transgenic Expression of a Human Glycosyltransferase. *J. Biol. Chem.* **1995**, *270*, 29515–29519.
143. Kelder, B.; Erney, R.; Kopchick, J.; Cummings, R.; Prieto, P. Glycoconjugates in Human and Transgenic Animal Milk. *Adv. Exp. Med. Biol.* **2001**, *501*, 269–278.
144. Appert, H. E.; Rutherford, T. J.; Tarr, G. E.; Thomford, N. R.; McCorquodale, D. J. Isolation of Galactosyltransferase from Human Milk and the Determination of Its N-Terminal Amino Acid Sequence. *Biochem. Biophys. Res. Commun.* **1986**, *138*, 224–229.
145. Endo, T.; Amano, J.; Berger, E. G.; Kobata, A. Structure Identification of the Complex-Type, Asparagine-Linked Sugar Chains of β-D-Galactosyl-Transferase Purified from Human Milk. *Carbohydr. Res.* **1986**, *150*, 241–263.
146. Prieels, J. P.; Monnom, D.; Dolmans, M.; Beyer, T. A.; Hill, R. L. Co-Purification of the Lewis Blood Group *N*-Acetylglucosaminide α1 Goes to 4 Fucosyltransferase and an *N*-Acetylglucosaminide α1 Goes to 3 Fucosyltransferase from Human Milk. *J. Biol. Chem.* **1981**, *256*, 10456–10463.
147. Eppenberger-Castori, S.; Lotscher, H.; Finne, J. Purification of the *N*-Acetylglucosaminide α(1-3/4)Fucosyltransferase of Human Milk. *Glycoconjugate J.* **1989**, *6*, 101–114.
148. Johnson, P. H.; Watkins, W. M. Purification of the Lewis Blood-Group Gene Associated α-3/4-Fucosyltransferase from Human Milk: An Enzyme Transferring Fucose Primarily to Type 1 and Lactose-Based Oligosaccharide Chains. *Glycoconjugate J.* **1992**, *9*, 241–249.

149. Johnson, P. H.; Donald, A. S.; Feeney, J.; Watkins, W. M. Reassessment of the Acceptor Specificity and General Properties of the Lewis Blood-Group Gene Associated α3/4-Fucosyltransferase Purified from Human Milk. *Glycoconjugate J.* **1992**, *9*, 251–264.
150. Hosomi, O.; Takeya, A. The Relationship Between the (β 1-3) *N*-Acetylglucosaminyltransferase and the Presence of Oligosaccharides Containing Lacto-*N*-triose II Structure in Bovine and Human Milk. *Nippon Juigaku Zasshi* **1989**, *51*, 1–6.
151. Wiederschain, G. Y.; Newburg, D. S. Compartmentalization of Fucosyltransferase and α-L-Fucosidase in Human Milk. *Biochem. Mol. Med.* **1996**, *58*, 211–220.
152. Palcic, M. M.; Venot, A. P.; Ratcliffe, R. M.; Hindsgaul, O. Enzymic Synthesis of Oligosaccharides Terminating in the Tumor-Associated Sialyl-Lewis[a] Determinant. *Carbohydr. Res.* **1989**, *190*, 1–11.
153. Stangier, K.; Palcic, M. M.; Bundle, D. R.; Hindsgaul, O.; Thiem, J. Fucosyltransferase-Catalyzed Formation of L-Galactosylated Lewis Structures. *Carbohydr. Res.* **1997**, *305*, 511–515.
154. Du, M.; Hindsgaul, O. Recognition of β-D-Gal*p*-(1→3)-β-D-Glc*p*NAc-OR Acceptor Analogues by the Lewis α(1→3/4)-Fucosyltransferase from Human Milk. *Carbohydr. Res.* **1996**, *286*, 87–105.
155. Lubineau, A., Auge, C.; Le Goff, N.; Le Narvor, C. Chemoenzymatic Synthesis of a 3IV,6III-Disulfated Lewis[X] Pentasaccharide, a Candidate Ligand for Human L-Selectin. *Carbohydr. Res.* **1997**, *305*, 501–509.
156. Chiu, M. H.; Thomas, V. H.; Stubbs, H. J.; Rice, K. G. Tissue Targeting of Multivalent Le[X]-Terminated N-Linked Oligosaccharides in Mice. *J. Biol. Chem.* **1995**, *270*, 24024–24031.
157. Thomas, V. H.; Elhalabi, J.; Rice, K. G. Enzymatic Synthesis of N-Linked Oligosaccharides Terminating in Multiple Sialyl-Lewis[X] and GalNAc-Lewis[X] Determinants: Clustered Glycosides for Studying Selectin Interactions. *Carbohydr. Res.* **1998**, *306*, 387–400.
158. de Vries, T.; Norberg, T.; Lonn, H.; Van den Eijnden, D. H. The use of Human Milk Fucosyltransferase in the Synthesis of Tumor-Associated Trimeric X Determinants. *Eur. J. Biochem.* **1993**, *216*, 769–777.
159. de Vries, T.; van den Eijnden, D. H. Biosynthesis of Sialyl-Oligomeric-Lewis[X] and VIM-2 Epitopes: Site Specificity of Human Milk Fucosyltransferase. *Biochemistry* **1994**, *33*, 9937–9944.
160. Nikrad, P. V.; Kashem, M. A.; Wlasichuk, K. B.; Alton, G.; Venot, A. P. Use of Human-Milk Fucosyltransferase in the Chemoenzymic Synthesis of Analogues of the Sialyl Lewis[a] and Sialyl Lewis[X] Tetrasaccharides Modified at the C-2 Position of the Reducing Unit. *Carbohydr. Res.* **1993**, *250*, 145–160.
161. Natunen, J.; Niemela, R.; Penttila, L.; Seppo, A.; Ruohtula, T.; Renkonen, O. Enzymatic Synthesis of two Lacto-*N*-neohexaose-Related Lewis[X] Heptasaccharides and Their Separation by Chromatography on Immobilized Wheat Germ Agglutinin. *Glycobiology* **1994**, *4*, 577–583.
162. Niemela, R.; Natunen, J.; Brotherus, E.; Saarikangas, A.; Renkonen, O. Alpha 1,3-Fucosylation of Branched Blood Group I-Type Oligo-(*N*-Acetyllactosamino)Glycans by Human Milk Transferases Is Restricted to Distal *N*-Acetyllactosamine Units: The Resulting Isomers Are Separated by WGA-Agarose Chromatography. *Glycoconjugate J.* **1995**, *12*, 36–44.
163. Natunen, J.; Aitio, O.; Helin, J.; Maaheimo, H.; Niemela, R.; Heikkinen, S.; Renkonen, O. Human α3-Fucosyltransferases Convert Chitin Oligosaccharides to Products Containing a GlcNAcβ1-4(Fucα1-3)GlcNAcβ1-4R Determinant at the Nonreducing Terminus. *Glycobiology* **2001**, *11*, 209–216.

164. Zeng, S.; Gallego, R. G.; Dinter, A.; Malissard, M.; Kamerling, J. P.; Vliegenthart, J. F.; Berger, E. G. Complete Enzymic Synthesis of the Mucin-Type Sialyl Lewis^X Epitope, Involved in the Interaction Between PSGL-1 and P-Selectin. *Glycoconjugate J.* **1999**, *16*, 487–497.
165. Di Virgilio, S.; Glushka, J.; Moremen, K.; Pierce, M. Enzymatic Synthesis of Natural and ^{13}C-Enriched Linear Poly-*N*-Acetyllactosamines as Ligands for Galectin-1. *Glycobiology* **1999**, *9*, 353–364.
166. Srivastava, G.; Alton, G.; Hindsgaul, O. Combined Chemical–Enzymic Synthesis of Deoxygenated Oligosaccharide Analogs: Transfer of Deoxygenated D-GlcpNAc Residues from Their UDP-GlcpNAc Derivatives Using *N*-Acetylglucosaminyltransferase I. *Carbohydr. Res.* **1990**, *207*, 259–276.
167. Alton, G.; Srivastava, G.; Kaur, K. J.; Hindsgaul, O. Use of *N*-Acetylglucosaminyltransferases I and II in the Synthesis of a Dideoxypentasaccharide. *Bioorg. Med. Chem.* **1994**, *2*, 675–680.
168. Bode, L.; Jantscher-Krenn, E. Structure–Function Relationships of Human Milk Oligosaccharides. *Adv. Nutr.* **2012**, *3*, 383S–391S.
169. Jantscher-Krenn, E.; Bode, L. Human Milk Oligosaccharides and Their Potential Benefits for the Breast-Fed Neonate. *Minerva Pediatr.* **2012**, *64*, 83–99.
170. Newburg, D. S.; Morelli, L. Human Milk and Infant Intestinal Mucosal Glycans Guide Succession of the Neonatal Intestinal Microbiota. *Pediatr. Res.* **2015**, 77, 115–120.
171. Rudloff, S.; Kunz, C. Milk Oligosaccharides and Metabolism in Infants. *Adv. Nutr.* **2012**, *3*, 398S–405S.
172. Kunz, C.; Rudloff, S. Potential Anti-Inflammatory and Anti-Infectious Effects of Human Milk Oligosaccharides. *Adv. Exp. Med. Biol.* **2008**, *606*, 455–465.
173. Zopf, D.; Roth, S. Oligosaccharide Anti-Infective Agents. *Lancet* **1996**, *347*, 1017–1021.
174. Harmsen, H. J.; Wildeboer-Veloo, A. C.; Raangs, G. C.; Wagendorp, A. A.; Klijn, N.; Bindels, J. G.; Welling, G. W. Analysis of Intestinal Flora Development in Breast-Fed and Formula-Fed Infants by Using Molecular Identification and Detection Methods. *J. Pediatr. Gastroenterol. Nutr.* **2000**, *30*, 61–67.
175. Garrido, D.; Barile, D.; Mills, D. A. A Molecular Basis for Bifidobacterial Enrichment in the Infant Gastrointestinal Tract. *Adv. Nutr.* **2012**, *3*, 415S–421S.
176. Bode, L.; Kunz, C.; Muhly-Reinholz, M.; Mayer, K.; Seeger, W.; Rudloff, S. Inhibition of Monocyte, Lymphocyte, and Neutrophil Adhesion to Endothelial Cells by Human Milk Oligosaccharides. *Thromb. Haemost.* **2004**, *92*, 1402–1410.
177. Chichlowski, M.; German, J. B.; Lebrilla, C. B.; Mills, D. A. The Influence of Milk Oligosaccharides on Microbiota of Infants: Opportunities for Formulas. *Annu. Rev. Food Sci. Technol.* **2011**, *2*, 331–351.
178. Bienenstock, J.; Buck, R. H.; Linke, H.; Forsythe, P.; Stanisz, A. M.; Kunze, W. A. Fucosylated but not Sialylated Milk Oligosaccharides Diminish Colon Motor Contractions. *PLoS One* **2013**, *8*, e76236.
179. György, P.; Jeanloz, R. W.; von Nicolai, H.; Zilliken, F. Undialyzable Growth Factors for *Lactobacillus bifidus* var. *pennsylvanicus*. Protective Effect of Sialic Acid Bound to Glycoproteins and Oligosaccharides Against Bacterial Degradation. *Eur. J. Biochem.* **1974**, *43*, 29–33.
180. Xiao, J. Z.; Takahashi, S.; Nishimoto, M.; Odamaki, T.; Yaeshima, T.; Iwatsuki, K.; Kitaoka, M. Distribution of in vitro Fermentation Ability of Lacto-*N*-biose I, a Major Building Block of Human Milk Oligosaccharides, in Bifidobacterial Strains. *Appl. Environ. Microbiol.* **2010**, *76*, 54–59.
181. Kiyohara, M.; Tachizawa, A.; Nishimoto, M.; Kitaoka, M.; Ashida, H.; Yamamoto, K. Prebiotic Effect of Lacto-*N*-biose I on Bifidobacterial Growth. *Biosci. Biotechnol. Biochem.* **2009**, *73*, 1175–1179.

182. Asakuma, S.; Hatakeyama, E.; Urashima, T.; Yoshida, E.; Katayama, T.; Yamamoto, K.; Kumagai, H.; Ashida, H.; Hirose, J.; Kitaoka, M. Physiology of Consumption of Human Milk Oligosaccharides by Infant Gut-Associated Bifidobacteria. *J. Biol. Chem.* **2011**, *286*, 34583–34592.
183. Katayama, T.; Sakuma, A.; Kimura, T.; Makimura, Y.; Hiratake, J.; Sakata, K.; Yamanoi, T.; Kumagai, H.; Yamamoto, K. Molecular Cloning and Characterization of *Bifidobacterium bifidum* 1,2-α-L-Fucosidase (AfcA), a Novel Inverting Glycosidase (Glycoside Hydrolase Family 95). *J. Bacteriol.* **2004**, *186*, 4885–4893.
184. Ashida, H.; Miyake, A.; Kiyohara, M.; Wada, J.; Yoshida, E.; Kumagai, H.; Katayama, T.; Yamamoto, K. Two Distinct α-L-Fucosidases from *Bifidobacterium bifidum* Are Essential for the Utilization of Fucosylated Milk Oligosaccharides and Glycoconjugates. *Glycobiology* **2009**, *19*, 1010–1017.
185. Kiyohara, M.; Tanigawa, K.; Chaiwangsri, T.; Katayama, T.; Ashida, H.; Yamamoto, K. An *exo*-α-Sialidase from Bifidobacteria Involved in the Degradation of Sialyloligosaccharides in Human Milk and Intestinal Glycoconjugates. *Glycobiology* **2011**, *21*, 437–447.
186. Wada, J.; Ando, T.; Kiyohara, M.; Ashida, H.; Kitaoka, M.; Yamaguchi, M.; Kumagai, H.; Katayama, T.; Yamamoto, K. *Bifidobacterium bifidum* Lacto-*N*-biosidase, a Critical Enzyme for the Degradation of Human Milk Oligosaccharides with a Type 1 Structure. *Appl. Environ. Microbiol.* **2008**, *74*, 3996–4004.
187. Suzuki, R.; Wada, J.; Katayama, T.; Fushinobu, S.; Wakagi, T.; Shoun, H.; Sugimoto, H.; Tanaka, A.; Kumagai, H.; Ashida, H.; Kitaoka, M.; Yamamoto, K. Structural and Thermodynamic Analyses of Solute-Binding Protein from *Bifidobacterium longum* Specific for Core 1 Disaccharide and Lacto-*N*-biose I. *J. Biol. Chem.* **2008**, *283*, 13165–13173.
188. Miwa, M.; Horimoto, T.; Kiyohara, M.; Katayama, T.; Kitaoka, M.; Ashida, H.; Yamamoto, K. Cooperation of β-galactosidase and β-*N*-Acetylhexosaminidase from Bifidobacteria in Assimilation of Human Milk Oligosaccharides with Type 2 Structure. *Glycobiology* **2010**, *20*, 1402–1409.
189. Marcobal, A.; Barboza, M.; Sonnenburg, E. D.; Pudlo, N.; Martens, E. C.; Desai, P.; Lebrilla, C. B.; Weimer, B. C.; Mills, D. A.; German, J. B.; Sonnenburg, J. L. Bacteroides in the Infant Gut Consume Milk Oligosaccharides via Mucus-Utilization Pathways. *Cell Host Microbe* **2011**, *10*, 507–514.
190. Sela, D. A.; Garrido, D.; Lerno, L.; Wu, S.; Tan, K.; Eom, H. J.; Joachimiak, A.; Lebrilla, C. B.; Mills, D. A. Bifidobacterium longum Subsp. infantis ATCC 15697 α-Fucosidases Are Active on Fucosylated Human Milk Oligosaccharides. *Appl. Environ. Microbiol.* **2012**, *78*, 795–803.
191. Garrido, D.; Ruiz-Moyano, S.; Mills, D. A. Release and Utilization of *N*-Acetyl-D-glucosamine from Human Milk Oligosaccharides by *Bifidobacterium longum* Subsp. *infantis*. *Anaerobe* **2012**, *18*, 430–435.
192. Sela, D. A.; Li, Y.; Lerno, L.; Wu, S.; Marcobal, A. M.; German, J. B.; Chen, X.; Lebrilla, C. B.; Mills, D. A. An Infant-Associated Bacterial Commensal Utilizes Breast Milk Sialyloligosaccharides. *J. Biol. Chem.* **2011**, *286*, 11909–11918.
193. Garrido, D.; Kim, J. H.; German, J. B.; Raybould, H. E.; Mills, D. A. Oligosaccharide Binding Proteins from *Bifidobacterium longum* Subsp. *infantis* Reveal a Preference for Host Glycans. *PLoS One* **2011**, *6*, e17315.
194. Jantscher-Krenn, E.; Lauwaet, T.; Bliss, L. A.; Reed, S. L.; Gillin, F. D.; Bode, L. Human Milk Oligosaccharides Reduce *Entamoeba histolytica* Attachment and Cytotoxicity In Vitro. *Br. J. Nutr.* **2012**, *108*, 1839–1846.
195. Terrazas, L. I.; Walsh, K. L.; Piskorska, D.; McGuire, E.; Harn, D. A., Jr. The Schistosome Oligosaccharide Lacto-*N*-neotetraose Expands Gr1(+) Cells That Secrete Anti-Inflammatory Cytokines and Inhibit Proliferation of Naive CD4(+) Cells:

A Potential Mechanism for Immune Polarization in Helminth Infections. *J. Immunol.* **2001**, *167*, 5294–5303.

196. Tong, H. H.; McIver, M. A.; Fisher, L. M.; DeMaria, T. F. Effect of Lacto-*N*-neotetraose, Asialoganglioside-GM1 and Neuraminidase on Adherence of Otitis Media-Associated Serotypes of *Streptococcus pneumoniae* to Chinchilla Tracheal Epithelium. *Microb. Pathog.* **1999**, *26*, 111–119.
197. Bode, L.; Kuhn, L.; Kim, H. Y.; Hsiao, L.; Nissan, C.; Sinkala, M.; Kankasa, C.; Mwiya, M.; Thea, D. M.; Aldrovandi, G. M. Human Milk Oligosaccharide Concentration and Risk of Postnatal Transmission of HIV Through Breastfeeding. *Am. J. Clin. Nutr.* **2012**, *96*, 831–839.
198. Idanpaan-Heikkila, I.; Simon, P. M.; Zopf, D.; Vullo, T.; Cahill, P.; Sokol, K.; Tuomanen, E. Oligosaccharides Interfere with the Establishment and Progression of Experimental Pneumococcal Pneumonia. *J. Infect. Dis.* **1997**, *176*, 704–712.
199. Yu, Z. T.; Chen, C.; Kling, D. E.; Liu, B.; McCoy, J. M.; Merighi, M.; Heidtman, M.; Newburg, D. S. The Principal Fucosylated Oligosaccharides of Human Milk Exhibit Prebiotic Properties on Cultured Infant Microbiota. *Glycobiology* **2013**, *23*, 169–177.
200. Huang, P.; Farkas, T.; Marionneau, S.; Zhong, W.; Ruvoen-Clouet, N.; Morrow, A. L.; Altaye, M.; Pickering, L. K.; Newburg, D. S.; LePendu, J.; Jiang, X. Noroviruses Bind to Human ABO, Lewis, and Secretor Histo-Blood Group Antigens: Identification of 4 Distinct Strain-Specific Patterns. *J. Infect. Dis.* **2003**, *188*, 19–31.
201. Cravioto, A.; Tello, A.; Villafan, H.; Ruiz, J.; del Vedovo, S.; Neeser, J. R. Inhibition of Localized Adhesion of Enteropathogenic *Escherichia coli* to HEp-2 Cells by Immunoglobulin and Oligosaccharide Fractions of Human Colostrum and Breast Milk. *J. Infect. Dis.* **1991**, *163*, 1247–1255.
202. Newburg, D. S.; Pickering, L. K.; McCluer, R. H.; Cleary, T. G. Fucosylated Oligosaccharides of Human Milk Protect Suckling Mice from Heat-Stabile Enterotoxin of *Escherichia coli*. *J. Infect. Dis.* **1990**, *162*, 1075–1080.
203. Chessa, D.; Winter, M. G.; Jakomin, M.; Baumler, A. J. *Salmonella enterica* Serotype *Typhimurium* Std Fimbriae Bind Terminal α(1,2)Fucose Residues in the Cecal Mucosa. *Mol. Microbiol.* **2009**, *71*, 864–875.
204. Ruiz-Palacios, G. M.; Cervantes, L. E.; Ramos, P.; Chavez-Munguia, B.; Newburg, D. S. *Campylobacter jejuni* Binds Intestinal H(O) Antigen (Fuc α1, 2Gal β1, 4GlcNAc), and Fucosyloligosaccharides of Human Milk Inhibit Its Binding and Infection. *J. Biol. Chem.* **2003**, *278*, 14112–14120.
205. Morrow, A. L.; Ruiz-Palacios, G. M.; Altaye, M.; Jiang, X.; Guerrero, M. L.; Meinzen-Derr, J. K.; Farkas, T.; Chaturvedi, P.; Pickering, L. K.; Newburg, D. S. Human Milk Oligosaccharides Are Associated with Protection Against Diarrhea in Breast-Fed Infants. *J. Pediatr.* **2004**, *145*, 297–303.
206. Brassart, D.; Woltz, A.; Golliard, M.; Neeser, J. R. In Vitro Inhibition of Adhesion of *Candida albicans* Clinical Isolates to Human Buccal Epithelial Cells by Fucα1→2Gal β-Bearing Complex Carbohydrates. *Infect. Immun.* **1991**, *59*, 1605–1613.
207. Hong, P.; Ninonuevo, M. R.; Lee, B.; Lebrilla, C.; Bode, L. Human Milk Oligosaccharides Reduce HIV-1-gp120 Binding to Dendritic Cell-Specific ICAM3-Grabbing Non-Integrin (DC-SIGN). *Br. J. Nutr.* **2009**, *101*, 482–486.
208. Kuhn, L.; Kim, H. Y.; Hsiao, L.; Nissan, C.; Kankasa, C.; Mwiya, M.; Thea, D. M.; Aldrovandi, G. M.; Bode, L. Oligosaccharide Composition of Breast Milk Influences Survival of Uninfected Children Born to HIV-Infected Mothers in Lusaka, Zambia. *J. Nutr.* **2015**, *145*, 66–72.
209. Xu, H. T.; Zhao, Y. F.; Lian, Z. X.; Fan, B. L.; Zhao, Z. H.; Yu, S. Y.; Dai, Y. P.; Wang, L. L.; Niu, H. L.; Li, N.; Hammarstrom, L.; Boren, T.; Sjostrom, R. Effects of

Fucosylated Milk of Goat and Mouse on *Helicobacter pylori* Binding to Lewis^{b} Antigen. *World J. Gastroenterol.* **2004**, *10*, 2063–2066.
210. Miyazaki, T.; Sato, T.; Furukawa, K.; Ajisaka, K. Enzymatic Synthesis of Lacto-*N*-difucohexaose I Which Binds to *Helicobacter pylori. Methods Enzymol.* **2010**, *480*, 511–524.
211. Atochina, O.; Harn, D. LNFPIII/Le^{X}-Stimulated Macrophages Activate Natural Killer Cells via CD40-CD40L Interaction. *Clin. Diagn. Lab. Immunol.* **2005**, *12*, 1041–1049.
212. Bode, L.; Rudloff, S.; Kunz, C.; Strobel, S.; Klein, N. Human Milk Oligosaccharides Reduce Platelet-Neutrophil Complex Formation Leading to a Decrease in Neutrophil β2 Integrin Expression. *J. Leukoc. Biol.* **2004**, *76*, 820–826.
213. Lewis, Z. T.; Totten, S. M.; Smilowitz, J. T.; Popovic, M.; Parker, E.; Lemay, D. G.; Van Tassell, M. L.; Miller, M. J.; Jin, Y. S.; German, J. B.; Lebrilla, C. B.; Mills, D. A. Maternal Fucosyltransferase 2 Status Affects the Gut Bifidobacterial Communities of Breastfed Infants. *Microbiome* **2015**, *3*, 13.
214. Newburg, D. S.; Ruiz-Palacios, G. M.; Altaye, M.; Chaturvedi, P.; Meinzen-Derr, J.; Guerrero Mde, L.; Morrow, A. L. Innate Protection Conferred by Fucosylated Oligosaccharides of Human Milk Against Diarrhea in Breastfed Infants. *Glycobiology* **2004**, *14*, 253–263.
215. Morrow, A. L.; Meinzen Derr, J.; Huang, P.; Schibler, K. R.; Cahill, T.; Keddache, M.; Kallapur, S. G.; Newburg, D. S.; Tabangin, M.; Warner, B. B.; Jiang, X. Fucosyltransferase 2 Non-Secretor and Low Secretor Status Predicts Severe Outcomes in Premature Infants. *J. Pediatr.* **2011**, *158*, 745–751.
216. Coppa, G. V.; Zampini, L.; Galeazzi, T.; Facinelli, B.; Ferrante, L.; Capretti, R.; Orazio, G. Human Milk Oligosaccharides Inhibit the Adhesion to Caco-2 Cells of Diarrheal Pathogens: *Escherichia coli*, *Vibrio cholerae*, and *Salmonella fyris*. *Pediatr. Res.* **2006**, *59*, 377–382.
217. Matrosovich, M. N.; Gambaryan, A. S.; Tuzikov, A. B.; Byramova, N. E.; Mochalova, L. V.; Golbraikh, A. A.; Shenderovich, M. D.; Finne, J.; Bovin, N. V. Probing of the Receptor-Binding Sites of the H1 and H3 Influenza A and Influenza B Virus Hemagglutinins by Synthetic and Natural Sialosides. *Virology* **1993**, *196*, 111–121.
218. Eiwegger, T.; Stahl, B.; Schmitt, J.; Boehm, G.; Gerstmayr, M.; Pichler, J.; Dehlink, E.; Loibichler, C.; Urbanek, R.; Szepfalusi, Z. Human Milk-derived Oligosaccharides and Plant-Derived Oligosaccharides Stimulate Cytokine Production of Cord Blood T-Cells In Vitro. *Pediatr. Res.* **2004**, *56*, 536–540.
219. Idota, T.; Kawakami, H.; Murakami, Y.; Sugawara, M. Inhibition of Cholera Toxin by Human Milk Fractions and Sialyllactose. *Biosci. Biotechnol. Biochem.* **1995**, *59*, 417–419.
220. Bouchara, J. P.; Sanchez, M.; Chevailler, A.; Marot-Leblond, A.; Lissitzky, J. C.; Tronchin, G.; Chabasse, D. Sialic Acid-Dependent Recognition of Laminin and Fibrinogen by *Aspergillus fumigatus conidia. Infect. Immun.* **1997**, *65*, 2717–2724.
221. Devaraj, N.; Sheykhnazari, M.; Warren, W. S.; Bhavanandan, V. P. Differential Binding of Pseudomonas aeruginosa to Normal and Cystic Fibrosis Tracheobronchial Mucins. *Glycobiology* **1994**, *4*, 307–316.
222. Kuntz, S.; Rudloff, S.; Kunz, C. Oligosaccharides from Human Milk Influence Growth-Related Characteristics of Intestinally Transformed and Non-Transformed Intestinal Cells. *Br. J. Nutr.* **2008**, *99*, 462–471.
223. Kavanaugh, D. W.; O'Callaghan, J.; Butto, L. F.; Slattery, H.; Lane, J.; Clyne, M.; Kane, M.; Joshi, L.; Hickey, R. M. Exposure of Subsp. to Milk Oligosaccharides Increases Adhesion to Epithelial Cells and Induces a Substantial Transcriptional Response. *PLoS One* **2013**, *8*, e67224.

224. Stehle, T.; Yan, Y.; Benjamin, T. L.; Harrison, S. C. Structure of Murine Polyomavirus Complexed with an Oligosaccharide Receptor Fragment. *Nature* **1994**, *369*, 160–163.
225. Stins, M. F.; Prasadarao, N. V.; Ibric, L.; Wass, C. A.; Luckett, P.; Kim, K. S. Binding Characteristics of S Fimbriated *Escherichia coli* to Isolated Brain Microvascular Endothelial Cells. *Am. J. Pathol.* **1994**, *145*, 1228–1236.
226. Virkola, R.; Parkkinen, J.; Hacker, J.; Korhonen, T. K. Sialyloligosaccharide Chains of Laminin as an Extracellular Matrix Target for S Fimbriae of *Escherichia coli*. *Infect. Immun.* **1993**, *61*, 4480–4484.
227. Mysore, J. V.; Wigginton, T.; Simon, P. M.; Zopf, D.; Heman-Ackah, L. M.; Dubois, A. Treatment of *Helicobacter pylori* Infection in Rhesus Monkeys Using a Novel Antiadhesion Compound. *Gastroenterology* **1999**, *117*, 1316–1325.
228. Evans, D. G.; Evans, D. J., Jr.; Moulds, J. J.; Graham, D. Y. *N*-Acetylneuraminyllactose-Binding Fibrillar Hemagglutinin of *Campylobacter pylori*: A Putative Colonization Factor Antigen. *Infect. Immun.* **1988**, *56*, 2896–2906.
229. Korhonen, T. K.; Valtonen, M. V.; Parkkinen, J.; Vaisanen-Rhen, V.; Finne, J.; Orskov, F.; Orskov, I.; Svenson, S. B.; Makela, P. H. Serotypes, Hemolysin Production, and Receptor Recognition of *Escherichia coli* Strains Associated with Neonatal Sepsis and Meningitis. *Infect. Immun.* **1985**, *48*, 486–491.
230. Van Niekerk, E.; Autran, C. A.; Nel, D. G.; Kirsten, G. F.; Blaauw, R.; Bode, L. Human Milk Oligosaccharides Differ Between HIV-Infected and HIV-Uninfected Mothers and Are Related to Necrotizing Enterocolitis Incidence in Their Preterm Very-Low-Birth-Weight Infants. *J. Nutr.* **2014**, *144*, 1227–1233.
231. Neu, J.; Walker, W. A. Necrotizing Enterocolitis. *N. Engl. J. Med.* **2011**, *364*, 255–264.
232. Holman, R. C.; Stoll, B. J.; Clarke, M. J.; Glass, R. I. The Epidemiology of Necrotizing Enterocolitis Infant Mortality in the United States. *Am. J. Pub. Health* **1997**, *87*, 2026–2031.
233. Dicken, B. J.; Sergi, C.; Rescorla, F. J.; Breckler, F.; Sigalet, D. Medical Management of Motility Disorders in Patients with Intestinal Failure: A Focus on Necrotizing Enterocolitis, Gastroschisis, and Intestinal Atresia. *J. Pediatr. Surg.* **2011**, *46*, 1618–1630.
234. Clark, R. H.; Gordon, P.; Walker, W. M.; Laughon, M.; Smith, P. B.; Spitzer, A. R. Characteristics of Patients Who Die of Necrotizing Enterocolitis. *J. Perinatol.* **2012**, *32*, 199–204.
235. Blakely, M. L.; Lally, K. P.; McDonald, S.; Brown, R. L.; Barnhart, D. C.; Ricketts, R. R.; Thompson, W. R.; Scherer, L. R.; Klein, M. D.; Letton, R. W.; Chwals, W. J.; Touloukian, R. J.; Kurkchubasche, A. G.; Skinner, M. A.; Moss, R. L.; Hilfiker, M. L.; Network, N.E. C. S. o. t. N. N. R. Postoperative Outcomes of Extremely Low Birth-Weight Infants with Necrotizing Enterocolitis or Isolated Intestinal Perforation: A Prospective Cohort Study by the NICHD Neonatal Research Network. *Ann. Surg.* **2005**, *241*, 984–989, discussion 989–994.
236. Jantscher-Krenn, E.; Zherebtsov, M.; Nissan, C.; Goth, K.; Guner, Y. S.; Naidu, N.; Choudhury, B.; Grishin, A. V.; Ford, H. R.; Bode, L. The Human Milk Oligosaccharide Disialyllacto-*N*-Tetraose Prevents Necrotising Enterocolitis in Neonatal Rats. *Gut* **2012**, *61*, 1417–1425.
237. Fernandez-Mayoralas, A.; Martin-Lomas, M. Synthesis of 3- and 2′-Fucosyl-lactose and 3,2′-Difucosyl-lactose from Partially Benzylated Lactose Derivatives. *Carbohydr. Res.* **1986**, *154*, 93–101.
238. Aly, M. R.; El Ibrahim, S. I.; El Ashry, E. S. H.; Schmidt, R. R. Synthesis of Lacto-*N*-neotetraose and Lacto-*N*-tetraose Using the Dimethylmaleoyl Group as Amino Protective Group. *Carbohydr. Res.* **1999**, *316*, 121–132.

239. Hsu, Y.; Lu, X. A.; Zulueta, M. M.; Tsai, C. M.; Lin, K. I.; Hung, S. C.; Wong, C. H. Acyl and Silyl Group Effects in Reactivity-Based One-Pot Glycosylation: Synthesis of Embryonic Stem Cell Surface Carbohydrates Lc4 and IV(2)Fuc-Lc4. *J. Am. Chem. Soc.* **2012**, *134*, 4549–4552.
240. Takamura, T.; Chiba, T.; Ishihara, H.; Tejima, S. Chemical Modification of Lactose. XIII. Synthesis of Lacto-*N*-tetraose. *Chem. Pharm. Bull.* **1980**, *28*, 1804–1809.
241. Aly, M. R. E.; El Ibrahim, S. I.; El Ashry, E. S. H.; Schmidt, R. R. Synthesis of Lacto-*N*-neohexaose and Lacto-*N*-neooctaose Using the Dimethylmaleoyl Moiety as an Amino Protective Group. *Eur. J. Org. Chem.* **2000**, 319–326.
242. Manzoni, L.; Lay, L.; Schmidt, R. R. Synthesis of Lewis^{a} and Lewis^{X} Pentasaccharides Based on *N*-Trichloroethoxycarbonyl Protection. *J. Carbohydr. Chem.* **1998**, *17*, 739–758.
243. Love, K. R.; Seeberger, P. H. Solution Syntheses of Protected Type II Lewis Blood Group Oligosaccharides: Study for Automated Synthesis. *J. Org. Chem.* **2005**, *70*, 3168–3177.
244. Zhang, Y.-M.; Esnault, J.; Mallet, J.-M.; Sinaÿ, P. Synthesis of the β-Methyl Glycoside of Lacto-*N*-fucopentaose III. *J. Carbohydr. Chem.* **1999**, *18*, 419–427.
245. Lay, L.; Manzoni, L.; Schmidt, R. R. Synthesis of *N*-Acetylglucosamine Containing Lewis^{a} and Lewis^{X} Building Blocks Based on *N*-Tetrachlorophthaloyl Protection—Synthesis of Lewis^{X} Pentasaccharide. *Carbohydr. Res.* **1998**, *310*, 157–171.
246. Cao, S.; Gan, Z.; Roy, R. Active-Latent Glycosylation Strategy Toward Lewis^{X} Pentasaccharide in a Form Suitable for Neoglycoconjugate Syntheses. *Carbohydr. Res.* **1999**, *318*, 75–81.
247. Sherman, A. A.; Yudina, O. N.; Mironov, Y. V.; Sukhova, E. V.; Shashkov, A. S.; Menshov, V. M.; Nifantiev, N. E. Study of Glycosylation with *N*-Trichloroacetyl-D-glucosamine Derivatives in the Syntheses of the Spacer-Armed Pentasaccharides Sialyl Lacto-*N*-neotetraose and Sialyl Lacto-*N*-tetraose, Their Fragments, and Analogues. *Carbohydr. Res.* **2001**, *336*, 13–46.
248. Mandal, P. K.; Misra, A. K. Concise Synthesis of Two Pentasaccharides Corresponding to the α-Chain Oligosaccharides of *Neisseria gonorrhoeae* and *Neisseria meningitidis*. *Tetrahedron* **2008**, *64*, 8685–8691.
249. Chernyak, A.; Oscarson, S.; Turek, D. Synthesis of the Lewis^{b} Hexasaccharide and Squarate Acid–HSA Conjugates Thereof with Various Saccharide Loadings. *Carbohydr. Res.* **2000**, *329*, 309–316.
250. Shimizu, H.; Ito, Y.; Kanie, O.; Ogawa, T. Solid Phase Synthesis of Polylactosamine Oligosaccharide. *Bioorg. Med. Chem. Lett.* **1996**, *6*, 2841–2846.
251. Takamura, T.; Chiba, T.; Tejima, S. Chemical Modification of Lactose. XVI. Synthesis of Lacto-*N*-Neohexaose. *Chem. Pharm. Bull.* **1981**, *29*, 2270–2276.
252. Roussel, F.; Takhi, M.; Schmidt, R. R. Solid-Phase Synthesis of a Branched Hexasaccharide Using a Highly Efficient Synthetic Strategy. *J. Org. Chem.* **2001**, *66*, 8540–8548.
253. Maranduba, A.; Veyrieres, A. Glycosylation of Lactose: Synthesis of Branched Oligosaccharides Involved in the Biosynthesis of Glycolipids Having Blood-Group I Activity. *Carbohydr. Res.* **1986**, *151*, 105–119.
254. Knuhr, P.; Castro-Palomino, J.; Grathwohl, M.; Schmidt, R. R. Complex Structures of Antennary Human Milk Oligosaccharides—Synthesis of a Branched Octasaccharide. *Eur. J. Org. Chem.* **2001**, 4239–4246.
255. Broder, W.; Kunz, H. Glycosyl Azides as Building Blocks in Convergent Syntheses of Oligomeric Lactosamine and Lewis^{X} Saccharides. *Bioorg. Med. Chem.* **1997**, *5*, 1–19.
256. Lee, J. C.; Wu, C. Y.; Apon, J. V.; Siuzdak, G.; Wong, C. H. Reactivity-Based One-Pot Synthesis of the Tumor-Associated Antigen N3 Minor Octasaccharide for the

Development of a Photocleavable DIOS-MS Sugar Array. *Angew. Chem. Int. Ed.* **2006**, *45*, 2753–2757.

257. Kim, H. M.; Kim, I. J.; Danishefsky, S. J. Total Syntheses of Tumor-Related Antigens N3: Probing the Feasibility Limits of the Glycal Assembly Method. *J. Am. Chem. Soc.* **2001**, *123*, 35–48.
258. Jennum, C. A.; Fenger, T. H.; Bruun, L. M.; Madsen, R. One-pot Glycosylations in the Synthesis of Human Milk Oligosaccharides. *Eur. J. Org. Chem.* **2014**, 3232–3241.
259. Han, N. S.; Kim, T. J.; Park, Y. C.; Kim, J.; Seo, J. H. Biotechnological Production of Human Milk Oligosaccharides. *Biotechnol. Adv.* **2012**, *30*, 1268–1278.
260. Monteiro, M. A.; Chan, K. H.; Rasko, D. A.; Taylor, D. E.; Zheng, P. Y.; Appelmelk, B. J.; Wirth, H. P.; Yang, M.; Blaser, M. J.; Hynes, S. O.; Moran, A. P.; Perry, M. B. Simultaneous Expression of Type 1 and Type 2 Lewis Blood Group Antigens by *Helicobacter pylori* Lipopolysaccharides. Molecular Mimicry Between *H. pylori* Lipopolysaccharides and Human Gastric Epithelial Cell Surface Glycoforms. *J. Biol. Chem.* **1998**, *273*, 11533–11543.
261. Smith, H.; Parsons, N. J.; Cole, J. A. Sialylation of Neisserial Lipopolysaccharide: A Major Influence on Pathogenicity. *Microb. Pathog.* **1995**, *19*, 365–377.
262. Chen, X.; Varki, A. Advances in the Biology and Chemistry of Sialic Acids. *ACS Chem. Biol.* **2010**, *5*, 163–176.
263. Johnson, K. F. Synthesis of Oligosaccharides by Bacterial Enzymes. *Glycoconjugate J.* **1999**, *16*, 141–146.
264. Chen, X. Fermenting Next Generation Glycosylated Therapeutics. *ACS Chem. Biol.* **2011**, *6*, 14–17.
265. Chen, X.; Kowal, P.; Wang, P. G. Large-Scale Enzymatic Synthesis of Oligosaccharides. *Curr. Opin. Drug. Discov. Dev.* **2000**, *3*, 756–763.
266. Ichikawa, Y.; Look, G. C.; Wong, C. H. Enzyme-Catalyzed Oligosaccharide Synthesis. *Anal. Biochem.* **1992**, *202*, 215–238.
267. Ichikawa, Y.; Wang, R.; Wong, C. H. Regeneration of Sugar Nucleotide for Enzymatic Oligosaccharide Synthesis. *Methods Enzymol.* **1994**, *247*, 107–127.
268. Tsai, T. I.; Lee, H. Y.; Chang, S. H.; Wang, C. H.; Tu, Y. C.; Lin, Y. C.; Hwang, D. R.; Wu, C. Y.; Wong, C. H. Effective Sugar Nucleotide Regeneration for the Large-Scale Enzymatic Synthesis of Globo H and SSEA4. *J. Am. Chem. Soc.* **2013**, *135*, 14831–14839.
269. Yu, H.; Lau, K.; Thon, V.; Autran, C. A.; Jantscher-Krenn, E.; Xue, M.; Li, Y.; Sugiarto, G.; Qu, J.; Mu, S.; Ding, L.; Bode, L.; Chen, X. Synthetic Disialyl Hexasaccharides Protect Neonatal Rats from Necrotizing Enterocolitis. *Angew. Chem. Int. Ed.* **2014**, *53*, 6687–6691.
270. Yu, H.; Chokhawala, H. A.; Huang, S.; Chen, X. One-Pot Three-Enzyme Chemoenzymatic Approach to the Synthesis of Sialosides Containing Natural and Non-Natural Functionalities. *Nat. Protoc.* **2006**, *1*, 2485–2492.
271. Yu, H.; Lau, K.; Li, Y.; Sugiarto, G.; Chen, X. One-Pot Multienzyme Synthesis of Lewis^X and Sialyl Lewis^X Antigens. *Curr. Protoc. Chem. Biol.* **2012**, *4*, 233–247.
272. Lau, K.; Yu, H.; Thon, V.; Khedri, Z.; Leon, M. E.; Tran, B. K.; Chen, X. Sequential Two-Step Multienzyme Synthesis of Tumor-Associated Sialyl T-Antigens and Derivatives. *Org. Biomol. Chem.* **2011**, *9*, 2784–2789.
273. Li, Y.; Chen, X. Sialic Acid Metabolism and Sialyltransferases: Natural Functions and Applications. *Appl. Microbiol. Biotechnol.* **2012**, *94*, 887–905.
274. Mehta, S.; Gilbert, M.; Wakarchuk, W. W.; Whitfield, D. M. Ready Access to Sialylated Oligosaccharide Donors. *Org. Lett.* **2000**, *2*, 751–753.
275. Yan, F.; Mehta, S.; Eichler, E.; Wakarchuk, W. W.; Gilbert, M.; Schur, M. J.; Whitfield, D. M. Simplifying Oligosaccharide Synthesis: Efficient Synthesis of

Lactosamine and Siaylated Lactosamine Oligosaccharide Donors. *J. Org. Chem.* **2003**, *68*, 2426–2431.
276. Hayashi, M.; Tanaka, M.; Itoh, M.; Miyauchi, H. A Convenient and Efficient Synthesis of SLe^X Analogs. *J. Org. Chem.* **1996**, *61*, 2938–2945.
277. Cao, H.; Huang, S.; Cheng, J.; Li, Y.; Muthana, S.; Son, B.; Chen, X. Chemical Preparation of Sialyl Lewis^X Using an Enzymatically Synthesized Sialoside Building Block. *Carbohydr. Res.* **2008**, *343*, 2863–2869.
278. Schmidt, D.; Thiem, J. Chemical Synthesis Using Enzymatically Generated Building Units for Construction of the Human Milk Pentasaccharides Sialyllacto-*N*-tetraose and Sialyllacto-*N*-neotetraose Epimer. *Beilstein J. Org. Chem.* **2010**, *6*, 18.
279. Yao, W.; Yan, J.; Chen, X.; Wang, F.; Cao, H. Chemoenzymatic Synthesis of Lacto-*N*-tetrasaccharide and Sialyl Lacto-*N*-tetrasaccharides. *Carbohydr. Res.* **2015**, *401*, 5–10.
280. Shaikh, F. A.; Withers, S. G. Teaching Old Enzymes New Tricks: Engineering and Evolution of Glycosidases and Glycosyl Transferases for Improved Glycoside Synthesis. *Biochem. Cell Biol.* **2008**, *86*, 169–177.
281. Williams, S. J.; Withers, S. G. Glycosyl Fluorides in Enzymatic Reactions. *Carbohydr. Res.* **2000**, *327*, 27–46.
282. Albert, M.; Repetschnigg, W.; Ortner, J.; Gomes, J.; Paul, B. J.; Illaszewicz, C.; Weber, H.; Steiner, W.; Dax, K. Simultaneous Detection of Different Glycosidase Activities by ^{19}F NMR Spectroscopy. *Carbohydr. Res.* **2000**, *327*, 395–400.
283. Zeuner, B.; Jers, C.; Mikkelsen, J. D.; Meyer, A. S. Methods for Improving Enzymatic Trans-Glycosylation for Synthesis of Human Milk Oligosaccharide Biomimetics. *J. Agric. Food Chem.* **2014**, *62*, 9615–9631.
284. Albermann, C.; Piepersberg, W.; Wehmeier, U. F. Synthesis of the Milk Oligosaccharide 2′-Fucosyllactose Using Recombinant Bacterial Enzymes. *Carbohydr. Res.* **2001**, *334*, 97–103.
285. Albermann, C.; Distler, J.; Piepersberg, W. Preparative Synthesis of GDP-β-L-Fucose by Recombinant Enzymes from Enterobacterial Sources. *Glycobiology* **2000**, *10*, 875–881.
286. Wang, G.; Boulton, P. G.; Chan, N. W.; Palcic, M. M.; Taylor, D. E. Novel *Helicobacter pylori* α1,2-Fucosyltransferase, a Key Enzyme in the Synthesis of Lewis Antigens. *Microbiology* **1999**, *145* (Pt 11), 3245–3253.
287. Lee, W. H.; Pathanibul, P.; Quarterman, J.; Jo, J. H.; Han, N. S.; Miller, M. J.; Jin, Y. S.; Seo, J. H. Whole Cell Biosynthesis of a Functional Oligosaccharide, 2′-Fucosyllactose, Using Engineered *Escherichia coli*. *Microb. Cell Fact.* **2012**, *11*, 48.
288. Chin, Y. W.; Kim, J. Y.; Lee, W. H.; Seo, J. H. Enhanced Production of 2′-Fucosyllactose in Engineered Escherichia coli BL21star(DE3) by Modulation of Lactose Metabolism and Fucosyltransferase. *J. Biotechnol.* **2015**, *210*, 107–115.
289. Baumgartner, F.; Seitz, L.; Sprenger, G. A.; Albermann, C. Construction of Escherichia coli Strains with Chromosomally Integrated Expression Cassettes for the Synthesis of 2′-Fucosyllactose. *Microb. Cell Fact.* **2013**, *12*, 40.
290. Prieto, P. A. Profiles of Human Milk Oligosaccharides and Production of Some Human Milk Oligosaccharides in Transgenic Animals. *Adv. Nutr.* **2012**, *3*, 456S–464S.
291. Nagae, M.; Tsuchiya, A.; Katayama, T.; Yamamoto, K.; Wakatsuki, S.; Kato, R. Structural Basis of the Catalytic Reaction Mechanism of Novel 1,2-α-L-Fucosidase from *Bifidobacterium bifidum*. *J. Biol. Chem.* **2007**, *282*, 18497–18509.
292. Wada, J.; Honda, Y.; Nagae, M.; Kato, R.; Wakatsuki, S.; Katayama, T.; Taniguchi, H.; Kumagai, H.; Kitaoka, M.; Yamamoto, K. 1,2-α-L-Fucosynthase: A Glycosynthase Derived from an Inverting α-Glycosidase with an Unusual Reaction Mechanism. *FEBS Lett.* **2008**, *582*, 3739–3743.

293. Yu, H.; Yu, H.; Karpel, R.; Chen, X. Chemoenzymatic Synthesis of CMP-Sialic Acid Derivatives by a One-Pot Two-Enzyme System: Comparison of Substrate Flexibility of Three Microbial CMP-Sialic Acid Synthetases. *Bioorg. Med. Chem.* **2004**, *12*, 6427–6435.
294. Yu, H.; Chokhawala, H.; Karpel, R.; Yu, H.; Wu, B.; Zhang, J.; Zhang, Y.; Jia, Q.; Chen, X. A Multifunctional *Pasteurella multocida* Sialyltransferase: A Powerful Tool for the Synthesis of Sialoside Libraries. *J. Am. Chem. Soc.* **2005**, *127*, 17618–17619.
295. Sugiarto, G.; Lau, K.; Li, Y.; Khedri, Z.; Yu, H.; Le, D. T.; Chen, X. Decreasing the Sialidase Activity of Multifunctional *Pasteurella multocida* α2-3-Sialyltransferase 1 (PmST1) by Site-Directed Mutagenesis. *Mol. Biosyst.* **2011**, 7, 3021–3027.
296. Sugiarto, G.; Lau, K.; Qu, J.; Li, Y.; Lim, S.; Mu, S.; Ames, J. B.; Fisher, A. J.; Chen, X. A Sialyltransferase Mutant with Decreased Donor Hydrolysis and Reduced Sialidase Activities for Directly Sialylating Lewis^X. *ACS Chem. Biol.* **2012**, 7, 1232–1240.
297. Endo, T.; Koizumi, S. Process for Producing alpha2,3/alpha2,8-Sialyltransferase and Sialic Acid-Containing Complex Sugar. Patent WO 2003027297 A1, April 3, 2004.
298. Schmolzer, K.; Czabany, T.; Luley-Goedl, C.; Pavkov-Keller, T.; Ribitsch, D.; Schwab, H.; Gruber, K.; Weber, H.; Nidetzky, B. Complete Switch from α-2,3- to α-2,6-Regioselectivity in *Pasteurella dagmatis* β-D-Galactoside Sialyltransferase by Active-Site Redesign. *Chem. Commun. (Cambridge)* **2015**, *51*, 3083–3086.
299. Schmolzer, K.; Ribitsch, D.; Czabany, T.; Luley-Goedl, C.; Kokot, D.; Lyskowski, A.; Zitzenbacher, S.; Schwab, H.; Nidetzky, B. Characterization of a Multifunctional α2,3-Sialyltransferase from *Pasteurella dagmatis*. *Glycobiology* **2013**, *23*, 1293–1304.
300. Guo, Y.; Jers, C.; Meyer, A. S.; Arnous, A.; Li, H.; Kirpekar, F.; Mikkelsen, J. D. A *Pasteurella multocida* Sialyltransferase Displaying Dual *trans*-Sialidase Activities for Production of 3′-Sialyl and 6′-Sialyl Glycans. *J. Biotechnol.* **2014**, *170*, 60–67.
301. Tanaka, H.; Ito, F.; Iwasaki, T. A System for Sialic Acid Transfer by Colominic Acid and a Sialidase That Preferentially Hydrolyzes Sialyl β-2,8 Linkages. *Biosci. Biotechnol. Biochem.* **1995**, *59*, 638–643.
302. Mcjarrow, P.; Garman, J.; Harvey, S.; Van Amelsfort, A. Diary Process and Product. Patent WO 2003049547 A2, June 19, 2003.
303. Pelletier, M.; Barker, W. A.; Hakes, D. J.; Zopf, D. A. Methods for Producing Sialyloligosaccharides in a Dairy Source. US Patent 6706492 B2, March 16, 2004.
304. Sallomons, E.; Wilbrink, M. H.; Sanders, P.; Kamerling, J. P.; Van Vuure, C. A.; Hage, J. A. Methods for Providing Sialylated Oligosaccharides. WO Patent 2013085384 A1, June 13, 2013.
305. Gilbert, M.; Bayer, R.; Cunningham, A. M.; DeFrees, S.; Gao, Y.; Watson, D. C.; Young, N. M.; Wakarchuk, W. W. The Synthesis of Sialylated Oligosaccharides Using a CMP-Neu5Ac Synthetase/Sialyltransferase Fusion. *Nat. Biotechnol.* **1998**, *16*, 769–772.
306. Endo, T.; Koizumi, S.; Tabata, K.; Ozaki, A. Large-Scale Production of CMP-NeuAc and Sialylated Oligosaccharides Through Bacterial Coupling. *Appl. Microbiol. Biotechnol.* **2000**, *53*, 257–261.
307. Priem, B.; Gilbert, M.; Wakarchuk, W. W.; Heyraud, A.; Samain, E. A New Fermentation Process Allows Large-Scale Production of Human Milk Oligosaccharides by Metabolically Engineered Bacteria. *Glycobiology* **2002**, *12*, 235–240.
308. Fierfort, N.; Samain, E. Genetic Engineering of *Escherichia coli* for the Economical Production of Sialylated Oligosaccharides. *J. Biotechnol.* **2008**, *134*, 261–265.
309. Yu, H.; Huang, S.; Chokhawala, H.; Sun, M.; Zheng, H.; Chen, X. Highly Efficient Chemoenzymatic Synthesis of Naturally Occurring and Non-Natural α-2,6-Linked Sialosides: A *P. damsela* α-2,6-Sialyltransferase with Extremely Flexible Donor–Substrate Specificity. *Angew. Chem. Int. Ed.* **2006**, *45*, 3938–3944.

310. Tsukamoto, H.; Takakura, Y.; Mine, T.; Yamamoto, T. Photobacterium sp. JT-ISH-224 Produces Two Sialyltransferases, α-/β-Galactoside α2,3-Sialyltransferase and β-Galactoside α2,6-Sialyltransferase. *J. Biochem.* **2008**, *143*, 187–197.
311. Drouillard, S.; Mine, T.; Kajiwara, H.; Yamamoto, T.; Samain, E. Efficient Synthesis of 6′-Sialyllactose, 6,6′-Disialyllactose, and 6′-KDO-Lactose by Metabolically Engineered *E. coli* Expressing a Multifunctional Sialyltransferase from the *Photobacterium* sp. JT-ISH-224. *Carbohydr. Res.* **2010**, *345*, 1394–1399.
312. Nyffenegger, C.; Nordvang, R. T.; Zeuner, B.; Lezyk, M.; Difilippo, E.; Logtenberg, M. J.; Schols, H. A.; Meyer, A. S.; Mikkelsen, J. D. Backbone Structures in Human Milk Oligosaccharides: Trans-Glycosylation by Metagenomic β-*N*-Acetylhexosaminidases. *Appl. Microbiol. Biotechnol.* **2015**, *99*, 7997–8009.
313. Murata, T.; Inukai, T.; Suzuki, M.; Yamagishi, M.; Usui, A. T. Facile Enzymatic Conversion of Lactose into Lacto-*N*-tetraose and Lacto-*N*-neotetraose. *Glycoconjugate J.* **1999**, *16*, 189–195.
314. Prieto, P. A.; Kleman-Leyer, K. M. Process for Synthesizing Oligosaccharides. US Patent 5,945,314, 1999.
315. Prieto, P. A. In Vitro and Clinical Experiences with a Human Milk Oligosaccharide, Lacto-*N*-neotetraose, and Fructooligosaccharides. *Foods Food Ingredients J. Jpn.* **2005**, *210*, 1018–1030.
316. Renaudie, L.; Daniellou, R.; Auge, C.; Le Narvor, C. Enzymatic Supported Synthesis of Lacto-*N*-neotetraose Using Dendrimeric Polyethylene Glycol. *Carbohydr. Res.* **2004**, *339*, 693–698.
317. Li, Y.; Yu, H.; Chen, Y.; Lau, K.; Cai, L.; Cao, H.; Tiwari, V. K.; Qu, J.; Thon, V.; Wang, P. G.; Chen, X. Substrate Promiscuity of *N*-Acetylhexosamine 1-Kinases. *Molecules* **2011**, *16*, 6396–6407.
318. Chen, Y.; Thon, V.; Li, Y.; Yu, H.; Ding, L.; Lau, K.; Qu, J.; Hie, L.; Chen, X. One-Pot Three-Enzyme Synthesis of UDP-GlcNAc Derivatives. *Chem. Commun. (Cambridge)* **2011**, *47*, 10815–10817.
319. Guan, W.; Ban, L.; Cai, L.; Li, L.; Chen, W.; Liu, X.; Mrksich, M.; Wang, P. G. Combining Carbochips and Mass Spectrometry to Study the Donor Specificity for the *Neisseria meningitidis* β1,3-*N*-Acetylglucosaminyltransferase LgtA. *Bioorg. Med. Chem. Lett.* **2011**, *21*, 5025–5028.
320. Muthana, M. M.; Qu, J.; Li, Y.; Zhang, L.; Yu, H.; Ding, L.; Malekan, H.; Chen, X. Efficient One-Pot Multienzyme Synthesis of UDP-Sugars Using a Promiscuous UDP-Sugar Pyrophosphorylase from *Bifidobacterium longum* (BLUSP). *Chem. Commun. (Cambridge)* **2012**, *48*, 2728–2730.
321. Lau, K.; Thon, V.; Yu, H.; Ding, L.; Chen, Y.; Muthana, M. M.; Wong, D.; Huang, R.; Chen, X. Highly Efficient Chemoenzymatic Synthesis of β1-4-Linked Galactosides with Promiscuous Bacterial β1-4-Galactosyltransferases. *Chem. Commun. (Cambridge)* **2010**, *46*, 6066–6068.
322. Sun, M.; Li, Y.; Chokhawala, H. A.; Henning, R.; Chen, X. N-Terminal 112 Amino Acid Residues Are not Required for the Sialyltransferase Activity of *Photobacterium damsela* α2,6-Sialyltransferase. *Biotechnol. Lett.* **2008**, *30*, 671–676.
323. Cheng, J.; Yu, H.; Lau, K.; Huang, S.; Chokhawala, H. A.; Li, Y.; Tiwari, V. K.; Chen, X. Multifunctionality of *Campylobacter jejuni* Sialyltransferase CstII: Characterization of GD3/GT3 Oligosaccharide Synthase, GD3 Oligosaccharide Sialidase, and Trans-Sialidase Activities. *Glycobiology* **2008**, *18*, 686–697.
324. Yu, H.; Cheng, J.; Ding, L.; Khedri, Z.; Chen, Y.; Chin, S.; Lau, K.; Tiwari, V. K.; Chen, X. Chemoenzymatic Synthesis of GD3 Oligosaccharides and Other Disialyl Glycans Containing Natural and Non-Natural Sialic Acids. *J. Am. Chem. Soc.* **2009**, *131*, 18467–18477.

325. Dumon, C.; Priem, B.; Martin, S. L.; Heyraud, A.; Bosso, C.; Samain, E. In Vivo Fucosylation of Lacto-*N*-neotetraose and Lacto-*N*-neohexaose by Heterologous Expression of *Helicobacter pylori* α1,3-Fucosyltransferase in Engineered *Escherichia coli*. *Glycoconjugate J.* **2001**, *18*, 465–474.
326. Drouillard, S.; Driguez, H.; Samain, E. Large-Scale Synthesis of H-Antigen Oligosaccharides by Expressing *Helicobacter pylori* α1,2-Fucosyltransferase in Metabolically Engineered *Escherichia coli* Cells. *Angew. Chem. Int. Ed.* **2006**, *45*, 1778–1780.
327. Dumon, C.; Samain, E.; Priem, B. Assessment of the Two *Helicobacter pylori* α-1,3-Fucosyltransferase Ortholog Genes for the Large-Scale Synthesis of Lewis^x Human Milk Oligosaccharides by Metabolically Engineered *Escherichia coli*. *Biotechnol. Prog.* **2004**, *20*, 412–419.
328. Baumgartner, F.; Conrad, J.; Sprenger, G. A.; Albermann, C. Synthesis of the Human Milk Oligosaccharide Lacto-*N*-tetraose in Metabolically Engineered, Plasmid-Free *E. coli*. *ChemBioChem* **2014**, *15*, 1896–1900.
329. Baumgartner, F.; Sprenger, G. A.; Albermann, C. Galactose-Limited Fed-Batch Cultivation of *Escherichia coli* for the Production of Lacto-*N*-tetraose. *Enzyme Microb. Technol.* **2015**, *75–76*, 37–43.
330. Sakurama, H.; Fushinobu, S.; Hidaka, M.; Yoshida, E.; Honda, Y.; Ashida, H.; Kitaoka, M.; Kumagai, H.; Yamamoto, K.; Katayama, T. 1,3-1,4-α-L-Fucosynthase that Specifically Introduces Lewis$^{a/X}$ Antigens into Type-1/2 Chains. *J. Biol. Chem.* **2012**, *287*, 16709–16719.
331. Koizumi, S.; Endo, T.; Tabata, K.; Ozaki, A. Large-Scale Production of UDP-Galactose and Globotriose by Coupling Metabolically Engineered Bacteria. *Nat. Biotechnol.* **1998**, *16*, 847–850.
332. Koizumi, S.; Endo, T.; Tabata, K.; Nagano, H.; Ohnishi, J.; Ozaki, A. Large-Scale Production of GDP-Fucose and Lewis^X by Bacterial Coupling. *J. Ind. Microbiol. Biotechnol.* **2000**, *25*, 213–217.
333. Dumon, C.; Bosso, C.; Utille, J. P.; Heyraud, A.; Samain, E. Production of Lewis^X Tetrasaccharides by Metabolically Engineered *Escherichia coli*. *ChemBioChem* **2006**, *7*, 359–365.
334. Antoine, T.; Priem, B.; Heyraud, A.; Greffe, L.; Gilbert, M.; Wakarchuk, W. W.; Lam, J. S.; Samain, E. Large-Scale in vivo Synthesis of the Carbohydrate Moieties of Gangliosides GM1 and GM2 by Metabolically Engineered *Escherichia coli*. *ChemBioChem* **2003**, *4*, 406–412.
335. Wang, Z.; Chinoy, Z. S.; Ambre, S. G.; Peng, W.; McBride, R.; de Vries, R. P.; Glushka, J.; Paulson, J. C.; Boons, G. J. A General Strategy for the Chemoenzymatic Synthesis of Asymmetrically Branched *N*-Glycans. *Science* **2013**, *341*, 379–383.
336. Li, L.; Liu, Y.; Ma, C.; Qu, J.; Calderon, A. D.; Wu, B.; Wei, N.; Wang, X.; Guo, Y.; Xiao, Z.; Song, J.; Sugiarto, G.; Li, Y.; Yu, H.; Chen, X.; Wang, P. G. Efficient Chemoenzymatic Synthesis of an *N*-Glycan Isomer Library. *Chem. Sci.* **2015**, *6*, 5652–5661.

AUTHOR INDEX

Note: Page numbers in Roman type indicate that the listed author is cited on that page of an article in this volume; numbers in italic denote the reference number, in the list of references for that article, where the literature citation is given.

A

Adav, S.S., 70, *43*, 90, *43*, 91, *43*, 91, *220*, 93, *43*, 95, *43*
Adibekian, A., 36, *101*, 36, *102*, 36, *103*
Adney, W.S., 65, *6*, 65, *7*, 66, *7*, 66, *17*, 66, *18*, 66, *19*, 66, *20*, 66, *24*, 68, *24*, 70, *20*, 71, *17*, 72, *18*, 72, *19*, 72, *20*, 72, *24*, 72, *51*, 73, *20*, 73, *24*, 73, *84*, 74, *18*, 74, *20*, 75, *17*, 75, *18*, 76, *126*, 77, *19*, 77, *126*, 79, *24*, 80, *24*, 84, *51*, 87, *18*, 87, *20*, 90, *17*, 90, *18*, 90, *20*
Ahlgren, S., 72, *53*, 83, *53*
Aibe, S., 43, *123*
Aitio, O., 152, *163*
Ajisaka, K., 156, *210*, 168, *210*
Akahori, M., 150, *116*, 150, *117*
Akao, T., 87, *190*
Akcapinar, G.B., 84, *173*, 85, *173*
Akira, N., 87, *197*
Akita, O., 87, *190*
Alais, J., 40, *110*, 41, *112*
Alapuranen, M., 72, *57*
Albermann, C., 160, *284*, 160, *285*, 161, *289*, 167, *328*, 168, *328*, 168, *329*
Albert, M., 160, *282*
Aldrovandi, G.M., 155, *197*, 156, *208*, 158, *197*
Alekozai, E.M., 72, *60*, 74, *60*
Al-Mafraji, K., 32, *73*
Alper, P.B., 43, *120*
Altaye, M., *10*, 116, *10*, 150, 155, *200*, 156, *205*, 156, *214*
Alton, G., 152, *160*, 152, *166*, 152, *167*
Aly, M.R.E., 158, *238*, 158, *241*
Amano, J., 121, *58*, 142, *58*, 143, *58*, 144, *58*, 145, *58*, 146, *58*, 147, *58*, 151, *125*, 152, *145*
Amaral, A.L., 73, *68*, 73, *69*
Ambre, S.G., 170, *335*
Ames, J.B., 162, *296*, 166, *296*
Amster, I.J., 32, *73*
An, H.J., 117, *48*, 150, *48*, 151, *48*
Ando, T., 73, 77, 73, *92*, 154, *186*
Andre, L., 72, *52*, 83, *52*
Angulo, J., 23, *14*, 28, *41*, 28, *42*
Annila, A., 73, *67*, 79, *67*
Anokhin, A.P., 117, *46*
Anraku, T., 151, *128*
Antoine, T., 169, *334*
Antonieto, A.C.C., 91, *211*, 93, *211*, 95, *211*
Antus, S., 29, *49*, 29, *50*, 30, *64*
Apon, J.V., 158, *256*
Appelmelk, B.J., 159, *260*
Appert, H.E., 152, *144*
Arai, I., 150, *116*, 150, *117*, 150, *121*, 151, *121*
Arantes, V., 73, *101*, 74, *116*, 81, *101*, 81, *148*
Archer, D.B., 87, *189*
Arco, S.D., 32, *74*, 32, *77*
Argyros, R., 88, *205*
Arnold, F.H., 75, *120*, 82, *120*, 83, *165*, 83, *120*, 83, *166*
Arnous, A., 162, *300*, 163, *300*
Aro, N., 95, *231*
Arungundram, S., 32, *73*
Arvas, M., 72, *59*, 95, *231*
Asada, Y., 87, *190*
Asakuma, S., 150, *116*, 150, *117*, 154, *182*, 155, *182*
Ashida, H., 93, *226*, 154, *181*, 154, *182*, 154, *184*, 154, *185*, 154, *186*, 154, *187*, 154, *188*, 155, *182*, 168, *330*
Aslam, N., 70, *44*, 76, *44*, 90, *44*, 91, *44*, 93, *44*
Asong, J., 32, *73*
Aspinall, G.O., 3, *1*, 3, *3*, 3, *4*, 3, *5*, 3, *6*
Athanasopoulos, V.I., 91, *213*
Atochina, O., 156, *211*

Atterbury, R., 87, *193*
Auer, S., 72, *61*, 74, *61*
Auge, C., 152, *155*, 165, *316*
Autran, C.A., 158, *230*, 159, *269*
Avizienyte, E., 39, *105*, 39, *106*, 39, *108*
Axelsson, J., 23, *5*

B

Baath, J.A., 95, *233*
Badsara, S.S., 41, *115*
Baer, H.H., 116, *24*, 116, *25*, 116, *26*, 120, *24*, 123, *25*, 124, *24*, 124, *26*, 124, *81*, 124, *83*
Baggett, N., 33, *89*, 44, *89*
Bai, Y., 93, *229*
Baier, W., 121, *57*, 141, *57*, 150, *57*, 153, *57*
Bailey, M., 72, *52*, 83, *52*
Baker, J.O., 66, *17*, 66, *18*, 71, *17*, 72, *18*, 72, *51*, 74, *18*, 75, *17*, 75, *18*, 82, *156*, 84, *51*, 87, *18*, 90, *17*, 90, *18*
Baker, S.E., 65, *7*, 66, *7*, 72, *59*, 95, *234*
Balan, V., 73, *78*
Baldassano, R.N., 117, *46*
Ballard, O., 150, *113*
Ballew, N., 86, *181*
Ban, L., 166, *319*
Banerjee, G., 65, *4*, 66, *4*
Bankowska, R., 88, *198*
Bansal, P., 81, *150*
Bao, X., 83, *167*, 83, *170*, 84, *167*, 84, *170*
Barabote, R., 72, *59*
Barath, M., 39, *104*, 39, *105*, 39, *109*
Barboza, M., 154, *189*, 155, *189*
Bardet, L., 116, *12*
Barile, D., 153, *175*, 154, *175*
Barker, W.A., 162, *303*
Barnhart, D.C., 158, *235*
Barone, G., 48, *138*, 48, *139*
Barroca, N., 29, *51*, 30, *51*
Barsberg, S., 73, *100*, 81, *100*
Bartolini, B., 23, *5*
Bartolozzi, A., 28, *40*, 28, *46*, 29, *46*
Bascou, A., 41, *112*
Basten, J., 30, *54*
Batt, C.A., 86, *179*
Bauer, S., 91, *216*
Baumgartner, F., 161, *289*, 167, *328*, 168, *328*, 168, *329*
Baumler, A.J., 155, *203*
Bayer, R., 162, *305*
Bayram Akcapinar, G., 82, *154*, 84, *174*, 85, *174*
Bazin, H.G., 46, *132*, 46, *133*
Beauchamp, J.L., 93, *79*
Becker, D., 74, *119*, 75, *119*
Beckham, G.T., 65, 7, 66, 7, 66, *18*, 66, *19*, 66, *23*, 66, *24*, 66, *25*, 68, *24*, 68, *25*, 72, *18*, 72, *19*, 72, *23*, 72, *24*, 72, *25*, 72, *58*, 72, *60*, 73, *23*, 73, *24*, 73, *25*, 73, *81*, 73, *82*, 73, *83*, 73, *84*, 73, *85*, 73, *86*, 73, *87*, 74, *18*, 74, *60*, 74, *117*, 75, *18*, 76, *25*, 76, *58*, 77, *19*, 77, *58*, 78, *23*, 78, *58*, 79, *24*, 79, *25*, 80, *24*, 80, *25*, 81, *25*, 81, *149*, 82, *25*, 87, *18*, 90, *18*, 90, *25*
Beetz, T., 30, *53*
Behrens, S.H., 81, *153*
Bellerose, R.J., 65, *1*
Benjamin, T.L., 157, *224*
Benkestock, K., 70, *45*
Benz, J.P., 91, *216*
Bereczky, Z., 30, *64*
Bergenstrahle, M., 66, *19*, 72, *19*, 77, *19*
Berger, E.G., 152, *145*, 152, *164*
Bergeron, J.M., 66, *13*
Bergmann, C., 87, *195*, 87, *196*
Berka, R.M., 72, *59*
Bernard, B.A., 76, *136*, 82, *136*
Bernard, T., 88, *206*
Bernhardt, P.V., 26, *35*
Bertino, E., 116, *11*, 121, *11*, 150, *11*, 151, *11*, 151, *133*
Bethell, G.S., 5, *66*, 5, *69*, 5, *73*, 5, *74*, 5, *78*, 5, *87*
Beyer, T.A., 152, *146*
Bhavanandan, V.P., 157, *221*
Bienenstock, J., 153, *178*, 156, *178*
Bier, J.L., 151, *134*
Bifidus Factor.V., 116, *20*
Bindels, J.G., 153, *174*
Bindschädler, P., 36, *102*, 36, *103*
Bird, K., 29, *48*
Blaauw, R., 158, *230*
Blakely, M.L., 158, *235*
Blanc, B., 116, *4*
Blanch, H.W., 73, *96*, 75, *125*, 81, *151*, 83, *125*, 83, *164*

Blanchard, S., 34, *99*, 35, *99*
Blaney, S., 117, *42*
Blaser, M.J., 159, *260*
Blattner, R., 5, *89*, 5, *105*, 5, *106*, 5, *114*, 5, *135*, 5, *140*, 5, *142*, 5, *146*, 5, *147*, 5, *159*, 5, *163*, 5, *168*, 5, *171*, 5, *172*, 5, *173*, 5, *175*, 5, *176*, 5, *179*, 5, *180*, 5, *181*, 5, *183*, 5, *187*, 5, *188*, 5, *191*, 7, *172*
Bliss, L.A., 155, *194*
Bloom, M., 82, *157*
Blunt, J.W., 5, *101*, 7, *101*
Bobrowicz, B., 86, *185*, 86, *186*, 86, *188*, 88, *188*
Bobrowicz, P., 86, *181*, 86, *185*, 86, *186*, 86, *187*, 86, *188*, 88, *188*
Bode, L., 117, *38*, 117, *44*, 118, *38*, 152, *38*, 153, *38*, 153, *44*, 153, *168*, 153, *169*, 153, *176*, 155, *194*, 155, *197*, 156, *176*, 156, *207*, 156, *208*, 156, *212*, 157, *38*, 157, *176*, 157, *212*, 158, *197*, 158, *230*, 158, *236*, 159, *269*
Bodenheimer, A.M., 77, *139*
Boehm, G., 150, *114*, 157, *218*
Boer, H., 66, *26*, 72, *26*, 72, *54*, 74, *118*, 74, *119*, 75, *118*, 75, *119*, 84, *26*, 90, *26*
Boisbrun, M., 35, *100*
Bok, J.W., 95, *234*
Bolam, D.N., 73, *64*, 79, *64*, 80, *64*
Bols, M., 50, *146*, 50, *147*, 51, *147*, 52, *146*
Bomble, Y.J., 66, *19*, 72, *19*, 77, *19*
Bommarius, A.S., 81, *150*, 81, *153*
Bond-Watts, B.B., 65, *1*, 65, *2*
Bonnaffé, D., 40, *110*, 41, *111*, 41, *112*
Bonnet, H., 116, *12*
Boolieris, D.S., 5, *77*
Boons, G.J.A, 30, *57*, 32, *73*, 170, *335*
Boonvitthya, N., 84, *175*, 85, *175*
Boraston, A.B., 73, *62*, 73, *63*, 73, *64*, 79, *64*, 80, *62*, 80, *63*, 80, *64*, 81, *62*, 81, *63*
Borbás, A., 29, *49*, 29, *50*, 30, *64*
Borch, K., 73, *80*, 73, *89*, 73, *90*, 74, *89*, 74, *90*
Boren, T., 156, *209*
Börjesson, J., 76, *129*
Bosso, C., 165, *325*, 166, *325*, 169, *333*
Bouchara, J.P., 157, *220*
Boulineau, F.P., 47, *134* , 47, *135*
Boulton, P.G., 161, *286*
Bouquelet, S., 122, *63*, 125, *86*
Bovin, N.V., 126, *89*, 157, *217*
Bozonnet, S., 84, *175*, 85, *175*
Brady, J.W., 65, *6*, 66, *19*, 72, *19*, 77, *19*
Branda, L.A., 5, *77*
Brassart, D., 156, *206*
Breckler, F., 158, *233*
Bretthauer, R.K., 86, *178*
Brettin, T.S., 72, *59*
Brevnova, E., 72, *53*, 83, *53*
Brew, K., 151, *139*
Brimacombe, J.S., 5, *64*, 5, *70*, 5, *71*, 5, *76*, 5, *82*, 5, *86*, 5, *97*, 5, *100*, 5, *103*
Brisson, J.-R., 28, *44*
Brito, N., 66, *14*, 71, *14*, 78, *14*
Broberg, K.R., 39, *105*
Brodbeck, U., 151, *138*
Broder, W., 158, *255*
Brossmer, R., 123, *78*
Brotherus, E., 152, *162*
Brown, K.L., 5, *119*, 6, *119*
Brown, N.A., 91, *211*, 93, *211*, 95, *211*
Brown, R.L., 158, *235*
Bruce, D., 72, *59*
Brumer, H., 70, *46*
Brunecky, R., 66, *18*, 72, *18*, 72, *58*, 74, *18*, 75, *18*, 76, *58*, 77, *58*, 78, *58*, 87, *18*, 90, *18*
Bruntz, R., 133, *100*, 134, *100*
Bruun, L.M., 158, *258*
Bryan, M.B., 75, *125*, 83, *125*
Bu, L., 66, *19*, 66, *24*, 68, *24*, 72, *19*, 72, *24*, 73, *24*, 73, *84*, 73, *85*, 77, *19*, 79, *24*, 80, *24*
Bubner, P., 73, *97*
Buck, R.H., 153, *178*, 156, *178*
Bundle, D.R., 152, *153*
Burapatana, V., 84, *175*, 85, *175*
Bush, C.A., 127, *93*
Busson, R., 96, *236*
Butto, L.F., 157, *223*
Byramova, N.E., 157, *217*

C

Cadirgi, H., 86, *184*
Cahill, P., 155, *198*
Cahill, T., 156, *215*
Cai, L., 166, *317*, 166, *319*

Cai, Y.W., 83, *169*, 88, *169*
Calderon, A.D., 170, *336*
Callewaert, N., 86, *183*, 86, *184*, 89, *208*, 90, *208*, 93, *208*
Calvert, N., 5, *71*
Calvin, M., 4, *12*
Camattari, A., 84, *176*, 85, *176*
Cantarel, B.L., 88, *206*
Cao, H., 160, *277*, 160, *279*, 166, *317*
Cao, S., 158, *246*
Cao, X., 32, *75*
Capila, I., 23, *6*, 30, *65*, 32, *65*
Caporaso, J.G., 117, *46*
Capretti, R., 157, *216*
Caputo, R., 49, *144*
Carloni, I., 116, *8*, 121, *8*, 150, *8*
Carlucci, A., 116, *8*, 121, *8*, 150, *8*
Carvalho, J., 73, *68*, 73, *70*, 76, *70*
Casadio, R., 82, *154*
Cassanas, G., 116, *12*
Castro-Palomino, J., 158, *254*
Casu, B., 23, *1*, 23, *3*, 23, *12*, 23, *13*
Catley, B.J., 5, *100*, 5, *103*
Ceppellini, R., 151, *141*
Cereghino, G.P.L., 84, *171*
Cereghino, J.L., 84, *171*
Cervantes, L.E., 155, *204*
Cha, J.K., 33, *85*
Chabasse, D., 157, *220*
Chadha, B.S., 91, *218*
Chaffey, P.K., 66, *25*, 68, *25*, 72, *25*, 73, *25*, 76, *25*, 79, *25*, 80, *25*, 81, *25*, 82, *25*, 90, *25*
Chai, W., 138, *105*, 139, *105*, 142, *107*
Chaiwangsri, T., 154, *185*
Chan, K.H., 159, *260*
Chan, N.W., 161, *286*
Chan, S.Y., 121, *60*, 150, *60*
Chandra, R.P., 73, *101*, 73, *104*, 73, *106*, 81, *101*, 81, *104*, 81, *106*
Chang, C.-H., 32, *74*
Chang, C.-L., 32, *74*
Chang, M.C., 65, *1*, 65, *2*
Chang, M.D.-T., 32, *70*
Chang, S.H., 159, *268*
Chang, S.K., 41, *116*
Chang, S.-W., 45, *127*, 45, *128*, 45, *131*
Chang, W., 32, *80*
Chao, L.T., 70, *43*, 90, *43*, 91, *43*, 93, *43*, 95, *43*
Chapleur, Y., 35, *100*
Chapman, J., 72, *59*
Chaturvedi, P., *10*, 116, *10*, 150, 156, *205*, 156, *214*
Chau, B.H., 91, *216*
Chavez-Munguia, B., 155, *204*
Chekkor, A., 125, *86*
Chen, C., 32, *82*, 155, *199*
Chen, C.-S., 44, *124*, 45, *124*, 45, *127*, 45, *131*
Chen, C.-Y., 32, *78*
Chen, G., 66, *22*, 72, *22*, 74, *22*, 75, *22*, 76, *22*, 90, *22*
Chen, H.L., 83, *169*, 88, *169*
Chen, J., 32, *72*
Chen, L., 66, *23*, 66, *25*, 68, *25*, 71, *48*, 72, *23*, 72, *25*, 73, *23*, 73, *25*, 76, *25*, 78, *23*, 79, *25*, 80, *25*, 81, *25*, 82, *25*, 90, *25*
Chen, M.T., 88, *205*
Chen, W., 166, *319*
Chen, X., 151, *142*, 154, *192*, 159, *262*, 159, *264*, 159, *265*, 159, *269*, 159, *270*, 159, *271*, 159, *272*, 159, *273*, 160, *277*, 160, *279*, 161, *142*, 161, *293*, 161, *294*, 162, *295*, 162, *296*, 163, *309*, 166, *293*, 166, *296*, 166, *309*, 166, *317*, 166, *318*, 166, *320*, 166, *321*, 166, *322*, 166, *323*, 166, *324*, 170, *336*
Chen, Y., 166, *317*, 166, *318*, 166, *321*, 166, *324*
Chen, Z.-G., 32, *78*
Cheng, G., 47, *134*
Cheng, J., 160, *277*, 166, *323*, 166, *324*
Cheng, X., 72, *60*, 74, *60*
Chernyak, A., 158, *249*
Chertkov, O., 72, *59*
Chessa, D., 155, *203*
Chester, M.A., 122, *62*, 148, *62*
Chevailler, A., 157, *220*
Chew, S., 5, *127*, 5, *137*, 5, *150*
Chi, F.-C., 45, *125*, 45, *127*, 45, *128*, 45, *129*
Chiba, T., 34, *90*, 34, *93*, 158, *240*, 158, *251*
Chiba, Y., 93, *226*
Chichlowski, M., 117, *43*, 153, *43*, 153, *176*
Chida, N., 33, *87*

Chin, S., 166, *324*
Chin, Y.W., 161, *288*
Chinoy, Z.S., 170, *335*
Chiu, M.H., 152, *156*
Chizhov, O.S., 5, *58*
Cho, H.Y., 83, *169*, 88, *169*
Choay, J., 23, *12*, 23, *13*, 23, *93*, 28, *39*, 29, *47*
Choi, B.K., 86, *181*, 86, *187*, 86, *188*, 88, *188*
Chokhawala, H.A., 83, *164*, 159, *270*, 161, *294*, 163, *309*, 166, *309*, 166, *322*, 166, *323*
Chou, Y.-C., 66, *17*, 66, *18*, 71, *17*, 72, *18*, 74, *18*, 75, *17*, 75, *18*, 87, *18*, 90, *17*, 90, *18*
Choudhury, B., 158, *236*
Christiansen, M.N., 71, *49*
Chu, J.-W., 74, *115*
Chulalaksananukul, W., 84, *175*, 85, *175*
Chundawat, S.P., 73, *78*, 81, *149*
Chwals, W.J., 158, *235*
Chyan, C.-L., 32, *77*
Ciment, D.M., 5, *34*, 5, *35*
Claeyssens, M., 67, *29*, 68, *29*, 68, *30*, 68, *33*, 68, *34*, 70, *29*, 70, *30*, 70, *33*, 70, *34*, 70, *37*, 70, *38*, 70, *42*, 70, *46*, 72, *30*, 72, *33*, 72, *34*, 72, *42*, 74, *119*, 75, *119*, 84, *29*, 84, *30*, 84, *33*, 84, *34*, 84, *37*, 84, *38*, 87, *33*, 87, *34*, 88, *33*, 88, *34*, 89, *33*, 89, *34*, 89, *37*, 89, *38*, 89, *42*, 89, *208*, 90, *33*, 90, *34*, 90, *38*, 90, *208*, 92, *30*, 92, *42*, 93, *33*, 93, *34*, 93, *208*
Clark, D.S., 73, *96*, 75, *125*, 81, *151*, 83, *125*, 83, *164*
Clark, R.H., 158, *234*
Clarke, M.J., 158, *232*
Cleary, T.G., 155, *202*
Clemens, B., 117, *42*
Clemente, J.C., 117, *46*
Clinch, K., 5, *171*, 5, *173*, 5, *175*, 5, *176*, 5, *179*, 5, *180*, 5, *183*, 5, *187*
Clore, G.M., 79, *147*
Clowers, B.H., 117, *30*, 118, *30*, 121, *30*, 150, *30*, 156, *30*
Clyne, M., 157, *223*
Codée, J.D.C., 24, *22*, 24, *24*, 25, *22*, 25, *24*, 30, *58*
Cole, C., 39, *106*, 39, *108*
Cole, J.A., 159, *261*
Collen, A., 70, *46*
Collins, P., 5, *174*, 7, *174*
Collins, P.M., 5, *68*, 7, *68*
Conrad, J., 167, *328*, 168, *328*
Contreras, H., 89, *208*, 90, *208*, 90, *209*, 93, *208*
Contreras, M., 117, *46*
Contreras, R., 67, *29*, 68, *29*, 68, *34*, 68, *36*, 70, *29*, 70, *34*, 70, *36*, 70, *40*, 72, *34*, 84, *29*, 84, *34*, 86, *184*, 87, *34*, 88, *34*, 89, *34*, 89, *208*, 90, *34*, 90, *208*, 90, *209*, 93, *34*, 93, *208*, 96, *236*
Cook, W.J., 86, *181*
Coombe, D.R., 23, *17*
Coorevits, A., 70, *42*, 72, *42*, 89, *42*, 92, *42*
Copeland, R.J., 66, *15*
Coppa, G.V., 116, *8*, 116, *11*, 121, *8*, 121, *11*, 150, *8*, 150, *11*, 151, *11*, 157, *216*
Corzana, F., 23, *14*
Costello, C.E., 121, *60*, 150, *60*
Coutinho, P.M., 72, *59*, 88, *206*
Cowan, P., 151, *127*
Cravioto, A., 155, *201*
Cregg, J.M., 84, *171*
Cremata, J.A., 70, *45*
Cresi, F., 151, *133*
Crowley, M.F., 66, *19*, 66, *23*, 66, *24*, 68, *24*, 72, *19*, 72, *23*, 72, *24*, 72, *58*, 72, *60*, 73, *23*, 73, *24*, 73, *81*, 73, *82*, 73, *83*, 73, *84*, 73, *85*, 73, *86*, 73, *87*, 74, *60*, 74, *117*, 76, *58*, 77, *19*, 77, *58*, 78, *23*, 78, *58*, 79, *24*, 80, *24*
Crucello, A., 91, *215*
Cruys-Bagger, N., 73, *80*, 73, *89*, 73, *90*, 74, *89*, 74, *90*
Csuk, R., 25, *34*, 26, *34*
Cukan, M., 86, *181*
Cullen, D., 72, *59*
Culyba, E.K., 75, *121*
Cummings, R., 151, *143*
Cuneo, M.J., 77, *139*
Cunningham, A.M., 162, *305*
Czabany, T., 162, *298*, 162, *299*, 163, *298*
Czechura, P., 32, *76*
Czinege, E., 31, *68*, 31, *69*
Czjzek, M., 77, *143*

D

Da'aboul, I., 5, *71*
Dabrowski, J., 123, *76*, 133, *100*, 134, *100*
Dabrowski, U., 123, *76*, 133, *100*, 134, *100*
Dai, D., 117, *31*, 118, *30*
Dai, H., 30, *63*
Dai, Y.P., 32, *71*, 156, *209*
Dai, Z., 65, 7, 66, 7
Dale, B.E., 73, *78*, 81, *149*
D'Alonzo, D., 25, *31*, 49, *141*, 49, *142*, 49, *143*
Dalton, C.E., 40, *109*
Daly, R., 84, *172*
Damasceno, L.M., 86, *179*
Dana, C.M., 75, *125*, 83, *125*, 83, *164*
Danchin, E.G., 72, *59*
Daniel, H., 117, *36*
Daniellou, R., 165, *316*
Danishefsky, S.J., 158, *257*
Darboe, M.K., 150, *75*
Davidson, B., 5, *117*, 5, *118*, 5, *125*, 5, *126*, 5, *134*, 5, *141*, 5, *145*, 5, *149*, 5, *152*, 5, *154*
Davidson, R.C., 86, *181*, 86, *185*, 86, *186*, 86, *187*, 86, *188*, 88, *188*
Davies, G.J., 73, *64*, 79, *64*, 80, *64*
Davies, N.W., 151, *122*
Davies, R.B., 5, *143*, 7, *143*
Dax, K., 25, *33*, 160, *282*
De Bruyn, A., 67, *29*, 68, *29*, 68, *36*, 70, *29*, 70, *36*, 84, *29*, 96, *236*
de Jong, A.J.M., 30, *53*
De Lartigue, G., 117, *43*, 153, *43*
de Leon, A.L., 72, *59*
De Lorenzo, F., 48, *138*
de Paz, J.-L., 23, *14*, 26, *38*, 28, *38*, 28, *41*, 28, *42*, 41, *113*
De Pourcq, K., 86, *183*
De Raadt, A., 5, *148*, 5, *162*
De Schutter, K., 86, *183*
de Vries, R.P., 170, *335*
de Vries, T., 152, *158*, 152, *159*
Decker, S.R., 66, *17*, 66, *19*, 66, *20*, 70, *20*, 71, *17*, 72, *19*, 72, *20*, 72, *51*, 73, *20*, 74, *20*, 75, *17*, 77, *19*, 82, *156*, 84, *51*, 87, *20*, 90, *17*, 90, *20*
DeFrees, S., 162, *305*
Deguchi, H., 28, *45*, 29, *45*
Dehareng, D., 78, *145*
Dehlink, E., 157, *218*
Dejgaard, S., 66, *13*
del Vedovo, S., 155, *201*
Delabona Pda, S., 91, *215*
Deleault, K., 72, *53*, 83, *53*
Dell, A., 117, *29*
DeMaria, T.F., 155, *196*
den Haan, R., 72, *53*, 82, *157*, 82, *158*, 83, *53*, 83, *159*, 83, *160*
Denis, F., 30, *65*, 32, *65*
Dennis, N.J., 5, *30*
Denton, W.L., 151, *138*
DePeters, E.J., 150, *118*, 150, *119*, 151, *118*, 151, *119*
Desai, P., 154, *189*, 155, *189*
Deshpande, N., 66, *10*, 68, *10*, 84, *10*, 85, *10*, 87, *10*
Desmadril, M., 126, *89*
Detter, C., 72, *59*
Devaraj, N., 157, *221*
Devreese, B., 68, *33*, 70, *33*, 70, *37*, 70, *38*, 70, *41*, 70, *42*, 72, *33*, 72, *41*, 72, *42*, 84, *33*, 84, *37*, 84, *38*, 87, *33*, 88, *33*, 89, *33*, 89, *37*, 89, *38*, 89, *41*, 89, *42*, 90, *33*, 90, *38*, 92, *41*, 92, *42*, 93, *33*
Dewaele, S., 89, *208*, 90, *208*, 93, *208*
Dewerte, I., 70, *40*, 89, *208*, 90, *208*, 93, *208*
Dhar, A., 75, *121*, 75, *123*
Di Nicola, P., 151, *133*
Di Virgilio, S., 152, *165*
Dicken, B.J., 158, *233*
Didierjean, C., 35, *100*
Diehl, J.W., 5, *119*, 6, *119*
Dienes, D., 76, *129*
Difilippo, E., 164, *312*
Dilhas, A., 41, *111*, 41, *112*
Ding, L., 159, *269*, 166, *318*, 166, *320*, 166, *321*, 166, *324*
Ding, S.-Y., 65, *6*, 66, *17*, 71, *17*, 75, *17*, 90, *17*
Dinkelaar, J., 24, *24*, 25, *24*
Dinter, A., 152, *164*
Distler, J., 160, *285*
Divne, C., 74, *119*, 75, *119*
Divne,C., 71, *50*, 72, *50*
Do Vale, L.H., 91, *214*

Dolmans, M., 152, *146*
Dominguez-Bello, M.G., 117, *46*
Donald, A.S., 126, *67*, 149, *67*, 150, *67*, 152, *149*
Dondoni, A., 41, *118*, 42, *118*
Dong, X., 34, *94*
Donovan, S.M., 150, *120*, 151, *120*
dos Santos Castro, L., 91, *211*, 93, *211*, 95, *211*
Dotson-Fagerstrom, A., 83, *164*
Drake, M.R., 66, *25*, 68, *25*, 72, *25*, 73, *25*, 76, *25*, 79, *25*, 80, *25*, 81, *25*, 82, *25*, 90, *25*
Drakenberg, T., 73, *67*, 79, *67*
Driguez, H., 74, *119*, 75, *119*, 165, *326*
Driguez, P.-A., 23, *19*, 30, *56*
Drouillard, S., 164, *311*, 165, *326*
Drzeniek, Z., 116, *5*
Du, M., 152, *154*
Du, Y.G., 26, *37*
Dua, V.K., 127, *93*
Dube, V.E., 127, *93*
Dubey, M.K., 88, *201*, 92, *201*, 92, *224*
Dubois, A., 158, *227*
Duchaussoy, P., 28, *39*, 29, *47*, 30, *55*, 30, *56*
Duffner, J., 30, *65*, 32, *65*
Dulaney, S.B., 24, *25*, 25, *25*, 30, *61*, 30, *63*
Dumon, C., 126, *89*, 165, *325*, 166, *325*, 166, *327*, 169, *333*
Dunn-Coleman, N., 72, *59*
Duverger, V., 41, *112*

E

Ebel, F., 88, *203*
Ebersold, A., 133, *100*, 134, *100*
Ebner, K.E., 151, *138*
Echeverria, B., 32, *76*
Echtenacher, B., 88, *203*
Edwards, I.R., 5, *143*, 7, *143*
Egge, H., 117, *29*, 123, *76*, 125, *86*, 133, *100*, 134, *100*
Eibinger, M., 73, *97*
Eichler, E., 159, *275*
Eigel, W.N., 151, *136*
Eiwegger, T., 157, *218*
Ekino, K., 83, *162*, 85, *162*
El Ashry, E.S.H., 158, *238*, 158, *241*
El Hadri, A., 30, *65*, 32, *65*
El Ibrahim, S.I., 158, *238*, 158, *241*
El-Dakdouki, M.H., 30, *62*, 30, *63*
Elhalabi, J., 152, *157*
Elmerdahl, J., 73, *89*, 74, *89*
Endo, T., 152, *145*, 162, *297*, 162, *306*, 169, *331*, 169, *332*
Eneyskaya, E.V., 76, *130*, 89, *207*, 95, *207*
Engel, J., 88, *203*
Engfer, M.B., 117, *36*
Enright, G., 28, *44*
Eom, H.J., 154, *190*
Eppenberger-Castori, S., 152, *147*
Erasmuson, A.F., 5, *88*, 5, *101*, 7, *101*
Eriksen, S.H., 87, *191*
Eriksson, T., 70, *46*
Erlansson, K., 123, *80*, 131, *80*, 132, *52*
Erney, R., 150, *70*, 151, *142*, 151, *143*, 161, *142*
Esko, J.D., 23, *9*
Esnault, J., 158, *244*
Esterbauer, H., 76, *138*
Etzold, S., 117, *44*, 153, *44*
Evans, B.R., 77, *139*, 77, *140*
Evans, D.G., 158, *228*
Evans, D.J., 158, *228*
Evans, J.D., 73, *112*, 81, *112*

F

Fabris, C., 116, *11*, 121, *11*, 150, *11*, 151, *11*, 151, *133*
Faca, V.M., 91, *211*, 93, *211*, 95, *211*
Facinelli, B., 157, *216*
Fairbanks, A.J., 93, *228*
Fan, B.L., 156, *209*
Fan, R., 47, *134*
Fan, T.-c., 32, *70*
Fang, S.-l., 32, *70*
Farkas, T., 155, *200*, 156, *205*
Farrell, M.P., 92, *222*
Feeney, J., 126, *67*, 149, *67*, 150, *67*, 152, *149*
Fekete, E., 95, *232*
Felby, C., 73, *100*, 73, *114*, 81, *100*, 81, *114*
Feller, G., 78, *145*
Fellman, J.H., 123, *64*, 150, *64*
Fenger, T.H., 158, *258*
Fergusson, D.M., 5, *143*, 7, *143*

Fernandez-Mayoralas, A., 158, *237*
Fernig, D.G., 23, *8*
Ferrante, L., 157, *216*
Ferreira, E.C., 73, *68*, 73, *69*
Ferrier, R.J., 3, *1*, 3, *2*, 3, *3*, 3, *4*, 3, *5*, 3, *6*, 4, 7, 4, *8*, 4, *9*, 4, *10*, 4, *11*, 4, *12*, 4, *13*, 4, *16*, 4, *17*, 4, *18*, 4, *19*, 4, *20*, 4, *21*, 4, *22*, 4, *23*, 4, *24*, 4, *25*, 4, *26*, 4, *27*, 4, *28*, 4, *29*, 4, *41*, 5, 174, 5, *13*, 5, *14*, 5, *15*, 5, *23*, 5, *25*, 5, *30*, 5, *31*, 5, *32*, 5, *33*, 5, *34*, 5, *35*, 5, *36*, 5, *37*, 5, *38*, 5, *39*, 5, *40*, 5, *41*, 5, *42*, 5, *43*, 5, *44*, 5, *45*, 5, *46*, 5, *47*, 5, *48*, 5, *49*, 5, *50*, 5, *51*, 5, *52*, 5, *53*, 5, *54*, 5, *55*, 5, *56*, 5, *57*, 5, *58*, 5, *59*, 5, *60*, 5, *61*, 5, *62*, 5, *63*, 5, *64*, 5, *65*, 5, *66*, 5, *67*, 5, *68*, 5, *69*, 5, *70*, 5, *71*, 5, *72*, 5, *73*, 5, *74*, 5, *75*, 5, *76*, 5, *77*, 5, *78*, 5, *79*, 5, *80*, 5, *81*, 5, *82*, 5, *83*, 5, *84*, 5, *85*, 5, *86*, 5, *87*, 5, *88*, 5, *89*, 5, *90*, 5, *91*, 5, *92*, 5, *93*, 5, *94*, 5, *95*, 5, *96*, 5, *97*, 5, *98*, 5, *99*, 5, *100*, 5, *101*, 5, *102*, 5, *103*, 5, *104*, 5, *105*, 5, *106*, 5, *107*, 5, *108*, 5, *109*, 5, *110*, 5, *111*, 5, *112*, 5, *113*, 5, *114*, 5, *115*, 5, *116*, 5, *117*, 5, *118*, 5, *119*, 5, *120*, 5, *121*, 5, *122*, 5, *123*, 5, *124*, 5, *125*, 5, *126*, 5, *127*, 5, *128*, 5, *129*, 5, *130*, 5, *131*, 5, *132*, 5, *134*, 5, *135*, 5, *136*, 5, *137*, 5, *138*, 5, *139*, 5, *140*, 5, *141*, 5, *142*, 5, *143*, 5, *144*, 5, *145*, 5, *146*, 5, *147*, 5, *148*, 5, *149*, 5, *150*, 5, *151*, 5, *152*, 5, *153*, 5, *154*, 5, *155*, 5, *156*, 5, *157*, 5, *158*, 5, *159*, 5, *160*, 5, *161*, 5, *162*, 5, *163*, 5, *164*, 5, *165*, 5, *166*, 5, *167*, 5, *168*, 5, *169*, 5, *170*, 5, *171*, 5, *172*, 5, *173*, 5, *175*, 5, *176*, 5, *177*, 5, *178*, 5, *179*, 5, *180*, 5, *181*, 5, *182*, 5, *183*, 5, *184*, 5, *185*, 5, *186*, 5, *187*, 5, *188*, 5, *189*, 5, *190*, 5, *191*, 5, *192*, 5, *193*, 5, *194*, 6, *102*, 6, *116*, 6, *119*, 6, *120*, 6, *121*, 6, *122*, 6, *123*, 6, *136*, 7, *68*, 7, *101*, 7, *143*, 7, *144*, 7, *164*, 7, *166*, 7, *172*, 7, *174*, 34, *91*, 34, *92*
Ferro, D.R., 23, *12*, 23, *13*
Ferro, V., 23, *18*, 26, *35*
Field, R.A., 5, *187*, 5, *188*, 5, *191*
Fierfort, N., 163, *308*
Fiers, W., 67, *29*, 68, *29*, 70, *29*, 84, *29*
Fievre, S., 120, *55*, 127, *55*, 128, *95*, 138, *55*, 139, *55*
Finke, B., 117, *36*, 121, *60*, 132, *52*, 134, *99*, 150, *60*
Finne, J., 152, *147*, 157, *217*, 158, *229*
Fisher, A.J., 162, *296*, 166, *296*
Fisher, L.M., 155, *196*
Flicker, K., 84, *176*, 85, *176*
Ford, H.R., 158, *236*
Forster, M.J., 23, *10*
Forsythe, P., 153, *178*, 156, *178*
Fournet, B., 122, *63*
Foust, T.D., 65, *6*
Fox, J.M., 73, *96*, 81, *151*
Franca, N.C., 5, *88*
Fransson, L., 73, *65*, 79, *65*
Freeman, S.L., 115, *1*, 151, *118*, 117, *1*, 117, *30*, 118, *30*, 121, *30*, 150, *118*, 150, *30*, 150, *75*, 153, *1*, 156, *30*
Freeze, H.H., 90, *210*
Freitag, M., 95, *234*
Freitas, S., 91, *215*
Frihed, T.G., 50, *147*, 52, *146*
Froehlich, A., 72, *53*, 83, *53*
Fügedi, P., 30, *60*, 31, *60*, 31, *68*, 31, *69*, 32, *60*, 32, *81*
Fujita, K., 93, *225*, 93, *227*
Fujita, S., 151, *127*
Fujita, H., 87, *190*
Fukase, K., 29, *48*
Fukuda, K., 121, *61*, 123, *61*, 124, *61*, 126, *61*, 128, *61*, 130, *61*, 132, *61*, 133, *28*, 134, *61*, 136, *61*, 138, *61*, 140, *61*, 142, *61*, 144, *61*, 146, *61*, 148, *61*, 150, *61*, 151, *122*, 151, *126*, 151, *127*, 151, *128*, 151, *129*, 153, *61*
Fukuda, R., 151, *129*
Fukui, S., 128, *94*, 129, *97*, 130, *97*, 136, *103*, 136, *104*, 140, *104*, 141, *94*, 149, *111*
Funakoshi, I., 87, *192*, 128, *94*, 129, *97*, 130, *97*, 136, *103*, 136, *104*, 140, *104*, 141, *94*, 149, *111*
Funaoka, M., 73, *107*, 81, *107*
Furneaux, R.H., 5, *85*, 5, *90*, 5, *91*, 5, *92*, 5, *93*, 5, *95*, 5, *104*, 5, *110*, 5, *119*, 5, *126*, 5, *134*, 5, *141*, 5, *145*, 5, *149*, 5, *152*, 5, *154*, 5, *159*, 5, *163*, 5, *164*, 5, *166*, 5, *168*, 5, *171*, 5, *172*, 5, *173*, 5, *175*, 5, *176*, 5, *179*, 5, *180*, 5, *181*, 5, *183*, 5, *187*, 5, *191*, 5, *188*, 5, *193*, 6, *119*, 7, *164*, 7, *166*, 7, *172*, 34, *92*

Furukawa, J., 34, *97*
Furukawa, K., 83, *162*, 85, *162*, 88, *204*, 156, *210*, 168, *210*
Furuta, T., 72, *61*, 74, *61*
Fushinobu, S., 154, *187*, 168, *330*

G

Gabe, E.J., 5, *131*
Gabrielli, O., 116, *8*, 116, *11*, 121, *8*, 121, *11*, 150, *8*, 150, *11*, 151, *11*
Gadiel, P.A., 151, *124*
Gagneux, P., 117, *48*, 150, *48*, 151, *48*
Gainsford, G.J., 5, *119*, 5, *120*, 5, *121*, 5, *131*, 5, *136*, 6, *119*, 6, *120*, 6, *121*, 6, *136*
Gaiotto, T., 73, *95*
Galanina, O., 126, *89*
Galcheva-Gargova, Z., 30, *65*, 32, *65*
Galeazzi, T., 116, *11*, 121, *11*, 150, *11*, 151, *11*, 157, *216*
Gallagher, T.C., 5, *173*, 5, *175*
Gallego, R.G., 152, *164*
Galoyan, A.A., 34, *98*, 35, *98*
Gama, M., 73, *68*, 73, *69*, 73, *70*, 73, *71*, 73, *114*, 76, *70*, 81, *114*
Gambaryan, A.S., 157, *217*
Gan, Z., 158, *246*
Ganner, T., 73, *97*
Gao, D., 73, *78*
Gao, J., 93, *79*
Gao, Y., 162, *305*
Garcia, R., 70, *45*
Gardiner, J.M., 5, *176*, 5, *179*, 5, *180*, 5, *183*, 5, *187*, 5, *188*, 39, *104*, 39, *105*, 39, *106*, 39, *107*, 39, *108*, 39, *109*
Garman, J., 162, *302*
Garrido, D., 153, *175*, 154, *175*, 154, *190*, 154, *191*, 154, *193*
Garrison, D., 87, *195*
Gatti, G., 23, *12*, 23, *13*
Gauhe, A., 116, *23*, 116, *25*, 116, *26*, 120, *50*, 123, *25*, 123, 77, 124, *26*, 124, *66*, 124, *81*, 124, *83*, 124, *84*, 125, *66*, 126, *50*, 126, *90*, 149, *66*, 150, *66*
Gautier, C., 126, *89*
Gavard, O., 40, *110*, 41, *112*
Gelb, M.H., 34, *99*, 35, *99*
Gerlach, J.Q., 92, *222*
German, J.B., 115, *1*, 117, *1*, 117, *30*, 117, *33*, 117, *43*, 117, *48*, 118, *30*, 118, *33*, 121, *30*, 121, *33*, 122, *33*, 123, *33*, 124, *33*, 126, *33*, 127, *33*, 128, *33*, 130, *33*, 132, *33*, 133, *28*, 134, *33*, 135, *33*, 136, *33*, 138, *33*, 139, *33*, 140, *33*, 142, *33*, 144, *33*, 145, *33*, 146, *33*, 147, *33*, 148, *33*, 149, *33*, 150, *30*, 150, *48*, 150, *75*, 150, *118*, 150, *119*, 150, *120*, 151, *48*, 151, *118*, 151, *119*, 151, *120*, 153, *1*, 153, *43*, 154, *189*, 154, *192*, 154, *193*, 155, *33*, 155, *189*, 156, *30*, 156, *213*
German, J.B., 97, *34*, 98, *34*, 100, *34*, 101, *34*, 102, *34*, 103, *34*, 104, *34*, 106, *34*, 108, *34*, 110, *34*, 112, *34*, 113, *28*, 114, *34*, 115, *34*, 116, *34*, 117, *34*, 118, *34*, 120, *34*, 122, *34*, 124, *34*, 126, *34*, 128, *34*, 129, *34*, 130, *34*, 108, *34*, 116, *34*, 127, *34*, 129, *34*, 129, *34*, 115, *1*, 117, *1*, 153, *1*, 117, *30*, 118, *30*, 121, *30*, 150, *30*, 156, *30*, 117, *33*, 118, *33*, 122, *33*, 123, *33*, 124, *33*, 126, *33*, 128, *33*, 130, *33*, 132, *33*, 133, *28*, 134, *33*, 135, *33*, 136, *33*, 138, *33*, 139, *33*, 140, *33*, 142, *33*, 144, *33*, 145, *33*, 146, *33*, 148, *33*, 149, *33*, 155, *33*, 121, *33*, 127, *33*, 127, *33*, 147, *33*, 149, *33*, 117, *43*, 153, *43*, 117, *48*, 150, *48*, 151, *48*, 150, *75*, 150, *118*, 151, *118*, 150, *119*, 151, *119*, 150, *120*, 151, *120*, 154, *189*, 155, *189*, 154, *192*, 154, *193*, 156, *213*
Gerngross, T.U., 86, *180*, 86, *181*, 86, *186*, 86, *187*, 86, *188*, 88, *188*
Gerstmayr, M., 157, *218*
Geysens, S., 68, *34*, 70, *34*, 70, *40*, 72, *34*, 84, *34*, 86, *184*, 87, *34*, 87, *189*, 88, *34*, 89, *34*, 90, *34*, 93, *34*
Ghattyvenkatakrishna, P.K., 72, *60*, 74, *60*
Gibson, G.R., 117, *47*, 153, *47*
Gilbert, H.J., 73, *64*, 79, *64*, 80, *64*
Gilbert, M., 159, *274*, 159, *275*, 162, *305*, 163, *307*, 165, *307*, 169, *334*
Gilboa-Garber, N., 126, *89*
Gilkes, N.R., 76, *134*
Gill, M., 28, *43*
Gillgren, T., 73, *110*, 78, *110*, 81, *110*
Gillin, F.D., 155, *194*
Ginburg, V., 116, *5*

Ginsburg, V., 120, *51*, 120, *52*, 124, *82*, 126, *51*, 126, *88*, 127, *51*, 128, *51*, 131, *51*, 131, *52*, 132, *52*, 149, *73*, 149, *74*, 150, *71*, 150, *72*, 150, *73*, 150, *74*, 151, *71*, 151, *74*, 151, *130*
Giuliani, F., 116, *11*, 121, *11*, 150, *11*, 151, *11*, 151, *133*
Gladkikh, O., 5, *182*
Glass, N.L., 91, *216*
Glass, R.I., 158, *232*
Glieder, A., 84, *176*, 85, *176*, 86, *182*
Glushka, J., 152, *165*, 170, *335*
Gnanakaran, S., 73, *95*
Gnoth, M.J., 117, *37*
Godbole, S., 72, *51*, 84, *51*
Golbraikh, A.A., 157, *217*
Goldman, G.H., 91, *211*, 93, *211*, 95, *211*
Golliard, M., 156, *206*
Golubev, A.M., 89, *207*, 95, *207*
Gomes, J., 160, *282*
Gomez-Mendoza, D.P., 91, *214*
Gong, B., 86, *181*
Gonzalez, C., 66, *14*, 71, *14*, 78, *14*
Gonzalez, D., 151, *142*, 161, *142*
Gonzalez, M., 66, *14*, 71, *14*, 78, *14*
Goodwin, P.M., 73, *95*
Gooley, A.A., 68, *32*, 71, *32*, 72, *32*, 78, *32*, 79, *32*, 84, *32*
Gordon, J.I., 117, *46*
Gordon, P., 158, *234*
G_orgens, J.F., 82, *157*
Gorka-Niec, W., 66, *16*, 88, *16*, 88, *198*, 88, *200*, 92, *16*
Gorka-Niec,W., 88, *199*, 96, *199*
Goso, K., 127, *93*
Goth, K., 158, *236*
Goto, M., 66, *11*, 68, *11*, 71, *11*, 83, *162*, 85, *162*, 88, *204*
Gottlieb, H.E., 5, *88*
Gotz, A.W., 73, *83*
Gourvenec, F., 30, *56*
Govindarajan, S., 83, *166*
Graham, D.Y., 158, *228*
Grand, E., 24, *28*, 25, *28*
Grange, D.C.L., 83, *159*
Grathwohl, M., 158, *254*
Greene, E.R., 66, *25*, 68, *25*, 72, *25*, 73, *25*, 76, *25*, 79, *25*, 80, *25*, 81, *25*, 82, *25*, 90, *25*
Greffe, L., 169, *334*
Griffiths, M., 151, *124*
Grigoriev, I.V., 72, *59*
Grimm, R., 97, *34*, 98, *34*, 100, *34*, 101, *34*, 102, *34*, 103, *34*, 104, *34*, 106, *34*, 108, *34*, 110, *34*, 112, *34*, 113, *28*, 114, *34*, 115, *34*, 116, *34*, 117, *34*, 118, *34*, 120, *34*, 122, *34*, 124, *34*, 126, *34*, 128, *34*, 129, *34*, 130, *34*, 108, *34*, 116, *34*, 127, *34*, 129, *34*, 129, *34*, 117, *30*, 118, *30*, 121, *30*, 150, *30*, 156, *30*, 117, *33*, 118, *33*, 122, *33*, 123, *33*, 124, *33*, 126, *33*, 128, *33*, 130, *33*, 132, *33*, 133, *28*, 134, *33*, 135, *33*, 136, *33*, 138, *33*, 139, *33*, 140, *33*, 142, *33*, 144, *33*, 145, *33*, 146, *33*, 148, *33*, 149, *33*, 155, *33*, 121, *33*, 127, *33*, 127, *33*, 147, *33*, 149, *33*, 150, *118*, 151, *118*, 150, *119*, 151, *119*
Grimmonprez, L., 116, *12*, 125, *85*, 151, *131*
Grishin, A.V., 158, *236*
Grollman, E.F., 149, *74*, 150, *71*, 150, *72*, 150, *74*, 151, *71*, 151, *74*, 151, *130*
Gronberg, G., 123, *80*, 125, *87*, 129, *96*, 130, *96*, 131, *80*, 131, *87*, 132, *52*, 149, *96*
Gronenborn, A.M., 79, *147*
Gruber, F., 66, *12*
Gruber, K., 84, *176*, 85, *176*, 162, *298*, 163, *298*
Gruebele, M., 75, *121*, 75, *123*
Grulee, C., 117, *40*
Grünstein, D., 36, *103*
Guan, W., 166, *319*
Guaragna, A., 25, *31*, 49, *141*, 49, *142*, 49, *143*, 49, *144*
Guedes, N., 32, *76*
Guerardel, Y., 133, *65*, 134, *65*, 150, *65*
Guerrero, M.L., 156, *205*
Guerrero Mde, L., 156, *214*
Gul, O., 84, *173*, 84, *174*, 85, *173*, 85, *174*
Guner, Y.S., 158, *236*
Guo, Y., 162, *300*, 163, *300*, 170, *336*
Gusina, N.B., 34, *98*, 35, *98*
Gustafsson, R., 23, *5*
Gustavsson, M., 73, *65*, 79, *65*
Guthrie, R.D., 5, *46*, 5, *55*, 5, *59*, 5, *64*, 5, *70*, 5, *76*, 5, *82*

GyÇorgy, P., 154, *179*
Gyorgy, P., 116, *20*, 116, *21*, 116, *22*, 116, *23*

H

Haab, D., 76, *131*, 76, *133*
Hacker, J., 157, *226*
Haeuw-Fievre, S., 121, *56*, 131, *56*, 139, *56*, 140, *56*, 141, *56*
Hage, J.A., 162, *304*, 164, *304*
Hägglund, P., 76, *129*
Hagspiel, K., 76, *131*, 76, *133*
Haines, S.R., 5, *13*, 5, *129*, 5, *130*, 5, *131*, 5, *132*, 5, *140*
Hakes, D.J., 162, *303*
Hakkinen, M., 95, *231*
Haliloglu, T., 76, *128*, 77, *128*
Hall, D.W., 5, *165*, 5, *169*
Hall, M., 81, *150*, 81, *153*
Hallen-Adams, H.E., 70, *44*, 76, *44*, 90, *44*, 91, *44*, 93, *44*
Hallgren, P., 148, *108*
Hamaguchi, T., 92, *221*
Hamilton, C., 5, *191*
Hamilton, S.R., 86, *180*, 86, *181*, 86, *185*, 86, *186*, 86, *187*, 86, *188*, 88, *188*
Hammarstrom, L., 156, *209*
Han, C.S., 72, *59*
Han, J.J., 73, *95*
Han, N.S., 158, *259*, 161, *287*
Hannaford, A.J., 4, *27*
Hansen, S.U., 39, *104*, 39, *105*, 39, *106*, 39, *107*, 39, *108*, 39, *109*
Hanson, S.R., 75, *121*
Hansson, H., 70, *47*
Harle, R., 74, *119*, 75, *119*
Harmsen, H.J., 153, *174*
Harn, D.A., Jr., 155, *195*, 156, *195*, 156, *211*
Harris, M., 74, *119*, 75, *119*
Harris, P., 72, *59*
Harrison, M.J., 68, *32*, 71, *32*, 72, *32*, 78, *32*, 79, *32*, 84, *32*
Harrison, S.C., 157, *224*
Hart, G.W., 66, *15*
Hartner, F.S., 86, *182*
Harvey, S., 162, *302*
Hassan, H.H.A.M., 25, *30*
Hatakeyama, E., 154, *182*, 155, *182*
Hatton, L.R., 5, *38*, 5, *42*, 5, *43*, 5, *44*
Hay, R.W., 5, *72*
Hayashi, M., 160, *276*
Haymond, M.W., 151, *134*, 151, *135*
Hayn, M., 76, *138*
He, J., 77, *139*
Hearn, M.T., 84, *172*
Heath, A.C., 117, *46*
Heesemann, J., 88, *203*
Heidtman, M., 155, *199*
Heikkinen, H., 73, *113*, 81, *113*
Heikkinen, S., 152, *163*
Heinzelman, P., 83, *165*, 83, *166*
Helin, J., 152, *163*
Heller, W.T., 77, *140*
Helliwell, M., 39, *105*
Heman-Ackah, L.M., 158, *227*
Hemming, F.W., 87, *193*, 87, *194*
Hendrix, M., 43, *120*
Henker, J., 116, *14*, 151, *14*
Henning, R., 166, *322*
Henrissat, B., 72, *59*, 77, *143*, 88, *206*
Herbert, J.-M., 30, *56*
Herczeg, M., 29, *49*, 29, *50*, 30, *64*
Herdewijn, P., 67, *29*, 68, *29*, 68, *36*, 70, *29*, 70, *36*, 84, *29*, 96, *236*
Hernández-Torres, J.M., 47, *134*
Hersant, Y., 41, *112*
Hertzog, M., 117, *42*
Heyraud, A., 163, *307*, 165, *307*, 165, *325*, 166, *325*, 169, *333*, 169, *334*
Hickey, R.M., 157, *223*
Hidaka, M., 73, *76*, 78, *76*, 87, *197*, 168, *330*
Hidalgo, G., 117, *46*
Hie, L., 166, *318*
Higelin, J., 49, *145*
Higuchi, T., 123, *64*, 150, *64*
Hilfiker, M.L., 158, *235*
Hill, R.L., 151, *139*, 152, *146*
Hillenkamp, F., 150, *112*
Hilty, M., 150, *70*
Himmel, M.E., 65, *6*, 65, *7*, 66, *7*, 66, *17*, 66, *18*, 66, *19*, 66, *20*, 66, *23*, 66, *24*, 66, *25*, 68, *24*, 68, *25*, 70, *20*, 71, *17*, 72, *18*, 72, *19*, 72, *20*, 72, *23*, 72, *24*, 72, *25*, 72, *51*, 72, *58*, 73, *20*, 73, *23*, 73, *24*, 73, *25*,

73, *81*, 73, *82*, 73, *84*, 73, *85*, 73, *86*, 73, *87*, 74, *18*, 74, *20*, 74, *117*, 75, *17*, 75, *18*, 76, *25*, 76, *58*, 76, *126*, 77, *19*, 77, *58*, 77, *126*, 78, *23*, 78, *58*, 79, *24*, 79, *25*, 80, *24*, 80, *25*, 81, *25*, 81, *149*, 82, *25*, 82, *156*, 84, *51*, 87, *18*, 87, *20*, 90, *17*, 90, *18*, 90, *20*, 90, *25*
Hinde, K., 117, *39*, 117, *48*, 150, *48*, 151, *48*
Hindsgaul, O., 152, *152*, 152, *153*, 152, *154*, 152, *166*, 152, *167*
Hinou, H., 43, *121*, 43, *122*
Hiraiwa, N., 129, *97*, 130, *97*
Hiratake, J., 154, *183*
Hirayama, H., 92, *224*
Hirose, J., 154, *182*, 155, *182*
Hitomi, Y., 33, *84*, 33, *86*
Ho, I., 72, *59*
Ho, P.C., 83, *169*, 88, *169*
Hobdey, S.E., 82, *156*
Hoberg, J.O., 5, *188*, 5, *189*, 5, *191*
Hodosi, G., 45, *130*
Högermeier, J.A., 41, *114*
Hoj, P.B., 68, *30*, 70, *30*, 72, *30*, 84, *30*, 92, *30*
Holden, S.G., 5, *182*
Holman, R.C., 158, *232*
Holte, K., 30, *65*, 32, *65*
Honda, Y., 161, *292*, 168, *330*
Hong, J., 82, *155*
Hong, P., 156, *207*
Hooman, S., 72, *57*
Höonig, H., 25, *34*, 26, *34*
Hoopes, J.P., 86, *181*, 86, *187*
Hoorelbeke, K., 70, *42*, 72, *42*, 89, *42*, 92, *42*
Hoover, J.R., 116, *23*
Hopkins, D., 86, *187*
Horimoto, T., 154, *188*
Horta, M.A., 91, *215*
Hosomi, A., 92, *224*
Hosomi, O., 152, *150*
Hossain, T.J., 92, *224*
Hou, J., 83, *167*, 83, *170*, 84, *167*, 84, *170*
Houston-Cummings, N.R., 86, *181*
How, M.J., 5, *46*, 5, *55*
Hsiao, L., 155, *197*, 156, *208*, 158, *197*
Hsieh-Wilson, L.C., 41, *116*
Hsu, Y., 158, *239*
Hu, F., 73, *102*, 81, *102*
Hu, J., 74, *116*
Hu, Y.-P., 24, *26*, 25, *26*, 32, 77, 32, *78*, 32, *80*
Huang, C.J., 83, *169*, 86, *179*, 88, *169*, 12, 77
Huang, C.-Y., 32, *80*
Huang, H., 93, *229*
Huang, P., 155, *200*, 156, *215*
Huang, R., 166, *321*
Huang, S., 159, *270*, 160, *277*, 163, *309*, 166, *309*, 166, *323*
Huang, T.-Y., 32, *74*
Huang, X.F., 24, *25*, 25, *25*, 30, *61*, 30, *62*, 30, *63*
Hubschwerlen, C., 49, *145*
Hughes, N.A., 5, *82*, 5, *86*, 5, *97*
Hui, J.P.M., 68, *31*, 68, *35*, 71, *31*, 71, *35*, 84, *31*, 95, *31*, 95, *35*
Hung, K.Y., 83, *169*, 88, *169*
Hung, S.-C., 24, *26*, 24, *27*, 25, *26*, 25, *27*, 31, *27*, 31, *67*, 32, *70*, 32, *74*, 32, *77*, 32, *78*, 32, *79*, 32, *80*, 44, *124*, 45, *124*, 45, *125*, 45, *126*, 45, *127*, 45, *128*, 45, *129*, 45, *131*, 158, *239*
Hunt, D.K., 41, *114*
Hurford, J.R., 5, *79*
Hwang, D.R., 159, *268*
Hynes, S.O., 159, *260*

I

Iadonisi, A., 48, *138*, 48, *139*
Ichikawa, M., 90, *210*
Ichikawa, Y., 159, *266*, 159, *267*
Idanpaan-Heikkila, I., 155, *198*
Idota, T., 149, *68*, 150, *68*, 157, *219*
Igarashi, K., 72, *61*, 73, *72*, 73, *73*, 73, 77, 73, *91*, 73, *92*, 74, *61*
Iikura, H., 87, *197*
Ikeda, T., 34, *97*
Ikegami, S., 33, *84*, 33, *86*
Ilgen, C., 84, *171*
Ilia, S., 126, *89*
Illaszewicz, C., 160, *282*
Ilmen, M., 72, *53*, 83, *53*
Imaeda, T., 83, *168*
Imberty, A., 126, *89*

Inagaki, F., 128, *94*, 136, *103*, 136, *104*, 140, *104*, 141, *94*, 149, *111*
Inamori, H., 151, *126*
Inch, T.D., 5, *59*, 5, *64*, 5, *70*, 5, *76*
Inoue, Y., 92, *221*
Inukai, T., 164, *313*, 167, *313*
Iozzo, R.V., 23, *7*
Irene, D., 32, *77*
Ishihara, H., 158, *240*
Ishii, S., 43, *123*
Ito, F., 162, *301*
Ito, K., 92, *221*
Ito, T., 92, *221*
Ito, Y., 158, *250*
Itoh, M., 160, *276*
Ivanova, C., 65, *8*, 67, *8*, 95, *233*
Iwahara, S., 91, *212*
Iwai, Y., 33, *84*, 33, *86*
Iwamatsu, A., 93, *225*
Iwasaki, T.A, 162, *301*
Iwatsuki, K., 154, *180*

J

Jackson, M., 72, *59*
Jacobsen, I., 88, *203*
Jacquinet, J.-C., 23, *12*, 23, *13*, 28, *39*, 29, *47*, 29, *51*, 34, *93*, 42, *119*
Jakomin, M., 155, *203*
Jalak, J., 73, *93*, 73, *98*, 74, *93*, 74, *98*
Jambusaria, R.B., 81, *151*
Jantscher-Krenn, E., 153, *168*, 153, *169*, 155, *194*, 158, *236*, 159, *269*
Jardine, D.R., 68, *32*, 71, *32*, 72, *32*, 78, *32*, 79, *32*, 84, *32*
Jarrell, H.C., 28, *44*
Jaurand, G., 30, *54*, 30, *56*
Jayson, G.C., 39, *104*, 39, *105*, 39, *106*, 39, *107*, 39, *108*, 39, *109*
Jeanloz, R.W., 154, *179*
Jenness, R., 116, *3*
Jennum, C.A., 158, *258*
Jensen, B., 87, *191*
Jensen, D.F., 88, *201*, 92, *201*, 92, *223*, 92, *224*
Jensen, P.H., 71, *49*
Jensen, R.G., 116, *4*
Jeoh, T., 66, *18*, 66, *20*, 70, *20*, 72, *18*, 72, *20*, 73, *20*, 73, *95*, 74, *18*, 74, *20*, 75, *18*, 87, *18*, 87, *20*, 90, *18*, 90, *20*
Jers, C., 160, *283*, 162, *300*, 163, *300*
Jess, P., 81, *151*
Jia, Q., 161, *294*
Jiang, B., 88, *205*
Jiang, W., 73, *81*, 74, *117*
Jiang, X., 117, *41*, 155, *200*, 156, *205*, 156, *215*
Jiang, Y., 86, *181*, 86, *187*
Jikibara, T., 91, *212*
Jin, C., 88, *202*
Jin, Y.S., 156, *213*, 161, *287*
Jo, J.H., 161, *287*
Joachimiak, A., 154, *190*
Johansson, L.S., 73, *103*, 81, *103*
Johansson, M.H., 33, *88*
Johnson, D.K., 65, *6*
Johnson, K.F., 159, *263*
Johnson, P.H., 152, *148*, 152, *149*
Jones, J.K.N., 5, *30*
Jones,T.A., 71, *50*, 72, *50*, 79, *147*
Jonsson, L.J., 73, *110*, 78, *110*, 81, *110*
Joshi, L., 92, *222*, 157, *223*
Jung, J., 73, *95*
Jung, S., 73, *102*, 81, *102*

K

Kaariainen, L., 70, *39*
Kadentsev, V.I., 5, *58*
Kadla, J.F., 73, *104*, 81, *104*
Kajiwara, H., 164, *311*
Kajtár-Peredy, M., 30, *60*, 31, *60*, 32, *60*
Kal, S.M., 75, *125*, 83, *125*, 83, *164*
Kalicheva, I.S., 34, *98*, 35, *98*
Kallapur, S.G., 156, *215*
Kallio, J., 72, *57*
Kalliola, A., 73, *112*, 81, *112*
Kamerling, J.P., 152, *164*, 162, *304*, 164, *304*
Kanaan, A., 83, *165*
Kane, M., 92, *222*, 157, *223*
Kania, A., 88, *199*, 96, *199*
Kanie, O., 158, *250*
Kankasa, C., 155, *197*, 156, *208*, 158, *197*
Kappelmayer, J., 30, *64*
Karaffa, L., 95, *232*

Karamanos, N., 23, *4*
Karas, M., 121, *60*, 132, *52*, 134, *99*, 150, *60*, 150, *112*
Karimi-Aghcheh, R., 95, *232*, 95, *234*
Karkehabadi, S., 70, *41*, 70, *47*, 72, *41*, 89, *41*, 92, *41*
Karlsson, M., 88, *201*, 92, *201*, 92, *223*, 92, *224*
Karpel, R., 161, *293*, 161, *294*, 166, *293*
Karst, N.A., 24, *21*, 25, *21*
Kartha, K.P.R., 5, *187*, 5, *188*, 5, *191*
Kashem, M.A., 152, *160*
Katayama, T., 154, *182*, 154, *183*, 154, *184*, 154, *185*, 154, *186*, 154, *187*, 154, *188*, 155, *182*, 161, *291*, 161, *292*, 168, *330*
Kato, R., 161, *291*, 161, *292*
Kaundinya, G.V., 30, *65*, 32, *65*
Kaur, K.J., 152, *167*
Kavanaugh, D.W., 157, *223*
Kavunja, H., 30, *63*
Kawahawa, K., 150, *121*, 151, *121*
Kawai, Y., 150, *121*, 151, *121*
Kawakami, H., 157, *219*
Kawasaki, T., 87, *192*, 128, *94*, 129, *97*, 130, *97*, 136, *103*, 136, *104*, 140, *104*, 141, *94*, 149, *111*
Ke, H.M., 83, *169*, 88, *169*
Ke, W., 28, *43*, 28, *44*
Keddache, M., 156, *215*
Keenan, T.W., 151, *136*
Kelder, B., 151, *142*, 151, *143*, 161, *142*
Keller, N.P., 95, *234*
Kelly, J.W., 75, *121*, 75, *122*, 75, *123*
Kemmitt, T., 5, *172*, 7, *172*
Kennedy, J.F., 5, *64*, 5, *70*, 5, *76*, 5, *82*, 5, *86*, 5, *97*, 5, *100*, 5, *103*, 5, *117*, 5, *118*
Kerns, R.J., 46, *132*
Kerry, K.R., 151, *123*
Kett, W.C., 23, *17*
Khan, R., 5, *134*
Khedri, Z., 159, *272*, 162, *295*, 166, *324*
Kilburn, D.G., 73, *62*, 73, *63*, 76, *134*, 80, *62*, 80, *63*, 81, *62*, 81, *63*
Kilcoyne, M., 92, *222*
Killeen, K., 117, *30*, 118, *30*, 121, *30*, 150, *30*, 156, *30*
Kim, H., 88, *204*
Kim, H.M., 158, *257*
Kim, H.Y., 155, *197*, 156, *208*, 158, *197*
Kim, I.J., 158, *257*
Kim, J., 117, *48*, 150, *48*, 151, *48*, 158, *259*
Kim, J.H., 154, *193*
Kim, J.Y., 161, *288*
Kim, M.S., 91, *214*
Kim, N., 86, *181*
Kim, S., 70, *41*, 70, *47*, 72, *41*, 89, *41*, 92, *41*
Kim, T.J., 158, *259*
Kim, Y.H., 87, *196*
Kimura, K., 150, *117*
Kimura, K.., 150, *116*
Kimura, S., 73, 77, 73, *91*, 73, *92*
Kimura, T., 154, *183*
King, D., 87, *195*, 87, *196*
Kinne-Saffran, E., 117, *37*
Kirpekar, F., 162, *300*, 163, *300*
Kirsten, G.F., 158, *230*
Kishi, Y., 33, *85*
Kitagawa, H., 128, *94*, 129, *97*, 130, *97*, 136, *103*, 136, *104*, 140, *104*, 141, *94*, 149, *111*
Kitagawa, T., 83, *168*
Kitaoka, M., 121, *61*, 123, *61*, 124, *61*, 126, *61*, 128, *61*, 130, *61*, 132, *61*, 133, *28*, 134, *61*, 136, *61*, 138, *61*, 140, *61*, 142, *61*, 144, *61*, 146, *61*, 148, *61*, 150, *61*, 153, *61*, 154, *180*, 154, *181*, 154, *182*, 154, *186*, 154, *187*, 154, *188*, 155, *182*, 161, *292*, 168, *330*
Kiyohara, M., 154, *181*, 154, *184*, 154, *185*, 154, *186*, 154, *188*
Klarskov, K., 68, *30*, 70, *30*, 72, *30*, 84, *30*, 92, *30*
Klein, M.D., 158, *235*
Klein, N., 121, *57*, 141, *57*, 150, *57*, 153, *57*, 156, *212*, 157, *212*
Kleman-Leyer, K.M., 164, *314*
Klijn, N., 153, *174*
Kling, D.E., 155, *199*
Knight, R., 117, *46*
Knights, D., 117, *46*
Knott, B.C., 73, *82*, 73, *83*, 73, *87*
Knowles, J.K.C., 70, *39*, 71, *50*, 72, *50*, 72, *52*, 79, *147*, 83, *52*
Knuhr, P., 158, *254*

Kobata, A., *53*, 116, 13, 116, *28*, 117, *28*, 118, *28*, 120, 120, *28*, 120, *51*, 120, *52*, 120, *54*, 121, *28*, 121, *59*, 121, *61*, 123, *28*, 123, *61*, 124, *28*, 124, *61*, 124, *82*, 126, *28*, 126, *51*, 126, *61*, 126, *88*, 126, *92*, 127, *51*, 127, *92*, 128, *28*, 128, *51*, 128, *61*, 129, *92*, 130, *28*, 130, *61*, 130, *98*, 131, *51*, 131, *52*, 132, *28*, 132, *52*, 132, *61*, 133, *28*, 133, *53*, 134, *28*, 134, *53*, 134, *61*, 135, *54*, 135, *102*, 136, *28*, 136, *54*, 136, *61*, 137, *54*, 138, *28*, 138, *54*, 138, *61*, 140, *28*, 140, *61*, 142, *28*, 142, *61*, 144, *28*, 144, *61*, 146, *28*, 146, *61*, 148, *28*, 148, *59*, 148, *61*, 149, *69*, 149, *73*, 149, *74*, 150, *59*, 150, *61*, 150, *69*, 150, *73*, 150, *74*, 151, *74*, 151, *125*, 151, *130*, 152, *145*, 153, *61*
Kobayashi, A., 73, *107*, 81, *107*
Kobayashi, K., 93, *225*
Kobayashi, R., 83, *163*, 85, *163*
Koga, T., 88, *204*
Kogelberg, H., 138, *105*, 139, *105*, 142, *107*
Kohda, K., 83, *168*
Koivula, A., 66, *26*, 72, *26*, 72, *53*, 72, *54*, 72, *55*, 72, *56*, 72, *57*, 72, *61*, 73, 77, 73, *91*, 73, *92*, 74, *61*, 74, *118*, 74, *119*, 75, *118*, 75, *119*, 83, *53*, 83, *55*, 83, *56*, 84, *26*, 90, *26*
Koizumi, S., 162, *297*, 162, *306*, 169, *331*, 169, *332*
Kokot, D., 162, *299*
Kolarich, D., 71, *49*
Komáromi, I., 29, *50*
Komor, R., 83, *165*
Kontteli, M., 73, *67*, 79, *67*
Kopchick, J., 151, *143*
Korhonen, T.K., 157, *226*, 158, *229*
Koshida, S., 29, *48*
Kotz, A., 88, *203*
Kovensky, J., 24, *28*, 25, *28*
Kövér, K.E., 30, *64*
Kowal, P., 159, *265*
Kozutsumi, Y., 87, *192*
Kraulis, P.J., 79, *147*
Krotkiewski, H., 88, *198*
Kroukamp, H., 82, *157*, 82, *158*
Kruszewska, J.S., 66, *16*, 88, *16*, 88, *198*, 88, *199*, 88, *200*, 92, *16*, 96, *199*
Kruus, K., 73, *103*, 73, *111*, 73, *112*, 73, *113*, 81, *103*, 81, *111*, 81, *112*, 81, *113*
Ku, C.-C., 32, *77*, 32, *78*
Kubicek, C.P., 66, *12*, 67, *27*, 72, *59*, 76, *131*, 76, *133*, 95, *232*, 95, *233*, 95, *234*
Kubota, T., 83, *163*, 85, *163*
Kuczynski, J., 117, *46*
Kuhn, L., 155, *197*, 156, *208*, 158, *197*
Kuhn, R., 116, *20*, 116, *22*, 116, *23*, 116, *24*, 116, *25*, 116, *26*, 120, *24*, 120, *50*, 123, *25*, 123, 77, 123, *78*, 123, *79*, 124, *24*, 124, *26*, 124, *66*, 124, *81*, 124, *83*, 124, *84*, 125, *66*, 126, *50*, 126, *90*, 149, *66*, 150, *66*
Kulkarni, S.S., 31, *67*, 32, *79*, 45, *125*, 45, *127*
Kull, A., 88, *205*
Kulminskaya, A.A., 76, *130*
Kulmiskaya, A.A., 89, *207*, 95, *207*
Kumagai, H., 93, *225*, 154, *182*, 154, *183*, 154, *184*, 154, *186*, 154, *187*, 155, *182*, 161, *292*, 168, *330*
Kumar Kolli, V.S., 87, *196*
Kumar, L., 73, *101*, 81, *101*
Kumar, R., 73, *102*, 81, *102*
Kumazaki, T., 151, *140*
Kunkel, G., 66, *17*, 71, *17*, 75, *17*, 90, *17*
Kuntz, S., 157, *222*
Kunz, C., 117, *37*, 121, *57*, 141, *57*, 150, *57*, 153, *57*, 153, *171*, 153, *172*, 153, *176*, 156, *176*, 156, *212*, 157, *176*, 157, *212*, 157, *222*
Kunz, H., 158, *255*
Kunze, W.A., 153, *178*, 156, *178*
Kurasin, M., 73, *93*, 73, *99*, 74, *93*
Kurita, K., 43, *123*
Kurkchubasche, A.G., 158, *235*
Kurosaka, A., 129, *97*, 130, *97*, 149, *111*
Kurosawa, H., 43, *121*
Kuroshu, R.M., 91, *215*
Kuske, C.R., 72, *59*
Kusumoto, S., 29, *48*
Kuszmann, J., 45, *130*
Kuzuhara, H., 43, *121*
Kwan, G., 40, *109*
Kwon, I., 83, *162*, 85, *162*

L

la Grange, D.C., 72, *53*, 83, *53*
Laborda, P., 91, *215*
Lage, J.L., 78, *145*
Laine, J., 73, *103*, 81, *103*
Lally, K.P., 158, *235*
Lam, J.S., 169, *334*
Lane, J., 157, *223*
Langsford, M.L., 76, *134*
Lanthier, P., 68, *31*, 71, *31*, 84, *31*, 95, *31*
Lappalainen, A., 72, *57*
Lappas, A., 73, *111*, 81, *111*
Larocque, S., 28, *43*
Laroy, W., 86, *184*
Larrondo, L.F., 72, *59*
Lassaletta, J.M., 26, *38*, 28, *38*
Lau, K., 159, *269*, 159, *271*, 159, *272*, 162, *295*, 162, *296*, 166, *296*, 166, *317*, 166, *318*, 166, *321*, 166, *323*, 166, *324*
Lauber, C., 117, *46*
Laughon, M., 158, *234*
Lauwaet, T., 155, *194*
Lavigne, J., 77, *141*, 78, *141*
Lawson, A.M., 138, *105*, 139, *105*, 142, *107*
Lay, L., 24, *20*, 25, *20*, 158, *242*, 158, *245*
Lázár, L., 29, *49*, 29, *50*, 30, *64*
Le Costaouec, T., 81, *152*
Le, D.T., 162, *295*
Le Goff, N., 152, *155*
Le Narvor, C., 152, *155*, 165, *316*
Le, P.Y., 5, *121*, 6, *121*
Leach, F.E., III, 32, *73*
Lebrilla, C.B., 115, *1*, 117, *1*, 117, *30*, 117, *33*, 117, *34*, 117, *48*, 118, *30*, 118, *33*, 118, *34*, 120, *34*, 121, *30*, 121, *33*, 121, *34*, 122, *33*, 122, *34*, 123, *33*, 123, *34*, 124, *33*, 124, *34*, 126, *33*, 126, *34*, 127, *33*, 128, *33*, 128, *34*, 130, *33*, 130, *34*, 132, *33*, 132, *34*, 133, *28*, 134, *33*, 134, *34*, 135, *33*, 135, *34*, 136, *33*, 136, *34*, 137, *34*, 138, *33*, 138, *34*, 139, *33*, 140, *33*, 140, *34*, 142, *33*, 142, *34*, 144, *33*, 144, *34*, 145, *33*, 146, *33*, 146, *34*, 147, *33*, 147, *34*, 148, *33*, 148, *34*, 149, *33*, 149, *34*, 150, *30*, 150, *34*, 150, *48*, 150, *75*, 150, *118*, 150, *119*, 150, *120*, 151, *48*, 151, *118*, 151, *119*, 151, *120*, 153, *1*, 153, *176*, 154, *189*, 154, *190*, 154, *192*, 155, *33*, 155, *189*, 156, *30*, 156, *207*, 156, *213*
Lederman, I., 28, *39*, 29, *47*, 30, *54*, 30, *56*
Lee, B., 156, *207*
Lee, C.K., 5, *160*
Lee, H.Y., 159, *268*
Lee, J.-C., 31, *67*, 32, *70*, 45, *127*, 45, *128*, 158, *256*
Lee, J.H., 81, *150*
Lee, P.-Y., 32, *78*
Lee, T.M., 75, *120*, 82, *120*, 83, *120*
Lee, W.H., 161, *287*, 161, *288*
Lehtio, J., 73, *65*, 78, *146*, 79, *65*
Lehtovaara, P., 72, *52*, 83, *52*
Leitao, A.F., 73, *114*, 81, *114*
Lemay, D.G., 156, *213*
Lemoine, J., 128, *95*, 133, *65*, 134, *65*, 150, *65*
Leon, M.E., 159, *272*
LePendu, J., 155, *200*
Lerno, L., 154, *190*, 154, *192*
Lespagnol, A., 116, *16*, 116, *17*
Letton, R.W., 158, *235*
Levene, P.A., 120, *49*, 123, *49*
Lever, R., 23, *15*
Levine, S.E., 73, *96*
Levy, Y., 75, *123*
Lewis, M.D., 33, *85*
Lewis, Z.T., 117, *39*, 156, *213*
Lezyk, M., 164, *312*
Li, H., 86, *181*, 86, *185*, 86, *186*, 86, *187*, 86, *188*, 88, *188*, 162, *300*, 163, *300*
Li, J., 91, *217*, 91, *219*
Li, L., 166, *319*, 170, *336*
Li, N., 156, *209*
Li, T., 32, *75*, 73, *88*
Li, W.H., 83, *169*, 88, *169*
Li, Y., 151, *142*, 154, *192*, 159, *269*, 159, *271*, 159, *273*, 160, *277*, 161, *142*, 162, *295*, 162, *296*, 166, *296*, 166, *317*, 166, *318*, 166, *320*, 166, *322*, 166, *323*, 170, *336*
Lian, G., 32, *83*
Lian, Z.X., 156, *209*
Liao, C.-C., 45, *127*
Liao, Y.-J., 41, *115*
Lichius, A., 95, *234*
Lico, L.S., 32, *74*, 45, *126*
Lide'n, G., 76, *129*
Lim, L.-H., 32, 77
Lim, S., 162, *296*, 166, *296*

Lima, L.H.F., 77, *142*
Limpaseni, T., 92, *221*
Lin, F., 32, *72*, 32, *83*
Lin, J.H., 66, *21*, 72, *21*, 84, *21*, 86, *21*, 90, *21*, 26, *37*
Lin, K.I., 158, *239*
Lin, S.-Y., 24, *26*, 24, *27*, 25, *26*, 25, *27*, 31, *27*, 32, *74*, 32, *77*, 32, *80*
Lin, Y.C., 159, *268*
Lin, Y.Y., 123, *64*, 150, *64*
Lin, Y.-J., 41, *115*
Lin, Y.-T., 32, *77*
Lindahl, U., 23, *1*
Lindeberg, G., 73, *66*, 73, *67*, 79, *66*, 79, *67*
Linder, M.B., 72, *54*, 73, *65*, 73, *66*, 73, *67*, 73, *75*, 73, *94*, 78, *146*, 79, *65*, 79, *66*, 79, *67*, 79, *94*
Lindh, F., 123, *80*, 125, *87*, 129, *96*, 130, *96*, 131, *80*, 131, *87*, 132, *52*, 149, *96*
Linger, J.G., 82, *156*
Linhardt, R.J., 23, *6*, 24, *21*, 25, *21*, 26, *36*, 26, *37*, 46, *132*, 46, *133*
Linke, H., 153, *178*, 156, *178*
Liphardt, J., 81, *151*
Lipniunas, P., 123, *80*, 125, *87*, 129, *96*, 130, *96*, 131, *80*, 131, *87*, 132, *52*, 149, *96*
Lipták, A., 29, *49*, 29, *50*, 30, *64*
Lissitzky, J.C., 157, *220*
Litjens, R.E.J.N., 24, *24*, 25, *24*, 28, *46*, 29, *46*
Liu, B., 155, *199*
Liu, D., 91, *217*
Liu, G., 91, *219*
Liu, J.-Y., 30, *61*, 32, *80*
Liu, Q., 32, *75*
Liu, R., 30, *61*
Liu, W., 66, *21*, 66, *22*, 72, *21*, 72, *22*, 74, *22*, 75, *22*, 76, *22*, 84, *21*, 86, *21*, 90, *21*, 90, *22*
Liu, X., 83, *161*, 96, *235*, 166, *319*
Liu, Y.-H., 32, *75*, 45, *127*, 170, *336*
Logtenberg, M.J., 164, *312*
Lohman, G.J.S., 41, *114*
Loibichler, C., 157, *218*
Lombard, V., 88, *206*
Lomino, J.V., 93, *230*
Lonn, H., 152, *158*
Look, G.C., 159, *266*
López-Prados, J., 28, *42*
Losfeld, M.E., 90, *210*
Lotscher, H., 152, *147*
Louie, K., 151, *134*
Love, K.R., 158, *243*
Lozupone, C.A., 117, *46*
Lu, F.-C., 45, *126*
Lu, L.-D., 32, *79*
Lu, X.-A., 31, *67*, 32, *70*, 32, *79*, 158, *239*
Lubineau, A., 40, *110*, 152, *155*
Lucas, S.M., 72, *59*
Luley-Goedl, C., 162, *298*, 162, *299*, 163, *298*
Lundblad, A., 122, *62*, 148, *62*, 148, *108*, 148, *110*
Lundgren, T., 123, *80*, 125, *87*, 129, *96*, 130, *96*, 131, *80*, 131, *87*, 132, *52*, 149, *96*
Luo, H., 93, *229*
Luo, S.-Y., 41, *115*
Lynd, L.R., 67, *28*, 83, *159*, 83, *160*
Lyskowski, A., 162, *299*

M

Ma, B., 76, *128*, 77, *128*
Ma, C., 170, *336*
Ma, L., 91, *219*
Maaheimo, H., 152, *163*
Maccarana, M., 23, *5*
Macchione, G., 28, *41*, 28, *42*
Macher, I., 25, *33*
Machetto, F., 30, *55*
Mackenzie, L.F., 73, *83*
Madsen, R., 158, *258*
Magnuson, J.K., 72, *59*
Magris, M., 117, *46*
Mahdi, S., 74, *119*, 75, *119*
Makarov, D.E., 76, *127*, 77, *127*, 78, *127*
Makarow, M., 70, *39*
Makela, P.H., 158, *229*
Makimura, Y., 154, *183*
Malavaki, C., 23, *4*
Malekan, H., 166, *320*
Malissard, M., 152, *164*
Mallet, J.-M., 30, *55*, 158, *244*
Malmstrom, A., 23, *5*
Manary, M.J., 117, *46*
Manavalan, A., 91, *220*
Mandal, P.K., 158, *248*

Mándi, A., 29, *50*
Mangoni, L., 48, *139*
Maningat, P., 151, *135*
Mansfield, R., 86, *181*
Manzoni, L., 158, *242*, 158, *245*
Mar Kayser, M., 28, *42*
Maranduba, A., 158, *253*
Maras, M., 67, *29*, 68, *29*, 68, *36*, 70, *29*, 70, *36*, 84, *29*, 89, *208*, 90, *208*, 93, *208*, 96, *236*
Marcobal, A.M., 154, *189*, 155, *189*, 154, *192*
Margolles-Clark, E., 78, *146*
Marionneau, S., 155, *200*
Marjamaa, K., 73, *103*, 73, *111*, 73, *112*, 73, *113*, 81, *103*, 81, *111*, 81, *112*, 81, *113*
Marot-Leblond, A., 157, *220*
Marques, D., 29, *48*
Marra, A., 34, *94*, 41, *118*, 42, *118*
Marsh, R.W., 4, *23*, 5, *23*
Marshall, R.D., 5, *82*, 5, *86*
Martano, C., 151, *133*
Martelli, P.L., 82, *154*
Martens, E.C., 154, *189*, 155, *189*
Martin, S.L., 165, *325*, 166, *325*
Martinet, W., 89, *208*, 90, *208*, 93, *208*
Martinez, D., 72, *59*
Martínez, L., 77, *142*
Martinez-Rossi, N.M., 91, *211*, 93, *211*, 95, *211*
Martín-Lomas, M., 26, *38*, 28, *38*, 32, *76*, 41, *113*, 158, *237*
Martin-Sampedro, R., 73, *103*, 73, *113*, 81, *103*, 81, *113*
Masaki, H., 73, *76*, 78, *76*, 87, *197*
Masri, N., 77, *141*, 78, *141*
Massi, A., 41, *118*, 42, *118*
Masuda, N., 43, *123*
Masuda, S., 48, *140*
Matrosovich, M.N., 157, *217*
Matsuoka, K., 43, *121*
Matthews, J.F., 65, 7, 66, 7, 66, *19*, 72, *19*, 77, *19*
Mattinen, M.-L., 73, *67*, 79, *67*
Mawatari, A., 28, *45*, 29, *45*
Mayer, K., 153, *176*, 156, *176*, 157, *176*
Mayrhofer, C., 73, *97*
Maza, S., 28, *41*, 28, *42*
McBride, J.E., 72, *53*, 83, *53*, 83, *159*
McBride, R., 170, *335*
McCabe, C., 66, *19*, 66, *24*, 68, *24*, 72, *19*, 72, *24*, 73, *24*, 73, *86*, 76, *126*, 77, *19*, 77, *126*, 79, *24*, 80, *24*
McCluer, R.H., 155, *202*
McCorquodale, D.J., 152, *144*
McCoy, J.M., 155, *199*
McCrumb, D.K., 126, *91*
McDonald, S., 158, *235*
McGaughey, J., 77, *140*
McGuire, E., 155, *195*, 156, *195*
McHugh, S.G., 68, *31*, 71, *31*, 84, *31*, 95, *31*
McIntosh, L.P., 77, *144*
McIver, M.A., 155, *196*
Mcjarrow, P., 162, *302*
McLennan, T.J., 5, *87*
Medaković, D., 34, *96*
Medve, J., 73, *74*
Mehta, S., 159, *274*, 159, *275*
Meilleur, F., 77, *139*
Meinzen-Derr, J.K., 156, *205*, 156, *214*, 156, *215*
Melin, P., 92, *223*
Mellon, M., 72, *53*, 83, *53*
Menshov, V.M., 158, *247*
Merighi, M., 155, *199*
Merino, S., 72, *59*
Mert, M., 82, *157*
Mertens, J.M.R., 30, *53*
Messer, M., 150, *121*, 151, *121*, 151, *123*, 151, *124*, 151, *125*, 151, *126*, 151, *127*, 151, *128*, 151, *129*
Messner, R., 66, *12*
Meyer, A.S., 160, *283*, 162, *300*, 163, *300*, 164, *312*
MIao, Y., 91, *217*
Michalski, J.C., 120, *55*, 121, *56*, 127, *55*, 128, *95*, 131, *56*, 133, 101, 138, *55*, 139, *55*, 139, *56*, 139, *106*, 140, *56*, 141, *56*
Michelena, O., 32, *76*
Michener, W., 66, *18*, 66, *20*, 70, *20*, 72, *18*, 72, *20*, 73, *20*, 74, *18*, 74, *20*, 75, *18*, 87, *18*, 87, *20*, 90, *18*, 90, *20*
Middleton, S., 5, *170*
Miele, R.G., 86, *185*

Mikander, S., 73, *111*, 73, *112*, 81, *111*, 81, *112*
Miki, S., 91, *212*
Mikkelsen, J.D., 160, *283*, 162, *300*, 163, *300*, 164, *312*
Milgate, S.M., 5, *128*
Miller, G.J., 39, *105*, 39, *106*, 39, *107*, 39, *108*, 40, *109*
Miller, I.J., 5, *164*, 5, *166*, 7, *164*, 7, *166*
Miller, M.J., 156, *213*, 161, *287*
Miller, R.C., Jr., 76, *134*
Mills, D.A., 115, *1*, 117, *1*, 117, *43*, 153, *1*, 153, *43*, 153, *175*, 153, *176*, 154, *175*, 154, *189*, 154, *190*, 154, *191*, 154, *192*, 154, *193*, 155, *189*, 156, *213*
Min Jou, W., 86, *184*
Mine, T., 164, *310*, 164, *311*
Minshull, J., 83, *166*
Mironov, Y.V., 158, *247*
Misra, A.K., 158, *248*
Misra, M., 72, *59*
Mitchell, E.P., 126, *89*
Mitchell, T., 86, *185*, 86, *186*, 86, *187*, 86, *188*, 88, *188*
Mitchinson, C., 70, *47*
Mitrovic, A., 84, *176*, 85, *176*
Miwa, M., 154, *188*
Miyake, A., 154, *184*
Miyama, N., 33, *86*
Miyauchi, H.A, 160, *276*
Miyazaki, T., 156, *210*, 168, *210*
Mizumoto, S., 23, *4*
Mochalova, L.V., 157, *217*
Mohamed, S., 26, *35*
Mohammad, M.A., 151, *135*
Mohler, S., 83, *165*
Momany, M., 65, *7*, 66, *7*
Momeni, M.H., 73, *83*
Monnom, D., 152, *146*
Monteiro, M.A., 5, *194*, 159, *260*
Montesino, R., 70, *45*
Montreuil, J., 116, *9*, 116, *18*, 116, *27*, 120, *55*, 121, *56*, 122, *63*, 123, *27*, 125, *85*, 125, *86*, 127, *55*, 128, *95*, 131, *56*, 133, *101*, 138, *55*, 139, *55*, 139, *56*, 139, *106*, 140, *56*, 141, *56*., 151, *131*, 154, *9*
Montruil, J., 148, *109*
Moo, G.M., 81, *151*
Moran, A.P., 159, *260*
Moreira, S., 73, *71*, 73, *114*, 81, *114*
Morelle, W., 133, *65*, 134, *65*, 150, *65*
Morelli, L., 153, *170*
Moremen, K.W., 151, *142*, 152, *165*, 161, *142*
Morikawa, Y., 83, *163*, 85, *163*
Morrison, I.M., 5, *117*, 5, *118*
Morrow, A.L., *10*, 116, *10*, 116, *15*, 117, *15*, 117, *41*, 121, *15*, 150, 150, *113*, 153, *15*, 155, *200*, 156, *205*, 156, *214*, 156, *215*, 170, *15*
Moser, B., 76, *134*
Moses, V., 4, *11*, 4, *12*
Moss, R.L., 158, *235*
Mota, M., 73, *68*, 73, *69*, 73, *70*, 73, *71*, 76, *70*
Motallebi, M., 84, *177*, 85, *177*
Moulds, J.J., 158, *228*
Mrksich, M., 166, *319*
Mu, S., 159, *269*, 162, *296*, 166, *296*
Muhly-Reinholz, M., 153, *176*, 156, *176*, 157, *176*
Mukerji, P., 151, *142*, 161, *142*
Muller-Werner, B., 150, *115*
Mullet, S., 116, *9*, 154, *9*
Mulloy, B., 23, *10*, 23, *15*
Muñoz-García, J.C., 23, *14*
Munro, M.H.G., 5, *101*, 7, *101*
Murakami, K., 43, *123*
Murakami, S., 93, *226*
Murakami, Y., 157, *219*
Murata, T., 164, *313*, 167, *313*
Murray, P.G., 72, *55*, 83, *55*
Muthana, M.M., 166, *320*, 166, *321*
Muthana, S., 160, *277*
Mutsaers, J.H.G.M., 87, *192*
Mwiya, M., 155, *197*, 156, *208*, 158, *197*
Myles, D.A., 77, *139*
Mysore, J.V., 158, *227*

N

Nagae, M., 161, *291*, 161, *292*
Nagano, H., 169, *332*
Nagendran, S., 70, *44*, 76, *44*, 90, *44*, 91, *44*, 93, *44*
Naggi, A., 23, *3*
Naidu, N., 158, *236*

Nakada, H., 128, *94*, 129, *97*, 130, *97*, 136, *103*, 136, *104*, 140, *104*, 141, *94*, 149, *111*
Nakagame, S., 73, *104*, 73, *105*, 73, *106*, 81, *104*, 81, *105*, 81, *106*
Nakamura, A., 72, *61*, 73, *76*, 74, *61*, 78, *76*
Nakamura, T., 150, *116*, 150, *117*, 150, *121*, 151, *121*, 151, *127*, 151, *129*
Nakao, Y., 87, *192*
Nakazawa, H., 83, *163*, 85, *163*
Namiki, M., 150, *121*, 151, *121*
Nanthkumar, N.N., 117, *31*, 118, *30*
Napolitano, C., 49, *141*, 49, *142*, 49, *143*
Nardelli, C., 151, *142*, 161, *142*
Narimatsu, H., 93, *226*
Natunen, J., 152, *161*, 152, *162*, 152, *163*
Nazir, A., 91, *218*
Neeser, J.R., 155, *201*, 156, *206*
Nel, D.G., 158, *230*
Nelson, B., 72, *59*
Nelson, M.A., 72, *59*
Nelson, S., 88, *205*
Nemeth, Z., 95, *232*
Nett, J.H., 86, *181*, 86, *185*, 86, *186*, 86, *187*, 86, *188*, 88, *188*
Neu, J., 158, *231*
Neubauer, S.H., 117, *32*, 118, *32*, 123, *32*, 124, *32*, 126, *32*, 128, *32*, 130, *32*, 132, *32*, 133, *28*, 134, *32*, 136, *32*, 138, *32*, 140, *32*, 142, *32*, 144, *32*, 146, *32*, 148, *32*
Neustroev, K.N., 76, *130*, 89, *207*, 95, *207*
Nevalainen, H., 66, *10*, 68, *10*, 68, *32*, 71, *32*, 71, *49*, 72, *32*, 78, *32*, 79, *32*, 84, *10*, 84, *32*, 85, *10*, 87, *10*
Newburg, D.S., *10*, 116, *2*, 116, *10*, 116, *15*, 117, *15*, 117, *31*, 117, *32*, 117, *35*, 117, *41*, 117, *42*, 118, *30*, 118, *32*, 121, *15*, 123, *32*, 124, *32*, 126, *32*, 128, *32*, 130, *32*, 132, *32*, 133, *28*, 134, *32*, 136, *32*, 138, *32*, 140, *32*, 142, *32*, 144, *32*, 146, *32*, 148, *32*, 150, 151, *132*, 152, *151*153, *2*, 153, *15*, 153, *35*, 153, *170*, 155, *199*, 155, *200*, 155, *202*, 156, *205*, 156, *214*, 156, *215*, 170, *15*
Newton, S.A., 76, *136*, 82, *136*
Nguyen, C., 83, *166*
Ngyuen, U., 150, *75*
Nicaise, M., 126, *89*
Nicol, S.C., 151, *122*
Nicolay, J., 66, *13*
Nidetzky, B., 73, *97*, 162, *298*, 162, *299*, 163, *298*
Niemela, R., 152, *161*, 152, *162*, 152, *163*
Nieto, P.M., 23, *14*, 28, *41*, 28, *42*
Nieves, R.A., 72, *51*, 84, *51*
Nifantiev, N.E., 158, *247*
Nikrad, P.V., 152, *160*
Nilges, M., 79, *147*
Nilsson, B., 122, *62*, 123, *80*, 125, *87*, 129, *96*, 130, *96*, 131, *80*, 131, *87*, 132, *52*, 148, *62*, 149, *96*
Nimlos, M.R., 34, *95*, 65, *6*, 73, *84*, 73, *85*
Nimpf, J., 25, *34*, 26, *34*
Ninonuevo, M.R., 117, *30*, 118, *30*, 121, *30*, 150, *30*, 156, *30*, 156, *207*
Niranjan, K., 91, *213*
Nishimoto, M., 154, *180*, 154, *181*
Nishimura, S.-I., 43, *123*
Nishiyama, K., 33, *86*
Nissan, C., 155, *197*, 156, *208*, 158, *197*, 158, *236*
Niu, H.L., 156, *209*
Nonaka, H., 73, *107*, 73, *108*, 81, *107*, 81, *108*
Norberg, T., 152, *158*
Nordvang, R.T., 164, *312*
Norris, R.F., 116, *21*
Noti, C., 24, *23*, 25, *23*, 36, *102*, 41, *23*
Nouwens, A.S., 68, *32*, 71, *32*, 72, *32*, 78, *32*, 79, *32*, 84, *32*
Numata, Y., 129, *97*, 130, *97*, 149, *111*
Nurmi-Rantala, S., 72, *56*, 83, *56*
Nussinov, R., 76, *128*, 77, *128*
Nyffenegger, C., 164, *312*

O

Obayashi, H.., 150, *116*
Oberoi, H.S., 91, *218*
O'Callaghan, J., 157, *223*
Occhi, L., 151, *133*
Ochiai, N., 83, *163*, 85, *163*
Ochonicky, K.L., 150, *120*, 151, *120*
Odamaki, T., 154, *180*

O'Donohue, M.J., 84, *175*, 85, *175*
Oftedal, O.T., 151, *122*, 151, *126*
Ogasawara, W., 83, *163*, 85, *163*
Ogawa, S., 33, *87*
Ogawa, T., 158, *250*
Oh, Y.I., 41, *116*
Ohmachi, T., 87, *190*
Ohnishi, J., 169, *332*
Ohnishi, M., 121, *61*, 123, *61*, 124, *61*, 126, *61*, 128, *61*, 130, *61*, 132, *61*, 133, *28*, 134, *61*, 136, *61*, 138, *61*, 140, *61*, 142, *61*, 144, *61*, 146, *61*, 148, *61*, 150, *61*, 153, *61*
Ohno, M., 73, *76*, 78, *76*
Oja, M., 95, *231*
Ojeda, R., 26, *38*, 28, *38*, 28, *41*, 41, *113*
Oka, T., 88, *204*
Okada, H., 83, *163*, 85, *163*
Okada, K., 83, *163*, 85, *163*
Okamoto, T., 73, *77*
Oldberg, Å., 23, *5*
Olden, K., 76, *136*, 82, *136*
Oliveira Neto, M., 77, *142*
Olsen, J., 87, *191*
Omann, M., 95, *234*
Omata, N., 83, *163*, 85, *163*
O'Neill, H.M., 77, *139*, 77, *140*
Onodera, T., 83, *163*, 85, *163*
Orazio, G., 157, *216*
Orgueira, H.A., 28, *40*, 28, *46*, 29, *46*, 47, *137*
Ori, A., 23, *8*
Orita, T., 121, *58*, 142, *58*, 143, *58*, 144, *58*, 145, *58*, 146, *58*, 147, *58*
Orlando, R., 87, *195*, 87, *196*
Orlowski, J., 66, *16*, 88, *16*, 92, *16*
Orosz, A., 95, *232*
Orskov, F., 158, *229*
Orskov, I., 158, *229*
Ortner, J., 160, *282*
Osanaï, M., 121, *58*, 142, *58*, 143, *58*, 144, *58*, 145, *58*, 146, *58*, 147, *58*
Oscarson, S., 158, *249*
Ossowski, P., 47, *136*
Osumi, K., 121, *58*, 142, *58*, 143, *58*, 144, *58*, 145, *58*, 146, *58*, 147, *58*
Osztrovszky, G., 30, *60*, 31, *60*, 32, *60*
Overend, W.G., 4, *7*, 4, *8*, 4, *17*, 4, *19*, 4, *27*, 4, *41*, 5, *14*, 5, *30*, 5, *31*, 5, *32*, 5, *41*, 5, *43*, 5, *44*
Overkleeft, H.S., 24, *22*, 24, *24*, 25, *22*, 25, *24*, 30, *58*
Ozaki, A., 162, *306*, 169, *331*, 169, *332*

P

Packer, N., 66, *10*, 68, *10*, 68, *32*, 71, *32*, 71, *49*, 72, *32*, 78, *32*, 79, *32*, 84, *10*, 84, *32*, 85, *10*, 87, *10*
Padella, L., 116, *11*, 121, *11*, 150, *11*, 151, *11*
Page, C.P., 23, *15*
Paholcsek, M., 95, *232*
Pakarinen, A., 81, *152*
Pakula, T.M., 70, *40*, 95, *231*
Palamarczyk, G., 66, *16*, 88, *16*, 88, *198*, 88, *199*, 88, *200*, 92, *16*, 96, *199*
Palcic, M.M., 152, *152*, 152, *153*, 161, *286*
Palmacci, E.R., 28, *46*, 29, *46*
Palonen, H., 73, *75*, 73, *109*, 78, *109*, 81, *109*
Palumbo, G., 25, *31*, 49, *141*, 49, *142*, 49, *143*, 49, *144*
Pan, G.-R., 32, *79*
Panda, T., 66, *12*
Pandey, A., 91, *214*
Panine, P., 77, *143*
Paolella, C., 49, *143*
Paper, J.M., 70, *44*, 76, *44*, 90, *44*, 91, *44*, 93, *44*
Pareek, N., 73, *110*, 78, *110*, 81, *110*
Park, Y.C., 117, *30*, 118, *30*, 121, *30*, 150, *30*, 156, *30*, 158, *259*
Parker, E., 156, *213*
Parkkinen, J., 157, *226*, 158, *229*
Parsons, N.J., 159, *261*
Pathanibul, P., 161, *287*
Patton, S., 116, *4*
Paul, B.J., 160, *282*
Paulson, J.C., 170, *335*
Pavkov-Keller, T., 162, *298*, 163, *298*
Payne, C.M., 65, *7*, 66, *7*, 66, *23*, 72, *23*, 72, *58*, 73, *23*, 73, *81*, 73, *86*, 73, *87*, 74, *117*, 76, *58*, 77, *58*, 78, *23*, 78, *58*
Peberdy, J.F., 87, *193*, 87, *194*
Pedatella, S., 49, *141*, 49, *144*

Pedersen, C.M., 50, *147*, 52, *146*
Pedersoli, W.R., 91, *211*, 93, *211*, 95, *211*
Peila, C., 116, *11*, 121, *11*, 150, *11*, 151, *11*, 151, *133*
Pelletier, M., 162, *303*
Pellissier, H., 25, *29*
Peng, B., 83, *167*, 84, *167*
Peng, W., 170, *335*
Penttila, L., 152, *161*
Penttila, M., 70, *40*, 72, *52*, 72, *53*, 72, *56*, 73, *77*, 73, *91*, 73, *92*, 76, *135*, 78, *135*, 83, *52*, 83, *53*, 83, *56*, 89, *208*, 90, *208*, 93, *208*, 95, *231*, 96, *236*
Pera, E., 23, *5*
Perlinska-Lenart, U., 66, *16*, 88, *16*, 88, *200*, 92, *16*
Perlinska-Lenart,U., 88, *199*, 96, *199*
Perret, S., 126, *89*
Perry, M.B., 159, *260*
Persinoti, G.F., 91, *211*, 93, *211*, 95, *211*
Peter-Katalinic, J., 125, *86*, 133, *100*, 134, *100*
Petersen, P.M., 5, *153*, 5, *158*, 5, *167*, 5, *169*
Petitou, M., 23, *12*, 23, *13*, 23, *16*, 28, *39*, 29, *47*, 30, *54*, 30, *55*, 30, *56*, 30, *65*, 30, *66*, 32, *65*, 34, *93*, 34, *94*
Petridis, L., 77, *140*
Pettersson, G., 73, *66*, 73, *67*, 79, *66*, 79, *67*, 79, *147*
Pettersson,G., 71, *50*, 72, *50*
Pfenninger, A., 121, *60*, 132, *52*, 134, *99*, 150, *60*
Phatale, P.A., 95, *234*
Pichler, J., 157, *218*
Pickering, L.K., *10*, 116, *10*, 150, *70*, 150, *200*, 155, *200*, 155, *202*, 156, *205*
Piens, K., 68, *30*, 70, *30*, 70, *47*, 72, *30*, 74, *119*, 75, *119*, 84, *30*, 89, *208*, 90, *208*, 92, *30*, 93, *208*
Piepersberg, W., 160, *284*, 160, *285*
Pierani, P., 116, *8*, 121, *8*, 150, *8*
Pierce, M., 151, *142*, 152, *165*, 161, *142*
Pilz, I., 76, *138*
Pingali, S.V., 77, *140*
Pinto, R., 73, *68*, 73, *69*, 73, *70*, 73, *71*, 76, *70*
Piskarev, V.E., 138, *105*, 139, *105*, 142, *107*
Piskorska, D., 155, *195*, 156, *195*
Plancke, Y., 121, *56*, 131, *56*, 133, *65*, 134, *65*, 139, *56*, 140, *56*, 141, *56*, 150, *65*
Plank, H., 73, *97*
Podányi, B., 45, *130*
Podkaminer, K., 82, *156*
Poe, R., 87, *196*
Pokorna, M., 126, *89*
Polak, J.J., 117, *42*
Polat, T., 30, *59*
Poletti, L., 24, *20*, 25, *20*
Polikarpov, I., 65, *9*, 77, *142*, 91, *215*
Polonowski, M., 116, *16*, 116, *17*, 116, *18*
Pongsawasdi, P., 92, *221*
Ponpipom, M.M., 5, *61*, 5, *62*
Poon, D.K., 77, *144*
Popovic, M., 156, *213*
Potgieter, T.I., 86, *185*
Potier, P., 23, *19*
Pourceau, G., 24, *28*, 25, *28*
Power, M.L., 117, *48*, 150, *48*, 151, *48*
Powers, D.L., 75, *122*
Powers, E.T., 75, *121*, 75, *122*, 75, *123*
Prabhu, A., 30, *57*
Pradella, J.G., 91, *215*
Praestgaard, E., 73, *89*, 74, *89*
Prasad, D., 4, *21*, 4, *22*, 4, *26*, 4, *28*, 4, *29*, 5, *128*
Prasad, N., 5, *39*, 5, *40*, 5, *45*, 5, *50*, 5, *51*, 5, *52*, 5, *53*, 5, *128*
Prasad, R.M., 5, *128*
Prasit, P., 5, *105*, 5, *113*, 5, *116*, 5, *119*, 5, *120*, 5, *121*, 5, *122*, 5, *123*, 5, *124*, 5, *137*, 6, *116*, 6, *119*, 6, *120*, 6, *121*, 6, *122*, 6, *123*
Prentice, A.M., 150, *75*
Pribowo, A., 74, *116*
Price, J.L., 75, *121*, 75, *122*, 75, *123*
Prieels, J.P., 152, *146*
Priem, B., 163, *307*, 165, *307*, 165, *325*, 166, *325*, 166, *327*, 169, *334*
Prieto, P., 150, *70*, 151, *143*
Prieto, P.A., 116, *6*, 116, *7*, 126, *91*, 151, *142*, 161, *142*, 161, *290*, 164, *290*, 164, *314*, 164, *315*
Prinz, B., 86, *181*
Pritchard, R.G., 39, *104*
Provasoli, A., 23, *12*, 23, *13*

Pudlo, N., 154, *189*, 155, *189*
Puranen, T., 72, *54*, 72, *57*, 73, *112*, 81, *112*, 81, *152*
Puranik, R., 45, *127*, 45, *129*
Putnam, N., 72, *59*

Q

Qasba, P.K., 151, *137*
Qi, F., 66, *22*, 72, *22*, 74, *22*, 75, *22*, 76, *22*, 90, *22*
Qin, Y., 83, *161*, 91, *219*
Qu, J., 159, *269*, 162, *296*, 166, *296*, 166, *317*, 166, *318*, 166, *320*, 170, *336*
Qu, Y., 66, *21*, 72, *21*, 83, *161*, 84, *21*, 86, *21*, 90, *21*, 91, *219*
Quarterman, J., 161, *287*
Quintero, O., 70, *45*

R

Raangs, G.C., 153, *174*
Rabenstein, D.L., 23, *2*
Rafferty, G.A., 4, *19*, 4, *41*, 5, *41*
Raftery, J., 39, *105*, 40, *109*
Ragauskas, A.J., 73, *102*, 81, *102*
Ragazzi, M., 23, *12*, 23, *13*
Rahikainen, J.L., 73, *103*, 73, *112*, 73, *113*, 81, *103*, 81, *111*, 81, *112*, 81, *113*
Rajgarhia, V., 72, *53*, 83, *53*
Ralston, G.B., 151, *124*
Ramakrishnan, B., 151, *137*
Raman, R., 23, *11*
Ramos, P., 155, *204*
Rancurel, C., 88, *206*
Rasko, D.A., 159, *260*
Rastall, R.A., 91, *213*
Ratcliffe, R.M., 152, *152*
Rausch, S., 86, *185*, 86, *186*
Rauvala, H., 70, *39*
Raybould, H.E., 117, *43*, 153, *43*, 154, *193*
Realff, M.J., 81, *150*
Receveur, V., 77, *143*
Re'czey, K., 76, *129*
Reed, S.L., 155, *194*
Reeder, J., 117, *46*
Reeves, R.D., 5, *143*, 7, *143*
Reichardt, N.-C., 32, *76*, 41, *113*
Reinikainen, T., 71, *50*, 72, *50*, 73, *66*, 73, *67*, 76, *135*, 78, *135*, 78, *146*, 79, *66*, 79, *67*
Reisinger, C., 84, *176*, 85, *176*
Rempe, C.S., 77, *140*
Ren, G.R., 73, *90*, 74, *90*
Renaudie, L., 165, *316*
Renfer, E., 86, *185*
Renkonen, O., 152, *161*, 152, *162*, 152, *163*
Renner, R., 5, *146*
Repetschnigg, W., 160, *282*
Resch, M.G., 66, *19*, 66, *23*, 66, *25*, 68, *25*, 72, *19*, 72, *23*, 72, *25*, 73, *23*, 73, *25*, 76, *25*, 77, *19*, 78, *23*, 79, *25*, 80, *25*, 81, *25*, 82, *25*, 90, *25*
Rescorla, F.J., 158, *233*
Rey, F.E., 117, *46*
Rhee, C., 95, *234*
Riazi-Farzad, T., 122, *63*
Ribitsch, D., 162, *298*, 162, *299*, 163, *298*
Ricart, C.A., 91, *214*
Rice, K.G., 152, *156*, 152, *157*
Richard, S., 30, *65*, 32, *65*
Richardson, A.C., 5, *118*, 5, *125*
Richardson, P., 72, *59*
Ricketts, R.R., 158, *235*
Rignall, T.R., 76, *126*, 77, *126*
Rinas, U., 87, *193*
Rios, S., 86, *181*, 86, *187*
Rittenhour, A., 86, *188*, 88, *188*
Robbertse, B., 72, *59*
Robinson, W.T., 5, *87*
Roche, C.M., 83, *164*
Rochepeau-Jobron, L., 42, *119*
Rodehorst, T.K., 117, *42*
Rodrigues, A.C., 73, *114*, 81, *114*
Roehrig, S., 47, *137*
Rojas, O.J., 73, *103*, 73, *113*, 81, *103*, 81, *113*
Rokhsar, D.S., 72, *59*
Romero, P., 83, *165*
Rose, C.S., 116, *20*, 116, *21*, 116, *22*, 116, *23*
Rose, S.H., 83, *160*
Roser, J., 86, *187*
Rosseto, F.R., 77, *142*
Rossi, A., 91, *211*, 93, *211*, 95, *211*

Roth, S., 153, *173*, 158, *173*
Roussel, F., 158, *252*
Routier, F.H., 88, *203*
Rouvinen, J., 72, *54*
Rovio, S., 73, *113*, 81, *113*
Roy, R., 68, *31*, 71, *31*, 84, *31*, 95, *31*, 158, *246*
Roy, S., 30, *65*, 32, *65*
Ruan, S.K., 83, *169*, 88, *169*
Rubin, E.M., 72, *59*
Rubin, J., 81, *153*
Rudloff, S., 117, *37*, 121, *57*, 141, *57*, 150, *57*, 153, *57*, 153, *171*, 153, *172*, 153, *176*, 156, *176*, 156, *212*, 157, *176*, 157, *212*, 157, *222*
Rudmark, A., 148, *108*
Rudowski, A., 4, *21*, 4, *22*, 4, *26*
Ruel, K., 73, *69*
Ruelius, H.W., 116, *23*
Ruhaak, L.R., 150, *75*
Ruiz, A., 32, *76*
Ruiz, J., 155, *201*
Ruiz-Moyano, S., 154, *191*
Ruiz-Palacios, G.M., 116, *10*, 116, *15*, 117, *15*, 117, *41*, 121, *15*, 150, *10*, 150, *70*, 153, *15*, 155, *204*, 156, *205*, 156, *214*, 170, *15*
Ruohonen, L., 71, *50*, 72, *50*
Ruohtula, T., 152, *161*
Rushton, G., 39, *106*, 39, *108*
Rutherford, T.J., 152, *144*
Ruvoen-Clouet, N., 155, *200*
Ryan, A.E., 4, *17*, 5, *14*, 5, *31*

S

Saarikangas, A., 152, *162*
Sabatino, G., 151, *133*
Sabharwal, H., 122, *62*, 148, *62*
Sabin, C., 126, *89*
Saddler, J.N., 73, *101*, 73, *104*, 73, *106*, 74, *116*, 81, *101*, 81, *104*, 81, *106*, 81, *148*
Sadeghi, M., 84, *177*, 85, *177*
Sadilek, M., 34, *99*, 35, *99*
Saelens, X., 86, *184*
Saija, P., 75, *125*, 83, *125*
Saito, A., 28, *45*, 29, *45*
Saito, T., 150, *121*, 151, *121*, 151, *122*, 151, *126*, 151, *127*, 151, *128*, 151, *129*
Sakata, K., 154, *183*
Sakuma, A., 154, *183*
Sakurama, H., 168, *330*
Salameh, B.A.B., 39, *104*
Salamone, S., 35, *100*
Salamov, A.A., 72, *59*
Sallomons, E., 162, *304*, 164, *304*
Saloheimo, M., 72, *59*, 95, *231*
Salovuori, I., 70, *39*
Samain, E.A, 163, *307*, 163, *308*, 164, *311*, 165, *307*, 165, *325*, 165, *326*, 166, *325*, 166, *327*, 169, *333*, 169, *334*
Samejima, M., 72, *61*, 73, *72*, 73, *73*, 73, *77*, 73, *91*, 73, *92*, 74, *61*
Sameshima, Y., 88, *204*
Sammond, D.W., 72, *58*, 76, *58*, 77, *58*, 78, *58*
Samuelson, O., 33, *88*
Samyn, B., 70, *42*, 72, *42*, 89, *42*, 92, *42*
Sanai, Y., 150, *116*, 150, *117*
Sanchez, M., 157, *220*
Sandercock, L.E., 73, *62*, 80, *62*, 81, *62*
Sanders, P., 162, *304*, 164, *304*
Sandgren, M., 66, *23*, 70, *41*, 70, *47*, 72, *23*, 72, *41*, 73, *23*, 73, *83*, 73, *87*, 78, *23*, 88, *201*, 89, *41*, 92, *41*, 92, *201*
Sandra, K., 68, *33*, 68, *34*, 70, *33*, 70, *34*, 70, *37*, 70, *38*, 72, *33*, 72, *34*, 84, *33*, 84, *34*, 84, *37*, 84, *38*, 87, *33*, 87, *34*, 88, *33*, 88, *34*, 89, *33*, 89, *34*, 89, *37*, 89, *38*, 90, *33*, 90, *34*, 90, *38*, 93, *33*, 93, *34*
Sandra, P., 70, *38*, 84, *38*, 89, *38*, 90, *38*
Sanford, H., 117, *40*
Sangster, I., 4, *21*
Sankey, G.H., 5, *32*, 5, *36*, 5, *37*, 5, *45*, 5, *53*
Sannigrahi, P., 73, *102*, 81, *102*
Santoro, L., 116, *11*, 121, *11*, 150, *11*, 151, *11*
Sartori, G.R., 77, *142*
Sasaki, M., 151, *136*
Sasisekharan, R., 23, *11*
Sasisekharan, V., 23, *11*
Sato, T., 156, *210*, 168, *210*
Savel'ev, A.N., 76, *130*, 89, *207*, 95, *207*
Savel'ev, N.V., 76, *130*
Sawada, D., 33, *86*
Sawant, R.C., 41, *115*

Sawatzki, G., 116, *14*, 117, *36*, 150, *112*, 150, *115*, 151, *14*
Schell, P., 28, *40*, 28, *46*, 29, *46*, 47, *137*
Scheper, T., 65, *5*, 81, *5*
Scherer, L.R., 158, *235*
Schiattarella, M., 30, *58*
Schibler, K.R., 156, *215*
Schirrmacher, G., 84, *176*, 85, *176*
Schmidt, D., 160, *278*
Schmidt, P., 5, *138*
Schmidt, R.R., 158, *238*, 158, *241*, 158, *242*, 158, *245*, 158, *252*, 158, *254*
Schmitt, J., 157, *218*
Schmoll, M., 67, *27*, 72, *59*
Schmolzer, K., 162, *298*, 162, *299*, 163, *298*
Schmuck, M., 76, *138*
Schoch, C.L., 72, *59*
Schols, H.A., 164, *312*
Schonfeld, H., 116, *19*, 154, *19*
Schraml, J., 67, *29*, 68, *29*, 68, *36*, 70, *29*, 70, *36*, 84, *29*
Schreferl-Kunar, G., 66, *12*
Schulein, M., 77, *143*
Schultes, B., 30, *65*, 32, *65*
Schulz, R., 72, *60*, 74, *60*
Schur, M.J., 159, *275*
Schützenmeister, N., 36, *102*
Schwab, H., 162, *298*, 162, *299*, 163, *298*
Schwartz, H., 117, *40*
Schwöorer, R., 30, *52*
Scott, B.R., 77, *141*, 78, *141*
Scott, C.R., 34, *99*, 35, *99*
Scott, D.A., 90, *210*
Scott, P.J., 5, *143*, 7, *143*
Scott-Craig, J.S., 65, *4*, 66, *4*
Sears, P., 43, *120*
Seeberger, P.H., 24, *23*, 25, *23*, 28, *40*, 28, *46*, 29, *46*, 36, *101*, 36, *102*, 36, *103*, 41, *23*, 41, *114*, 47, *137*, 158, *243*
Seeger, W., 153, *176*, 156, *176*, 157, *176*
Seiboth, B., 65, *8*, 67, *8*, 95, *233*, 95, *234*
Seidl-Seiboth, V., 65, *8*, 67, *8*
Seitz, L., 161, *289*
Sekii, N., 151, *122*
Sela, D.A., 154, *190*, 154, *192*
Selig, M.J., 73, *100*, 81, *100*
Seo, J.H., 158, *259*, 161, *287*, 161, *288*
Seppo, A., 152, *161*
Sergeant, K., 70, *42*, 72, *42*, 89, *42*, 92, *42*
Sergi, C., 158, *233*
Serpa, V.I., 65, *9*, 77, *142*
Sethi, A., 73, *95*
Sethuraman, N., 86, *181*, 86, *187*
Severn, W.B., 5, *164*, 5, *166*, 7, *164*, 7, *166*
Sezerman, U.O., 82, *154*, 84, *173*, 84, *174*, 85, *173*, 85, *174*
Shabalin, K.A., 76, *130*, 89, *207*, 95, *207*
Shaikh, F.A., 160, *280*
Shang, B.Z., 74, *115*
Sharma, M., 91, *218*
Shashkov, A.S., 158, *247*
Shen, J., 32, *75*
Shen, L., 150, *71*, 151, *71*
Shen, Q., 91, *217*
Shen, Y., 83, *167*, 83, *170*, 84, *167*, 84, *170*, 91, *217*
Shenderovich, M.D., 157, *217*
Sheng, G.J., 41, *116*
Shental-Bechor, D., 75, *123*
Sherman, A.A., 158, *247*
Sheykhnazari, M., 157, *221*
Shi, P., 93, *229*
Shi, Z., 32, *77*, 32, *78*
Shida, T., 33, *86*
Shie, C.-R., 32, *79*
Shih, M.C., 83, *169*, 88, *169*
Shimada, I., 129, *97*, 130, *97*, 149, *111*
Shimizu, H., 158, *250*
Shirts, M.R., 73, *81*, 73, *84*, 73, *85*, 74, *117*
Shiyama, T., 29, *48*
Shoun, H., 154, *187*
Siegel, M., 116, *14*, 151, *14*
Sigalet, D., 158, *233*
Siika-Aho, M., 72, *53*, 72, *57*, 83, *53*
Silva, R.N., 91, *211*, 93, *211*, 95, *211*
Silva- Rocha, R., 91, *211*, 93, *211*, 95, *211*
Simon, P.M., 155, *198*, 158, *227*
Sina?, P., 158, *244*
Sinaÿ, P., 23, *12*, 23, *13*, 28, *39*, 29, *47*, 30, *55*, 34, *90*, 34, *93*, 34, *94*
Singh, B., 76, *134*
Singleton, M.F., 4, *18*
Sinkala, M., 155, *197*, 158, *197*
Sinnott, M.L., 74, *119*, 75, *119*
Sinnwell, V., 5, *150*
Siuzdak, G., 158, *256*

Sizun, P., 30, *56*
Sjoblad, S., 122, *62*, 148, *62*
Sjostrom, R., 156, *209*
Skinner, M.A., 158, *235*
Skropeta, D., 75, *124*
Slattery, H., 157, *223*
Smilowitz, J.T., 115, *1*, 117, *1*, 153, *1*, 156, *213*
Smith, B.C., 4, *27*
Smith, D.F., 116, *5*, 116, *6*, 116, *7*, 126, *91*, 151, *142*, 161, *142*
Smith, H., 159, *261*
Smith, J.C., 77, *140*
Smith, K.M., 95, *234*
Smith, P.B., 158, *234*
Smithson, A., 33, *89*, 44, *89*
Smolenska-Sym, G., 88, *199*, 96, *199*
Snow, C.D., 83, *165*, 83, *166*
Sobel, M., 28, *45*, 29, *45*, 29, *48*
Sobotka, H., 120, *49*, 123, *49*
Sohn, C.H., 93, *79*
Sokol, K., 155, *198*
Solere, M., 116, *12*
Somers, P.J., 5, *55*, 5, *59*
Somerville, C.R., 91, *216*
Somsák, L., 5, *161*, 34, *91*
Son, B., 160, *277*
Song, J., 170, *336*
Song, X., 82, *155*
Soni, R., 91, *218*
Sonnenburg, E.D., 154, *189*, 155, *189*
Sonnenburg, J.L., 154, *189*, 155, *189*
Sorlie, M., 73, *87*
Sousa, M.V., 91, *214*
Souza, A.P., 91, *215*
Specklin, J.L., 49, *145*
Spitzer, A.R., 158, *234*
Sprenger, G.A., 161, *289*, 167, *328*, 168, *328*, 168, *329*
Squire, M.A., 93, *228*
Srisodsuk, M., 76, *135*, 78, *135*, 78, *146*
Srivastava, G., 152, *166*, 152, *167*
Srivastava, V.K., 5, *94*
Stadheim, T.A., 86, *181*, 86, *185*, 86, *186*, 86, *187*, 86, *188*, 88, *188*, 88, *205*
Stagel, A., 95, *232*
Stahl, B., 117, *36*, 121, *60*, 132, *52*, 134, *99*, 150, *60*, 150, *112*, 150, *114*, 157, *218*
Stahlberg, J., 66, *23*, 68, *30*, 70, *30*, 70, *45*, 71, *50*, 72, *23*, 72, *30*, 72, *50*, 73, *23*, 73, *67*, 73, *74*, 73, *82*, 73, *83*, 73, *85*, 73, *87*, 78, *23*, 79, *67*, 84, *30*, 92, *30*
Ståhlberg, J., 70, *46*, 74, *119*, 75, *119*, 76, *129*
Stals, I., 66, *23*, 68, *33*, 68, *34*, 70, *33*, 70, *34*, 70, *37*, 70, *38*, 70, *41*, 70, *42*, 70, *46*, 72, *23*, 72, *33*, 72, *34*, 72, *41*, 72, *42*, 73, *23*, 78, *23*, 84, *33*, 84, *34*, 84, *37*, 84, *38*, 87, *33*, 87, *34*, 88, *33*, 88, *34*, 89, *33*, 89, *34*, 89, *37*, 89, *38*, 89, *41*, 89, *42*, 90, *33*, 90, *34*, 90, *38*, 92, *41*, 92, *42*, 93, *33*, 93, *34*
Stangier, K., 152, *153*
Stanisz, A.M., 153, *178*, 156, *178*
Stehle, T., 157, *224*
Steindorff, A.S., 91, *211*, 93, *211*, 95, *211*
Steiner, W., 160, *282*
Steinkellner, G., 84, *176*, 85, *176*
Stenlid, J., 92, *223*
Stepans, M.B., 117, *42*
Steup, M., 150, *112*
Stevenson, C.D., 5, *143*, 7, *143*
Stimpson, W.T., 39, *104*
Stoll, B.J., 158, *232*
St-Pierre, P., 77, *141*, 78, *141*
Strassel, J.-M., 30, *56*
Strawbridge, R.R., 86, *181*, 86, *187*
Strecker, G., 120, *55*, 121, *56*, 122, *63*, 125, *86*, 127, *55*, 128, *95*, 131, *56*, 133, *65*, 133, 101, 134, *65*, 138, *55*, 139, *55*, 139, *56*, 139, *106*, 140, *56*, 141, *56*, 148, *109*, 150, *65*
Strobel, S., 121, *57*, 141, *57*, 150, *57*, 153, *57*, 156, *212*, 157, *212*
Stubba, B., 30, *58*
Stubbs, H.J., 152, *156*
Stuetz, A.E., 5, *155*, 5, *157*
Sturgeon, C.M., 5, *117*, 5, *118*
Sturgeon, R.J., 5, *70*, 5, *76*, 5, *82*, 5, *86*, 5, *97*, 5, *100*, 5, *103*, 5, *117*, 5, *118*
Sturman, J.A., 123, *64*, 150, *64*
Suda, Y., 28, *45*, 29, *45*, 29, *48*
Sugahara, D., 121, *58*, 142, *58*, 143, *58*, 144, *58*, 145, *58*, 146, *58*, 147, *58*
Sugahara, K., 23, *4*
Sugawara, M., 149, *68*, 150, *68*, 157, *219*

Sugiarto, G., 159, *269*, 159, *271*, 162, *295*, 162, *296*, 166, *296*, 170, *336*
Sugimoto, H., 154, *187*
Sugimoto, N., 73, *72*, 73, *73*
Sugiyama, J., 73, *65*, 79, *65*
Sukhova, E.V., 158, *247*
Sun, B., 30, *61*
Sun, M., 163, *309*, 166, *309*, 166, *322*
Sunehag, A.L., 151, *134*, 151, *135*
Sung, H.M., 83, *169*, 88, *169*
Sungsuwan, S., 30, *63*
Suzuki, H., 83, *168*
Suzuki, M., 164, *313*, 167, *313*
Suzuki, R., 154, *187*
Suzuki, T., 92, *224*
Svahn, C.-M., 34, *95*
Svenson, S.B., 158, *229*
Svensson, S., 148, *108*, 148, *110*
Swartz, P.D., 77, *139*
Swift, R.J., 87, *193*, 87, *194*
Szakmary, K., 76, *131*
Sze, S.K., 70, *43*, 90, *43*, 91, *43*, 91, *220*, 93, *43*, 95, *43*
Szepfalusi, Z., 157, *218*

T

Tabangin, M., 156, *215*
Tabata, K., 162, *306*, 169, *331*, 169, *332*
Tabeur, C., 30, *55*
Tachibana, Y., *53*, 116, *13*, 120, 120, *54*, 126, *92*, 127, *92*, 129, *92*, 130, *98*, 132, *52*, 133, *53*, 134, *53*, 135, *54*, 135, *102*, 136, *54*, 137, *54*, 138, *54*
Tachizawa, A., 154, *181*
Taheri, M., 66, *13*
Takahashi, H., 33, *84*, 33, *86*
Takahashi, S., 154, *180*
Takakura, Y., 164, *310*
Takamura, T., 158, *240*, 158, *251*
Takaoka, M., 128, *94*, 141, *94*
Takaoka, Y., 93, *226*
Takashima, S., 73, *76*, 78, *76*, 87, *197*
Takegawa, K., 91, *212*
Takeuchi, M., 93, *225*
Takeya, A., 152, *150*
Takhi, M., 158, *252*
Talib, M.F., 66, *24*, 68, *24*, 72, *24*, 73, *24*, 79, *24*, 80, *24*
Tamminen, T., 73, *111*, 73, *112*, 73, *113*, 81, *111*, 81, *112*, 81, *113*
Tan, K., 154, *190*
Tan, Z., 66, *23*, 66, *25*, 68, *25*, 72, *23*, 72, *25*, 73, *23*, 73, *25*, 76, *25*, 78, *23*, 79, *25*, 80, *25*, 81, *25*, 82, *25*, 90, *25*
Tanahashi, N., 151, *138*
Tanaka, A., 154, *187*
Tanaka, H., 162, *301*
Tanaka, M., 160, *276*
Tang, H., 83, *167*, 83, *170*, 84, *167*, 84, *170*
Tanigawa, K., 154, *185*
Taniguchi, H., 161, *292*
Tao, N., 117, *33*, 117, *48*, 118, *33*, 121, *33*, 122, *33*, 123, *33*, 124, *33*, 126, *33*, 127, *33*, 128, *33*, 130, *33*, 132, *33*, 133, *28*, 134, *33*, 135, *33*, 136, *33*, 138, *33*, 139, *33*, 140, *33*, 142, *33*, 144, *33*, 145, *33*, 146, *33*, 147, *33*, 148, *33*, 149, *33*, 150, *48*, 150, *118*, 150, *119*, 150, *120*, 151, *48*, 151, *118*, 151, *119*, 151, *120*, 155, *33*
Tarr, G.E., 152, *144*
Tatai, J., 30, *60*, 31, *60*, 31, *68*, 31, *69*, 32, *60*, 32, *81*
Tate, S., 128, *94*, 136, *103*, 136, *104*, 140, *104*, 141, *94*
Tatsumi, H., 73, *80*, 73, *90*, 74, *90*
Taufik, E., 151, *122*, 151, *129*
Taylor, C.B., 66, *19*, 66, *24*, 68, *24*, 72, *19*, 72, *24*, 73, *24*, 73, *86*, 77, *19*, 79, *24*, 80, *24*
Taylor, D.E., 159, *260*, 161, *286*
Taylor, L.E., 66, *23*, 72, *23*, 73, *23*, 78, *23*, 82, *156*
Taylor, M.A., 5, *153*
Tedder, J.M., 3, *2*
Teeri, T.T., 66, *26*, 71, *50*, 72, *26*, 72, *50*, 72, *52*, 73, *65*, 73, *66*, 73, *94*, 74, *119*, 75, *119*, 76, *135*, 78, *135*, 78, *146*, 79, *65*, 79, *66*, 79, *94*, 83, *52*, 84, *26*, 90, *26*
Tejima, S., 158, *240*, 158, *251*
Tello, A., 155, *201*
Tenkanen, M., 73, *75*, 73, *109*, 78, *109*, 81, *109*
Terabayashi, T., 121, *61*, 123, *61*, 124, *61*, 126, *61*, 128, *61*, 130, *61*, 132, *61*, 133, *28*, 134, *61*, 136, *61*, 138, *61*, 140, *61*,

142, *61*, 144, *61*, 146, *61*, 148, *61*, 150, *61*, 153, *61*
Terrazas, L.I., 155, *195*, 156, *195*
Terry, A., 72, *59*
Terunuma, D., 43, *121*
Teugjas, H., 73, *93*, 74, *93*
Thayer, N., 72, *59*
Thea, D.M., 155, *197*, 156, *208*, 158, *197*
Thelin, M.A., 23, *5*
Thibault, P., 68, *31*, 68, *35*, 71, *31*, 71, *35*, 84, *31*, 95, *31*, 95, *35*
Thiem, J., 47, *136*, 152, *153*, 160, *278*
Thomas, D.A., 93, *79*
Thomas, D.Y., 66, *13*
Thomas, S.R., 72, *51*, 84, *51*
Thomas, V.H., 152, *156*, 152, *157*
Thomford, N.R., 152, *144*
Thompson, W.R., 158, *235*
Thon, V., 159, *269*, 159, *272*, 166, *317*, 166, *318*, 166, *321*
Thopate, S.R., 45, *128*
Thorngren, N., 72, *53*, 83, *53*
Thurl, S., 116, *14*, 150, *112*, 150, *115*, 151, *14*
Tiden, A.-K., 5, *172*, 7, *172*
Tigas, S., 151, *134*
Tilbrook, D.M.G., 5, *187*, 5, *188*
Timári, I., 30, *64*
Timmer, M.S.M., 36, *101*, 36, *102*
Ting, C.L., 76, *127*, 77, *127*, 78, *127*
Tiruchinapally, G., 30, *61*, 30, *62*
Tiwari, V.K., 166, *317*, 166, *323*, 166, *324*
Tjerneld, F., 70, *46*, 73, *74*, 73, *109*, 76, *129*, 78, *109*, 81, *109*
Tomabechi, Y., 93, *228*
Tomashek, J.J., 77, *141*, 78, *141*
Tong, H.H., 155, *196*
Torres, B.V., 116, *6*
Torri, G., 23, *3*, 23, *12*, 23, *13*, 28, *39*, 29, *47*
Totten, S.M., 150, *75*, 156, *213*
Touloukian, R.J., 158, *235*
Toumieux, S., 24, *28*, 25, *28*
Tovar, K., 116, *14*, 151, *14*
Tran, B.K., 159, *272*
Trappe, S., 87, *193*
Trehan, I., 117, *46*
Trentesaux-Chauvet, C., 122, *63*
Trinci, A.P.J., 87, *193*, 87, *194*
Tronchin, G., 157, *220*
Trouilleux, P., 23, *19*
Trudeau, D.L., 75, *120*, 82, *120*, 83, *120*
Tsai, C.J., 76, *128*, 77, *128*
Tsai, C.M., 116, *5*, 158, *239*
Tsai, T.-I., 32, 77, 159, *268*
Tsuchiya, A., 161, *291*
Tsuda, M., 149, *73*, 150, *73*
Tsukada, T., 72, *61*, 74, *61*
Tsukamoto, H., 164, *310*
Tsunemi, M., 150, *117*
Tu, Y.C., 159, *268*
Tucker, L.C.N., 5, *71*
Tuohy, M.G., 72, *55*, 83, *55*
Tuomanen, E., 155, *198*
Turecek, F., 34, *99*, 35, *99*
Turek, D., 158, *249*
Turnbull, J.E., 30, *52*, 32, *73*
Turner, M.J., 75, *123*
Tuzikov, A.B., 157, *217*
Tykesson, E., 23, *5*
Tyler, P.C., 5, *89*, 5, *93*, 5, *99*, 5, *107*, 5, *108*, 5, *109*, 5, *114*, 5, *119*, 5, *122*, 5, *136*, 5, *137*, 5, *138*, 5, *139*, 5, *154*, 5, *159*, 5, *163*, 5, *168*, 5, *171*, 5, *172*, 5, *173*, 5, *175*, 5, *176*, 5, *179*, 5, *180*, 5, *183*, 5, *187*, 5, *188*, 5, *191*, 6, *119*, 6, *122*, 6, *136*, 7, *172*, 30, *52*
Tzelepis, G.D., 92, *223*, 92, *224*

U

Uberbacher, E.C., 72, *60*, 74, *60*
Ubhayasekera, W., 88, *201*, 92, *201*
Uchihashi, T., 73, 77, 73, *92*
Uenishi, J.i., 48, *140*
Uozumi, T., 73, *76*, 78, *76*, 87, *197*
Uppugundla, N., 73, *78*
Urashima, T., 121, *61*, 123, *61*, 124, *61*, 126, *61*, 128, *61*, 130, *61*, 132, *61*, 133, *28*, 134, *61*, 136, *61*, 138, *61*, 140, *61*, 142, *61*, 144, *61*, 146, *61*, 148, *61*, 150, *61*, 150, *116*, 150, *117*, 151, *122*, 151, *126*, 151, *127*, 151, *128*, 151, *129*, 153, *61*, 154, *182*, 155, *182*
Urban, V.S., 77, *140*
Urbanek, R., 157, *218*
Usui, A.T., 164, *313*, 167, *313*

Utille, J.P., 169, *333*
Uusitalo, J., 70, *40*, 96, *236*

V

Vaisanen-Rhen, V., 158, *229*
Valjamae, P., 73, *93*, 73, *98*, 73, *99*, 74, *93*, 74, *98*
Valtonen, M.V., 158, *229*
Van Aelst, S.F., 30, *53*
Van Amelsfort, A., 162, *302*
Van Beeumen, J., 68, *30*, 68, *33*, 68, *34*, 70, *30*, 70, *33*, 70, *34*, 70, *37*, 70, *38*, 72, *30*, 72, *33*, 72, *34*, 84, *30*, 84, *33*, 84, *34*, 84, *37*, 84, *38*, 87, *33*, 87, *34*, 88, *33*, 88, *34*, 89, *33*, 89, *34*, 89, *37*, 89, *38*, 90, *33*, 90, *34*, 90, *38*, 90, *209*, 92, *30*, 93, *33*, 93, *34*
van Boeckel, C.A.A., 23, *16*, 24, *22*, 25, *22*, 30, *53*, 30, *54*, 30, *58*, 30, *66*
van Boom, J.H., 30, *58*
van den Bos, L.J., 24, *24*, 25, *24*
van den Bosch, R.H., 30, *53*
Van den Eijnden, D.H., 152, *158*, 152, *159*
van der Marel, G.A., 24, *22*, 24, *24*, 25, *22*, 25, *24*, 30, *58*
van der Vlugt, F.A., 30, *53*
Van Halbeek, H., 87, *192*
Van Landschoot, A., 70, *41*, 72, *41*, 89, *41*, 92, *41*
Van Niekerk, E., 158, *230*
Van Petegem, F., 90, *209*
Van Tassell, M.L., 156, *213*
Van Vuure, C.A., 162, *304*, 164, *304*
van Zyl, J.-H.D., 82, *158*
van Zyl, W.H., 67, *28*, 72, *53*, 82, *157*, 82, *158*, 83, *53*, 83, *159*, 83, *160*
Vander Wall, T., 82, *156*
Varki, A., 159, *262*
Varnai, A., 81, *152*
Vehmaanpera, J., 72, *57*, 72, *54*
Venot, A.P., 30, *57*, 32, *73*, 152, *152*, 152, *160*
Venturini, A., 82, *154*
Vervecken, W., 96, *236*
Vethaviyasar, N., 5, *56*, 5, *58*, 5, *63*, 5, *72*, 5, *75*, 5, *96*
Veyrieres, A., 158, *253*
Vicentini, R., 91, *215*
Viikari, L., 72, *57*, 73, *111*, 81, *111*, 81, *152*
Vilanova, M., 73, *69*
Villafan, H., 155, *201*
Villalobos, A., 83, *166*
Vinzant, T.B., 66, *17*, 71, *17*, 72, *51*, 75, *17*, 84, *51*, 90, *17*
Virkola, R., 157, *226*
Viverge, D., 116, *12*
Vlahov, I.R., 26, *36*
Vliegenthart, J.F.G., 87, *192*, 152, *164*
Vogl, T., 86, *182*
von Nicolai, H., 117, *29*, 154, *179*
von Ossowski, I., 74, *119*, 75, *119*
Vos, J.N., 30, *53*
Voutilainen, S.P., 72, *53*, 72, *54*, 72, *55*, 72, *56*, 72, *57*, 83, *53*, 83, *55*, 83, *56*
Voznyi, Y.V., 34, *98*, 35, *98*
Vullo, T., 155, *198*

W

Wada, J., 154, *184*, 154, *186*, 154, *187*, 161, *292*
Wada, M., 72, *61*, 73, *73*, 73, *77*, 73, *91*, 73, *92*, 74, *61*
Wagendorp, A.A., 24, *28*, 25, *28*, 153, *174*
Wagener, J., 88, *203*
Wagner, T.E., 151, *142*, 161, *142*
Wakabayashi, S., 48, *140*
Wakagi, T., 154, *187*
Wakao, M., 28, *45*, 29, *45*
Wakarchuk, W.W., 159, *274*, 159, *275*, 162, *305*, 163, *307*, 165, *307*, 169, *334*
Wakatsuki, S., 161, *291*, 161, *292*
Walker, W.A., 117, *31*, 118, *30*, 158, *231*
Walker, W.M., 158, *234*
Wall, H.M., 4, *19*
Wall, H.W., 4, *41*, 5, *41*
Wallis, G.L.F., 87, *193*, 87, *194*
Walsh, K.L., 155, *195*, 156, *195*
Walton, J.D., 65, *4*, 66, *4*, 70, *44*, 76, *44*, 90, *44*, 91, *44*, 93, *44*
Wanek, E., 25, *33*
Wang, B., 117, *45*, 153, *45*, 157, *45*
Wang, C., 71, *48*, 83, *170*, 84, *170*, 91, *219*
Wang, C.H., 159, *268*
Wang, C.-C., 32, *77*, 32, *78*, 45, *127*, 45, *128*, 45, *131*
Wang, F., 160, *279*
Wang, G., 161, *286*

Wang, I.L., 83, *169*, 88, *169*
Wang, J., 32, *75*
Wang, K., 93, *229*
Wang, L., 66, *21*, 72, *21*, 84, *21*, 86, *21*, 90, *21*
Wang, L.L., 156, *209*
Wang, L.X., 93, *230*
Wang, M., 91, *217*
Wang, P.G., 32, *75*, 159, *265*, 166, *317*, 166, *319*, 170, *336*
Wang, Q., 71, *48*
Wang, R., 159, *267*
Wang, S., 91, *219*
Wang, T.Y., 83, *161*, 83, *169*, 88, *169*, 96, *235*
Wang, W.C., 32, *75*, 126, *91*
Wang, X., 117, *47*, 153, *47*, 170, *336*
Wang, Y., 82, *155*
Wang, Z.-G., 30, *61*, 30, *62*, 76, *127*, 77, *127*, 78, *127*, 170, *335*
Ward, M., 70, *41*, 72, *41*, 72, *59*, 89, *41*, 92, *41*
Ward, R.E., 117, *30*, 118, *30*, 121, *30*, 150, *30*, 156, *30*
Warner, B.B., 117, *46*, 156, *215*
Warren, C.D., *10*, 116, *10*, 150
Warren, R.A.J., 73, *62*, 73, *63*, 76, *134*, 80, *62*, 80, *63*, 81, *62*, 81, *63*
Warren, W.S., 157, *221*
Watanabe, Y., 150, *116*, 150, *117*
Watkins, W.M., 152, *148*, 152, *149*
Watson, D.C., 162, *305*
Weber, H., 160, *282*, 162, *298*, 163, *298*
Wehmeier, U.F., 160, *284*
Wei, A., 47, *134*, 47, *135*
Wei, L., 66, *21*, 72, *21*, 84, *21*, 86, *21*, 90, *21*
Wei, N., 170, *336*
Wei, S., 96, *235*
Wei, W., 71, *48*
Wei, X., 83, *161*, 91, *219*
Weidmann, H., 25, *33*, 25, *34*, 26, *34*
Weimer, B.C., 154, *189*, 155, *189*
Weimer, P.J., 67, *28*
Welling, G.W., 153, *174*
Wells, L., 87, *195*, 87, *196*
Wen, M., 65, *2*
Wen, Y.-S., 31, *67*, 45, *127*, 45, *128*
Wenkert, E., 5, *88*
West, C.M., 76, *137*, 82, *137*
Westerholm-Parvinen, A., 72, *59*
Westh, P., 73, *80*, 73, *89*, 73, *90*, 74, *89*, 74, *90*
Westman, J., 34, *95*
White, T.C., 68, *31*, 68, *35*, 70, *42*, 71, *31*, 71, *35*, 72, *42*, 77, *141*, 78, *141*, 84, *31*, 89, *42*, 92, *42*, 95, *31*, 95, *35*
Whitelock, J.M., 23, 7
Whitfield, D.M., 28, *43*, 28, *44*, 159, *274*, 159, *275*
Whyteside, G., 87, *189*
Wiebe, M.G., 87, *193*
Wiederschain, G.Y., 152, *151*
Wieruszeski, J.M., 120, *55*, 121, *56*, 125, *86*, 127, *55*, 128, *95*, 131, *56*, 133, 101, 138, *55*, 139, *55*, 139, *56*, 139, *106*, 140, *56*, 141, *56*
Wigginton, T., 158, *227*
Wightman, R.H., 5, *145*, 5, *149*, 5, *152*, 5, *154*, 5, *159*, 5, *163*, 5, *168*, 5, *171*, 5, *173*, 5, *175*, 5, *176*, 5, *179*, 5, *180*, 5, *183*, 5, *187*, 5, *188*, 5, *191*
Wilbrink, M.H., 162, *304*, 164, *304*
Wildeboer-Veloo, A.C., 153, *174*
Wildt, S., 86, *181*, 86, *185*, 86, *186*, 86, *187*, 86, *188*, 88, *188*
Wilhelm, S.L., 117, *42*
Wilkins, M.R., 66, *10*, 68, *10*, 84, *10*, 85, *10*, 87, *10*
Wilkinson, M.C., 23, *8*
Williams, D.T., 5, *30*, 5, *47*
Williams, J.M., 5, *100*, 5, *103*
Williams, N.R., 4, *19*, 4, *20*, 4, *41*, 5, *41*, 5, *86*, 5, *97*, 5, *100*, 5, *103*, 5, *117*, 5, *118*, 5, *125*, 5, *126*, 5, *134*, 5, *141*, 5, *145*, 5, *149*, 5, *152*, 5, *154*, 5, *159*, 5, *163*, 5, *168*, 5, *171*, 5, *173*, 5, *175*, 5, *176*
Williams, S.J., 160, *281*
Wilson, D.B., 65, *3*
Wimmerova, M., 126, *89*
Winter, M.G., 155, *203*
Wirth, H.P., 159, *260*
Wischnewski, H., 86, *186*, 86, *187*
Wiswall, E., 72, *53*, 83, *53*
Withers, S.G., 73, *83*, 77, *144*, 160, *280*, 160, *281*

Wlasichuk, K.B., 152, *160*
Wohlert, J., 66, *19*, 72, *19*, 77, *19*
Wolff, M.W., 46, *133*
Woltz, A., 156, *206*
Wong, C.-H., 30, *59*, 32, *77*, 34, *97*, 43, *120*, 75, *121*, 158, *239*, 158, *256*, 159, *266*, 159, *267*, 159, *268*
Wong, D., 166, *321*
Wood, T.A., 5, *160*
Woosley, B.D., 87, *195*, 87, *196*
Wu, B., 161, *294*, 170, *336*
Wu, B.x, 161, *294*
Wu, C.Y., 158, *256*, 159, *268*
Wu, G., 66, *21*, 72, *21*, 84, *21*, 86, *21*, 90, *21*
Wu, I., 83, *166*
Wu, S., 97, *34*, 98, *34*, 100, *34*, 101, *34*, 102, *34*, 103, *34*, 104, *34*, 106, *34*, 108, *34*, 110, *34*, 112, *34*, 113, *28*, 114, *34*, 115, *34*, 116, *34*, 117, *33*, 117, *34*, 117, *48*, 118, *33*, 118, *34*, 120, *34*, 121, *33*, 122, *33*, 122, *34*, 123, *33*, 124, *33*, 124, *34*, 126, *33*, 126, *34*, 127, *33*, 127, *34*, 128, *33*, 128, *34*, 129, *34*, 130, *33*, 130, *34*, 132, *33*, 133, *28*, 134, *33*, 135, *33*, 136, *33*, 138, *33*, 139, *33*, 140, *33*, 142, *33*, 144, *33*, 145, *33*, 146, *33*, 147, *33*, 148, *33*, 149, *33*, 150, *48*, 150, *75*, 151, *48*, 154, *190*, 154, *192*, 155, *33*
Wu, W.-g, 28, *44*
Wyman, C.E., 73, *102*, 81, *102*

X

Xiao, J.Z., 154, *180*
Xiao, P., 96, *235*
Xiao, Z., 170, *336*
Xie, G., 72, *59*
Xie, M., 87, *195*
Ximenes, F.F.E., 91, *214*
Xu, D., 23, *9*
Xu, H.T., 156, *209*
Xu, L., 83, *167*, 83, *170*, 84, *167*, 84, *170*
Xu, P., 32, *71*
Xu, W., 32, *71*
Xu, Y., 30, *61*
Xue, M., 159, *269*
Xue, X., 93, *229*
Xun, L., 91, *219*

Y

Yaeshima, T., 154, *180*
Yaguchi, M., 68, *31*, 71, *31*, 84, *31*, 95, *31*
Yahara, A., 87, *190*
Yamada, E., 33, *87*
Yamada,O., 87, *190*
Yamagishi, M., 164, *313*, 167, *313*
Yamaguchi, M., 87, *190*, 154, *186*
Yamamoto, K., 93, *225*, 93, *226*, 93, *227*, 154, *181*, 154, *182*, 154, *183*, 154, *184*, 154, *185*, 154, *186*, 154, *187*, 154, *188*, 155, *182*, 161, *291*, 161, *292*, 168, *330*
Yamamoto, T., 164, *310*, 164, *311*
Yamanoi, T., 154, *183*
Yamaoka, K., 150, *121*, 151, *121*
Yamashina, I., 87, *192*, 128, *94*, 129, *97*, 130, *97*, 136, *103*, 136, *104*, 140, *104*, 141, *94*, 149, *111*
Yamashita, K., *53*, 116, 13, 120, 120, *54*, 124, *82*, 126, *92*, 127, *92*, 129, *92*, 130, *98*, 132, *52*, 133, *53*, 134, *53*, 135, *54*, 135, *102*, 136, *54*, 137, *54*, 138, *54*, 149, *69*, 150, *69*
Yan, F., 159, *275*
Yan, J., 160, *279*
Yan, S., 73, *88*
Yan, X., 71, *48*
Yan, Y., 157, *224*
Yang, B., 30, *61*, 30, *63*
Yang, M., 159, *260*
Yang, Y., 32, *71*
Yao, B., 93, *229*
Yao, J., 72, *59*
Yao, L., 82, *155*
Yao, W., 160, *279*
Yarbrough, J.M., 66, *19*, 72, *19*, 77, *19*
Yatsunenko, T., 117, *46*
Ye, H., 32, *75*
Yike, I., 76, *132*
Yin, H., 117, *30*, 118, *30*, 121, *30*, 150, *30*, 156, *30*
Yin, Z.J., 30, *62*, 30, *63*
Yoshida, A., 151, *140*
Yoshida, E., 154, *182*, 154, *184*, 155, *182*, 168, *330*
Yoshida, K., 30, *63*
Yoshida, T., 87, *190*
Yoshiuchi, K., 87, *190*

Young, N.M., 162, *305*
Yu, B., 32, *71*, 32, *72*, 32, *82*
Yu, F., 30, *65*, 32, *65*
Yu, H.N., 34, *97*, 159, *269*, 159, *270*, 159, *271*, 159, *272*, 161, *293*, 161, *294*, 162, *295*, 163, *309*, 166, *293*, 166, *309*, 166, *317*, 166, *318*, 166, *320*, 166, *321*, 166, *323*, 166, *324*, 170, *336*
Yu, S.Y., 156, *209*
Yu, S.-H., 28, *43*
Yu, X., 83, *165*
Yu, Z.T., 155, *199*
Yudina, O.N., 158, *247*
Yun, J.S., 151, *142*, 161, *142*

Z

Zacchi, G., 73, *109*, 78, *109*, 81, *109*
Zachara, N.E., 68, *32*, 71, *32*, 72, *32*, 78, *32*, 79, *32*, 84, *32*
Zahri, S., 84, *177*, 85, *177*
Zamani, M.R., 84, *177*, 85, *177*
Zampini, L., 116, *8*, 116, *11*, 121, *8*, 121, *11*, 150, *8*, 150, *11*, 151, *11*, 157, *216*
Zeilinger, S., 95, *234*
Zembek, P., 66, *16*, 88, *16*, 88, *200*, 92, *16*
Zeng, J., 150, *112*
Zeng, S., 152, *164*
Zeuner, B., 160, *283*, 164, *312*
Zha, D., 86, *181*
Zhang, F., 66, *22*, 72, *22*, 74, *22*, 75, *22*, 76, *22*, 90, *22*
Zhang, J., 71, *48*, 117, *30*, 118, *30*, 121, *30*, 150, *30*, 156, *30*, 161, *294*
Zhang, L., 91, *219*, 166, *320*
Zhang, R., 91, *217*
Zhang, S., 82, *155*
Zhang, W., 66, *22*, 72, *22*, 74, *22*, 75, *22*, 76, *22*, 90, *22*
Zhang, Y.-M., 82, *155*, 138, *105*, 139, *105*, 142, *107*, 158, *244*, 161, *294*
Zhao, G.P., 91, *219*
Zhao, S., 91, *217*
Zhao, W., 32, *75*
Zhao, X., 66, *19*, 72, *19*, 76, *126*, 77, *19*, 77, *126*
Zhao, Y.F., 74, *119*, 75, *119*, 156, *209*
Zhao, Z.H., 156, *209*
Zheng, D., 91, *216*
Zheng, H., 91, *219*, 163, *309*, 166, *309*
Zheng, P.Y., 159, *260*
Zherebtsov, M., 158, *236*
Zhong, W., 155, *200*
Zhong, Y.-Q., 25, *32*, 32, *70*, 32, *78*, 96, *235*
Zhou, L., 32, *75*
Zhou, Y., 32, *72*, 32, *83*
Zhou, Z., 71, *48*, 91, *219*
Zhuang, G., 66, *21*, 72, *21*, 84, *21*, 86, *21*, 90, *21*
Zilliken, F., 116, *20*, 116, *22*, 116, *23*, 154, *179*
Zitzenbacher, S., 162, *299*
Zivkovic, A.M., 150, *75*
Zolotarev, B.M., 5, *58*
Zopf, D.A., 116, *5*, 124, *82*, 153, *173*, 155, *198*, 158, *173*, 158, *227*, 162, *303*
Zou, G., 71, *48*, 91, *219*
Zubkov, O.A., 5, *190*
Zubkova, O.V., 30, *52*
Zulueta, M.M.L., 24, *26*, 24, *27*, 25, *26*, 25, *27*, 25, *32*, 31, *27*, 32, *70*, 32, 77, 32, *78*, 32, *80*, 158, *239*

SUBJECT INDEX

Note: Page numbers followed by "*f*" indicate figures, "*t*" indicate tables, and "*s*" indicate schemes.

A

N-Acetylglucosaminyltransferase I (GnT-I), 152
4-*O*-Acetyl-*N*-acetylneuraminic acid ($Neu4,5Ac_2$)-containing MOS, 150–151
Antiadhesive antimicrobials, 153
L-Arabinose, 36
Aspergillus species, 85*f*, 86–88

B

Bacillus circulans, 168
Baeyer–Villiger oxidation, 35–36
Bifidobacteria, 153
Bifidobacterium bifidum, 161, 168
Biofuels, 65–66

C

Campylobacter jejuni, 161, 163
Cellulases, 65–66
Cellulose, 65
Chemoenzymatic methods, 159–160
Corynebacterium ammoniagenes, 162, 169
Cyanohydrin, 36–40, 37*s*, 39*s*

D

6-Deoxy-L-hexoses, 50–52, 51*s*
4-Deoxypentenoside, 47
Diastereoselective cyanohydrin formation, 36–40, 37*s*, 39*s*
Diastereoselective Tishchenko reaction, 48–49, 49*s*
5,6-Dihydro-1,4-dithiin-2-yl [(4-methoxybenzyl)oxy]methane, 49–50, 50*s*
1,2:5,6-Di-*O*-isopropylidene-α-D-glucofuranose, 25, 28–29, 45
Disialyllacto-*N*-tetraose (DSLNT), 158, 165, 167*f*
Disialyl oligosaccharides, 164–165, 167*f*

E

Engineered living-cell strategy, 160–161, 169
Exo-glycals hydroboration, 42–46
Expression host engineering, 96

F

Fleming–Tamao oxidation, 50–52
Fucα1–2LNnT, 165
Fucosylated HMOS, 155–156
2′-Fucosyllactose (2′FL), 160–161
3-Fucosyllactose (3FL), 168
α1–2-Fucosyltransferase (FUT2), 116, 151
α1–3/4-Fucosyltransferase (FUT3), 116, 151–152
Fungal cellulase glycosylation
  Aspergillus species expression, 85*f*, 86–88
  glycan-active enzymes, 88–96
  N-glycans, 68–70, 69*f*, 72–76
  O-glycans, 68–70, 69*f*, 76–82
  Pichia pastoris expression, 84–86, 85*f*
  Saccharomyces cerevisiae expression, 82–84, 85*f*
  T.reesei (*see Trichoderma reesei* cellulase glycosylation)

G

β1–4-Galactosyltransferase 1 (β1–4GalT1), 151
$GlcMan_9GlcNAc_2$-dolichol phosphate, 66
Globotetraose oligosaccharides, 168–169
Globotriose oligosaccharides, 168–169
D-Glucuronic acid glycal, 47–48, 48*s*
N-Glycans, 68–70, 69*f*, 72–76
O-Glycans, 68–70, 69*f*, 76–82
N-Glycolylneuraminic acid (Neu5Gc)-containing MOS, 150–151
Glycosaminoglycans (GAGs), 23
  protein interactions, 52
  structure–activity studies, 23–25

Glycosylation
definition, 66, 96
eukaryotes, 66, 67*f*
fungal cellulases
Aspergillus species expression, 85*f*, 86–88
glycan-active enzymes, 88–96
N-glycans, 68–70, 69*f*, 72–76
O-glycans, 68–70, 69*f*, 76–82
Pichia pastoris expression, 84–86, 85*f*
Saccharomyces cerevisiae expression, 82–84, 85*f*
T.reesei (*see Trichoderma reesei* cellulase glycosylation)
Glycosyltransferase-catalyzed reactions, 159

H

Helicobacter pylori, 160–161, 165–166
Hemicellulose, 65
Heparin, 23
Hexose-5-ulose, 48–49, 49*s*
HMOS. *See* Human milk oligosaccharides (HMOS)
Host–strain glycoengineering, 96
Human milk oligosaccharides (HMOS)
antiadhesive antimicrobials, 153
biosynthesis, 151–152
core structures, 118–151, 123*t*
enzyme-catalyzed production processes
2′FL, 160–161
3FL, 168
3′SL, 161–163
6′SL, 163–164
disialyl oligosaccharides, 164–165
Fucα1–2LNnT, 165
globotriose and globotetraose oligosaccharides, 168–169
LDFT, 168
Le[a] tetrasaccharide, 168
Le[X] tetrasaccharide, 168
LNDFH I, 168
LNFP I, 168
LNFP II, 168
LNnD, 164–165
LNnDFH, 166
LNnFP IV, 166
LNnH, 164–165
LNnO, 164–165
LNnT, 164–165
LNT, 167–168
LNT2, 164–165
LSTd, 164–165
UDP-galactose, 169
fucosylated, 155–156
glycosidic linkages, 118, 122*t*
immunomodulators, 153
monosaccharide building blocks, 117–118, 119*t*
vs. MOS, 150–151
neutral non-fucosylated, 153–155
neutral oligosaccharides, 118, 120*t*
nutrient providers, 153
prebiotics, 153
sialylated, 156–158

I

L-Idose
diastereoselective Tishchenko reaction, 48–49, 49*s*
exo-glycals hydroboration, 42–46, 43*s*
structures, 24*f*
synthesis, 30, 31*s*
2-(trimethylsilyl)thiazole, stereoselective addition of, 41–42, 41*s*
L-Iduronic acid (IdoA)
chair conformations, 23, 24*f*
6-deoxy-L-hexoses, C–H activation of, 50–52, 51*s*
diastereoselective Tishchenko reaction, 48–49, 49*s*
5,6-dihydro-1,4-dithiin-2-yl [(4-methoxybenzyl)oxy]methane, homologation with, 49–50, 50*s*
D-gluco derivatives, epimerization of
C-5 epimerization, 33–36, 34–35*s*
Mitsunobu reaction, 32–33, 33*s*
sulfonate leaving group, nucleophilic displacement of, 25–32, 27*s*, 29*s*, 31*s*
skew-boat conformation, 23, 24*f*
structures, 23, 24*f*
tetroses and pentoses, homologation of
diastereoselective cyanohydrin formation, 36–40, 37*s*, 39*s*
Mukaiyama-type aldol reaction, 36, 37*s*
organometallic reagents, addition of, 40–41, 40*s*

2-(trimethylsilyl)thiazole, stereoselective addition of, 41–42
unsaturated sugars isomerization
4-deoxypentenoside, 47
exo-glycals hydroboration, 41–48
D-glucuronic acid glycal, 47–48
Δ^4-uronates, 46–47
Immunomodulators, 153
1,2-O-Isopropylidene-α-D-glucofuranurono-6,3-lactone, 25–26

L

Lactodifucotetraose (LDFT), 168
Lacto-*N*-difuco-hexoase I (LNDFH I), 168
Lacto-*N*-fucopentaose I (LNFP I), 168
Lacto-*N*-fucopentaose II (LNFP II), 168
Lacto-*N*-neodecaose (LNnD), 164–165
Lacto-*N*-neodifucohexaose (LNnDFH), 166
Lacto-*N*-neofucopentaose IV (LNnFP IV), 166
Lacto-*N*-neohexaose (LNnH), 152, 164–165
Lacto-*N*-neotetraos (LNnT), 155, 164–165
Lacto-*N*-neotetraose (LNT), 155, 167–168
Lacto-*N*-triose II (LNT2), 164–165
Lactose, 151
Lewis^a tetrasaccharide, 168
Lewis^X tetrasaccharide, 168–169
Lignin, 65
O-Linked glycosylation, 66

M

Methyl 2,3-*O*-isopropylidene-L-glycerate, 49–50, 50*s*
Methyl α-D-glucopyranoside, 35–36
Milk oligosaccharides (MOS), 150–151
Mitsunobu reaction, 32–33, 33*s*
Mukaiyama-type aldol reaction, 36

N

Necrotizing enterocolitis (NEC), 153, 158
Neisseria gonorrhoeae, 168–169
Neisseria meningitidis, 161–163
Neutral non-fucosylated HMOS, 153–155
N-linked glycosylation, 66, 67*f*
Nutrient providers, 153

O

One-pot glycosylation techniques, 158
One pot multienzyme (OPME) systems, 159, 165, 166–167*f*
Organometallic reagents, 40–41

P

Parikh–Doering oxidation, 36–38
Pasteurella multocida, 161–162
Photobacterium damselae, 163
Pichia pastoris, 84–86, 85*f*
Prebiotics, 153

R

Radical tandem decarboxylation–cyclization, 35–36, 35*s*
Recombinant bacterial glycosyltransferases, 158–159

S

Saccharification, 65
Saccharomyces cerevisiae, 82–84, 85*f*
Sialylated HMOS, 156–158
Sialyllacto-*N*-tetraose (LSTd), 164–165
Sialyllactose, 157–158
3′-Sialyllactose (3′SL), 161–163
6′-Sialyllactose (6′SL), 163–164
3′-Sialyl-*N*-acetyllactosamine (3′SLN), 161–163
6′-Sialyl-*N*-acetyllactosamine (6′SLN), 163–164
Streptococcus pneumoniae, 164
Sugar nucleotides, 159
Swern oxidation, 41–42

T

1-Thio-L-idoglycosides, 50–52, 51*s*
TrCel7A. *See Trichoderma reesei* cellulase glycosylation
α,α-Trehalose exo-glycals, 43–44, 43*s*
Trichoderma reesei cellulase glycosylation
ancestral hierarchy and mutagenesis source of, 68–70, 70*f*
Aspergillus species expression, 85*f*, 86–88
categories, 66–68

Trichoderma reesei cellulase glycosylation (*Continued*)
  glycan-modifying enzymes, secretomic and transcriptomic evidence of, 93, 94*t*
  N-glycans, 68–70, 69*f*
    catalytic domain, 72–76
  O-glycans, 68–70, 69*f*
    Family 1 CBMs, 79–82, 79*f*
    linker domain, 76–79
  molecular snapshot, 68*f*
  Pichia pastoris expression, 84–86, 85*f*
  Saccharomyces cerevisiae expression, 82–84, 85*f*
  secreted glycan-active α-mannosidases, 89–91, 89*f*, 90*t*
2-(Trimethylsilyl)thiazole, 41–42
Trypanosoma cruzi, 164

U

Δ^4-Uronates, 46–47

W

Whole-cell synthesis, 160, 169

CPI Antony Rowe
Eastbourne, UK
November 12, 2015